Marine Robot Autonomy

Mae L. Seto
Editor

Marine Robot Autonomy

Springer

Editor
Mae L. Seto
Department of Mechanical Engineering
Dalhousie University
Halifax, Nova Scotia, Canada

ISBN 978-1-4614-5658-2 ISBN 978-1-4614-5659-9 (eBook)
DOI 10.1007/978-1-4614-5659-9
Springer New York Heidelberg Dordrecht London

Library of Congress Control Number: 2012952931

Printed on acid-free paper

Springer is part of Springer Science+Business Media (www.springer.com)

Preface

This book provides an update to select underwater autonomy areas. It is intended for researchers and engineers who are new to the field of marine robot autonomy, at the same time, appealing to those with more experience. This book was inspired by: (1) quite a few researchers looking for a reference for graduate courses in marine autonomous robotics with emphasis on autonomy and (2) researchers and engineers who are new to the area with little or no formal training or experience in the area. It is hoped that the extensive references deliberately compiled by each chapter author provide a valuable starting point for further study.

The introductory chapter sets the background and provides definitions for subsequent chapters. It starts with a motivation for why autonomy is necessary and timely for marine robots. Then, it briefly reviews existing metrics and standards for marine autonomy and the components that exist in many intelligent autonomy architectures. The autonomy requirements of military and oceanographic users are touched upon and briefly contrasted to give the reader an appreciation for two different users of marine robots. Then, the fundamentals of what limits marine autonomy specifically, the underwater medium, underwater navigation, and critical enablers like energy are introduced. In-depth exploration is left to subsequent chapters; however, tie-ins and developments related to the following chapters are noted.

To start, two different intelligent autonomy architectures are highlighted, the behaviour-based MOOS-IvP (Benjamin et al., Chap. 2) and the deliberative T-REX (Rajan et al., Chap. 3). These architectures are implemented on actual systems and have received a fair amount of interest from the military and scientific users.

One of the basic functionalities of an intelligent autonomy architecture (for all environments) is motion or path planning. This occurs once a map of the environment exists. Plans optimized around constraints have to be generated. For reactive applications as in obstacle avoidance, plans are generated in near real time. For more deliberative search applications where a detailed digital map of the area exists this can take longer. Path planning is discussed in Chap. 4 (Paull et al.).

UUVs on long deployment require fault tolerance as the UUV itself will change if the mission is long enough. When unexpected hardware failures occur, the intelligent autonomy should allow the UUV to reconfigure itself to use alternative combinations of the remaining functioning resources (Lane et al., Chap. 5). This has been termed “autonomous embedded recoverability.” Chapter 5 describes work on a declarative goal-based solution for adaptive mission planning that builds in the ability to adapt and recover from failures.

The ability to scan or sense a wider area and to work cooperatively has the potential to vastly improve the efficiency and effectiveness of mission operations. However, given the complexity and difficulty of the underwater environment, cooperation between underwater vehicles faces many challenges. This is covered in Chap. 6 (Redfield).

In the very dynamic ocean environment where operators work with little or no a priori information the value of machine learning emerges. Reinforcement learning is a methodology in robot learning where a scalar evaluation (reward) of the performance of the algorithm is available from interaction with the environment. The objective in reinforcement learning is to maximize the expected reward through adjusting a value function. The role of machine learning in an intelligent autonomy architecture is highlighted in Chap. 7 (Carreras et al.).

SLAM is an example of a truly autonomous capability with little or no human intervention. With SLAM, a spatial map of the UUV’s environment is built for navigation purposes. The UUV uses its sensors (sonars, bathymetric sensors, etc.) to perceive the environment. The sensors are modelled as real with errors and finite ranges. The sensor measurements are assembled to create the map. There is no a priori environment information for the UUV to work with. With SLAM, beacons and networks do not have to be deployed or used making the technique useful in GPS-denied, underwater, or under-ice environments. Chapter 8 showcases this capability through three case studies.

I would like to acknowledge the efforts of the many reviewers who have given generously their time. Each chapter was meticulously peer reviewed. Lastly, I would like to thank all the authors who have contributed to this book. They are among the researchers who are expanding the envelope on the state of the art in autonomy for marine robots.

Halifax, NS, Canada Mae L. Seto

Contributors

J. Barreiro NASA Ames Research Center, Moffett Field, CA, USA

Michael R Benjamin Computer Science and Artificial Intelligence Laboratory, Massachusetts Institute of Technology, Cambridge, MA, USA

K. Brown Ocean Systems Laboratory, Heriot-Watt University, Edinburgh, Scotland, UK

M. Carreras Computer Vision and Robotics Group, University of Girona, Girona, Spain

A. El-fakdi Control Engineering and Intelligent Systems Group, University of Girona, Girona, Spain

B. Englot Department of Mechanical Engineering, Massachusetts Institute of Technology, Cambridge, MA, USA

Maurice F. Fallon Computer Science and Artificial Intelligence Laboratory, Massachusetts Institute of Technology, Cambridge, MA, USA

J. Folkesson Center for Autonomous Systems, Kungliga Tekniska Högskolan (KTH), Stockholm, Sweden

F. Hover Department of Mechanical Engineering, Massachusetts Institute of Technology, Cambridge, MA, USA

H. Johannsson Computer Science and Artificial Intelligence Laboratory, Massachusetts Institute of Technology, Cambridge, MA, USA

Michael Kaess Computer Science and Artificial Intelligence Laboratory, Massachusetts Institute of Technology, Cambridge, MA, USA

D. Lane Ocean Systems Laboratory, School of Engineering and Physical Sciences, Heriot-Watt University, Edinburgh, Scotland, UK

John J. Leonard Computer Science and Artificial Intelligence Laboratory, Massachusetts Institute of Technology, Cambridge, MA, USA

H. Li COBRA group, Department of Electrical and Computer Engineering, University of New Brunswick, Fredericton, NB, Canada

Hunter McClelland Robotics and Mechanisms Laboratory, Virginia Polytechnic Institute of Technology and State University (Virginia Tech), Blacksburg, VA, USA

Emilio Miguelanez SeeByte Ltd, Edinburgh, Scotland, UK

Paul M. Newman Department of Engineering Science, University of Oxford, Oxford, UK

Pedro Patron SeeByte Ltd, Edinburgh, Scotland, UK

L. Paull Department of Electrical and Computer Engineering, University of New Brunswick, Fredericton, NB, Canada

Y. Petillot Ocean Systems Laboratory, School of Engineering and Physical Sciences, Heriot-Watt University, Edinburgh, Scotland, UK

F. Py Monterey Bay Aquarium Research Institute, CA, USA

K. Rajan Monterey Bay Aquarium Research Institute, CA, USA

S. Redfield Office of Naval Research Global, London, UK

Naval Surface Warfare Center, Panama City Division, Panama City Division, FL, USA

P. Ridao Computer Vision and Robotics Group, University of Girona, Girona, Spain

S. Saeedi Department of Electrical and Computer Engineering, University of New Brunswick, Fredericton, NB, Canada

H. Schmidt Computer Sience and Artificial Intelligence Laboratory, Massachusetts Institute of Technology, Cambridge, MA, USA

M. L. Seto Intelligent Marine Systems Laboratory, Dalhousie University, Halifax, NS, Canada

Contents

1 Introduction to Autonomy for Marine Robots 1
M.L. Seto, L. Paull, and S. Saeedi

2 Autonomy for Unmanned Marine Vehicles with MOOS-IvP 47
Michael R. Benjamin, Henrik Schmidt, Paul M. Newman, and John J. Leonard

3 Towards Deliberative Control in Marine Robotics 91
Kanna Rajan, Frédéric Py, and Javier Barreiro

4 Path Planning for Autonomous Underwater Vehicles 177
Liam Paull, Sajad Saeedi, and Howard Li

5 An Ontology-Based Approach to Fault Tolerant Mission Execution for Autonomous Platforms 225
David Lane, Keith Brown, Yvan Petillot, Emilio Miguelanez, and Pedro Patron

6 Cooperation Between Underwater Vehicles 257
Signe Redfield

7 Behavior Adaptation by Means of Reinforcement Learning 287
Marc Carreras, Andrés El-fakdi, and Pere Ridao

8 Simultaneous Localization and Mapping in Marine Environments 329
Maurice F. Fallon, Hordur Johannsson, Michael Kaess, John Folkesson, Hunter McClelland, Brendan J. Englot, Franz S. Hover, and John J. Leonard

Index 373

Chapter 1
Introduction to Autonomy for Marine Robots

M.L. Seto, L. Paull, and S. Saeedi

Abstract An unmanned underwater vehicle (UUV) or autonomous underwater vehicle (AUV) is a marine robot [1] used for a wide range of oceanographic and military tasks including underwater surveys, inspection of submerged structures, tracking oceanographic features, undersea mapping, laying undersea cable, searching for downed aircraft, and finding naval sea mines, to name but a few. An unmanned surface vehicle (USV) is also categorized as a marine robot; however, most of this book is concerned with UUVs. In terms of physical shape, UUVs can be the traditional torpedo-shaped bodies as shown in Fig. 1.1 or an underwater glider such as that seen in Fig. 1.2. At the lowest level of control, UUVs operate with closed-loop control for basic actions such as maintaining depth, pitch, roll, and heading. As an extension to this capability, UUVs can also line-follow between a series of waypoints while logging data from mission sensors, as well as sensors that monitor the UUV status and functionality. Such basic actions can be grouped into *behaviours*.

1 Why More Autonomy?

An unmanned underwater vehicle (UUV) or autonomous underwater vehicle (AUV) is a marine robot [1] used for a wide range of oceanographic and military tasks including underwater surveys, inspection of submerged structures, tracking oceanographic features, undersea mapping, laying undersea cable, searching for

M.L. Seto (✉)
Intelligent Marine Systems Laboratory, Dalhousie University, 1360 Barrington Street, Halifax, NS, Canada B3J 1Z1
e-mail: mae.seto@dal.ca

L. Paull • S. Saeedi
Department of Electrical and Computer Engineering, University of New Brunswick, Fredericton, NB, Canada

M.L. Seto (ed.), *Marine Robot Autonomy*, DOI 10.1007/978-1-4614-5659-9_1,

Fig. 1.1 Surfaced Explorer UUV being recharged

downed aircraft, and finding naval sea mines, to name but a few. An unmanned surface vehicle (USV) is also categorized as a marine robot; however, most of this book is concerned with UUVs. In terms of physical shape, UUVs can be the traditional torpedo-shaped bodies as shown in Fig. 1.1 or an underwater glider such as that seen in Fig. 1.2. At the lowest level of control, UUVs operate with closed-loop control for basic actions such as maintaining depth, pitch, roll, and heading. As an extension to this capability, UUVs can also line-follow between a series of waypoints while logging data from mission sensors, as well as sensors that monitor the UUV status and functionality. Such basic actions can be grouped into *behaviours*.

A behaviour is a collection of actions or maneuvers that the UUV performs to achieve a goal. *Autonomous behaviours* (e.g., track a thermocline, follow waypoints) are on-board capabilities that provide actions for *known* situations and events. The behaviour-based subsumption architecture [2] for UUV control implements operator-scripted mission plans, based on autonomous behaviours, generated prior to a mission. These plans are then downloaded onto the UUV vehicle computer. The UUV performs its mission tasks at precise times as per this mission plan. However,

Fig. 1.2 Surfaced underwater glider

timing of tasks in this manner does not allow the UUV flexibility to adapt its mission plan in situ.

Adaptation is an attribute of the desired autonomy in marine robots. High energy density power sources and larger UUVs make long duration missions possible. Over long missions, goals likely evolve, the ocean environment changes, and the UUV itself changes. Unexpected UUV changes include instrument failure, loss of communications with a collaborative robot (or operator), inability to profile the ocean floor, problems with thrusters or control fins, greater than expected energy consumption, etc. The ocean environment can and will unexpectedly change due to unexpectedly strong currents, weather systems impacting underwater communications and propagation conditions, unknown topography, obstructions, islands, changes in salinity, etc. Until lately, unforeseen UUV or environment changes that adversely impact a mission resulted in a rescue ship recovering the UUV and rectifying the problem. This both interrupts the mission and requires ship support which detracts from the initial reason to use a robot. It also pre-supposes a ship can reach the UUV in a timely manner. For example, in an under-ice environment, ship support is possibly not feasible. If the UUV can adapt through its autonomy to solve its problem it increases the likelihood for mission success.

Intelligent behaviours are on-board capabilities that provide a structure to reason about unexpected or unknown situations or events and allow the UUV to adapt and complete a mission where possible. Intelligent behaviours can be the basis for *intelligent autonomy*. Intelligent autonomy provides a framework for deliberation and reasoning with intelligent behaviours. The intelligent autonomy framework can use autonomous agents (systems that perceive the environment through sensors and act upon it through actuators) to aid with intelligent behaviours like *decision-making*, *mission planning/re-planning*, and *fault tolerance*. Complex tasks require decision-making abilities rather than programmed choices to autonomously sense situations, make evaluations, and decide on follow-on actions. With mission planning, the UUV's motion or path planning is performed to satisfy constraints (e.g., percentage overlap for the sensor coverage). Autonomous fault tolerance monitors the UUV's condition and adapts to recover from UUV subsystem faults or failures where possible.

UUVs that employ autonomous agents appeared in the 1990s [3] with a growing body of research developing towards intelligent autonomy [4–12]. Alternatively, partial measures give the autonomous system limited autonomy that is shared with an operator [13]. Examples of intelligent autonomy are historically from spacecraft. NASA's Deep Space 1 spacecraft [14–18] used on-board autonomous agents to control the spacecraft for 2 days in May 1999 while it was 97 million km from Earth. The Mars Exploration Rover missions have autonomous agents [17,18] as well. Rockwell International's fault-tolerant autonomous agents safely landed an autonomous scale model F/A-18 aircraft after 60 % of one wing was deliberately blown off [19]. The application of autonomy to marine robots is at an earlier stage, despite its more advanced application to aviation and spacecraft. For marine robots, why more autonomy? Why are such high-level autonomy capabilities needed now? UUVs increasingly have requirements to:

- Improve endurance (longer missions)
- Respond to a complex, dynamic environment
- Perform more complex missions
- Perform missions only a UUV can achieve (e.g., under-ice bathymetric data collection)
- Implement multiple vehicle cooperation for force multiplication—a different instantiation than single UUV autonomy

An illustrative list of UUV capabilities that use intelligent autonomy include:

- Trim adjustments based on in situ salinity or temperature measurements
- Trim adjustments to compensate for changes in buoyancy or weight after a payload (e.g., sensor networks or cable) is deployed
- Maintaining a stand-off from a suspected mine (that it detected and identified) because the UUV's own magnetic signature can set off the mine
- Redistributing control authority on the control surfaces in order to compensate for an unexpected fresh water layer
- Re-planning a mission between communicating members because a member of the cooperative is no longer in communications with the rest
- Making opportunistic measurements for a rare interesting phenomena that was not the main focus of the mission
- Learning in situ (or a priori) to look for specific targets or to adapt to seabed types to optimize the on-board detection tools
- Devising adaptive sampling missions, based on in situ measured gradients of a quantity (e.g., water temperature), to a given resolution
- Tracking a chemical plume to its source
- Adaptively monitoring marine habitats

- Monitor and track incursions of select fish into a new water body
- Tracking a thermocline
- Mapping an oil spill
- Cooperatively navigating given poor underwater communications
- Building a map of an area with no a priori knowledge
- Monitoring and adapting to critical and non-critical system faults (e.g., from mission critical sensors to malfunctioning lights)
- Mission re-planning due to unexpected:
  - Changes in currents
  - Higher UUV energy consumption (e.g., UUV discovers it is off course and must expend additional energy to get back on course)
  - Doppler velocity log (DVL) not achieving sea bottom tracking
  - Loss of track on a mobile target (s)

To realize such capabilities UUVs need more intelligent autonomy than most currently possess.

Discussions of what constitutes autonomy for robots can vary and range from the philosophical to intentions and ethics [20–22]. One definition of "autonomous" from [23] is "Capable of operating without operator input for extended periods of time. Implicit in this description is the requirement that the UUV's sortie accomplishes its assigned goal and makes the appropriate rendezvous for a successful recovery."

For example, robotic autonomy can specifically include the ability to fulfill goals by: (1) making decisions given a set of automated planned options and, if these plans are inadequate, to make new plans; (2) analysing and interpreting in situ sensor data to create information; (3) diagnosing and recommending solutions for robot or mission problems detected through sensing and monitoring; and (4) learning and adaptation; and (5) collaborating with other UUVs, systems, or humans through point-to-point or network communications.

These abilities allow the UUV to adapt to the dynamic ocean environment, faults within the UUV, and changes to the mission. Such adaptation could be achieved through a *structure* to reason about follow-on actions related to unexpected events or developments. This structure is functionally a control architecture in the form of an autonomy framework (realized in software). The autonomy framework can be reactive [24], deliberative [25], or a hybrid [26–28]. Reactive architectures use the sense-plan-act schema and are suited to structured and predictable environments and missions. Reactive architectures are simpler and reasonably straightforward to implement; however, they do not deal well with deviations from the nominal mission, environment, or health of the UUV. Deliberative architectures base their planning on a world model. They take on-board sensor data and process it into information to build and update that world model. The world model is consulted to determine how the UUV could implement the operator's pre-defined mission goals and ultimately, the commands sent to sensors, motors, control surfaces, and other subsystems. Deliberative architectures can reason and make predictions about the environment. Hybrid architectures combine the advantage of fast reactive responses based on a behaviour-based layer coupled with a deliberative layer capable of decisions and planning. In some hybrid architectures, the deliberative layer oversees the sensing and the reactive behaviour-based layer by choosing and prioritizing behaviours. The behaviour-based layer oversees the low-level vehicle controllers and generates commands to the actuators and thrusters as needed. Two different autonomy architectures are presented in the subsequent chapters. One is behaviour-based (Benjamin et al., Chap. 2) and the other is more deliberative (Rajan et al., Chap. 3).

With intelligent autonomy the mission may also be specified as high-level objectives that are declarative or goal-based. Two examples are presented in this book (Lane et al., Chap. 5; Rajan et al., Chap. 3). The next section briefly discusses how and where such architectures could be implemented on board the marine robot.

1.1 Payload Autonomy Implementation

One implementation is to have the autonomy architecture reside on board a second computer, the backseat or *payload computer*, that interfaces to the original equipment manufacturer (OEM) *vehicle computer* through middleware. This is the backseat architecture [29] or payload autonomy used on some UUV systems. This decouples the vehicle control (and thus UUV hardware) for the basic UUV actions mentioned earlier (e.g., maintaining depth, speed, and heading) from the intelligent autonomy.

The backseat architecture's advantage is that it leaves the OEM vehicle computer untouched (though its interface has to be known). Changes in UUV intelligent capabilities are effected through the autonomy architecture, and the backseat computer itself can be ported to different platforms with minimal work, since it is relatively independent of the vehicle computer. In this control paradigm the *payload* computer issues high-level commands to the *vehicle* computer to execute *payload computer* decisions arrived at through the autonomy framework. This is covered in more detail in Chap. 2 (Benjamin et al.).

The PC-I04 and peripheral component interface (PCI) bus architectures are popular choices for some to implement payload autonomy. The advantages of the PC-104 interface is its compact size and low power consumption. However, a distinct disadvantage is its 16-bit data bus as opposed to the 32-bit bus of the PCI architecture. Embedded processors will continue to follow Moore's Law as commercial technology advances for low-power processors. Lately, advances are more due to multiple cores and rather than processor speeds. More complex autonomy frameworks and autonomous agents could be realized if more computationally powerful embedded processors could be applied. Having said that, the UUV autonomy framework does not have to send directives to the vehicle computer at high rates (compared to other robots), as UUVs have top speeds that range only from about 0.6–3 m/s.

Consideration for how complex the autonomy framework should be for a UUV should occur at the design stage [16]. Beyond the vehicle capabilities, the level of autonomy impacts the UUV concept of operations, survivability, serviceability, and affordability. The architecture should accommodate increasing levels of autonomy through software upgrades primarily. As the autonomy of an UUV increases, the number and complexity of its missions can also increase. As well, the UUV might perform missions that were not previously possible due to complexity, risk, or low connectivity with an operator. As an example, with intelligent autonomy on board a fleet of UUVs, they may require less operators, be more rapidly deployable, and require less operator directives. Semi-quantitative measures of autonomy exist [23] to assess the level of autonomy a UUV should have for design or specification purposes.

2 Levels of Autonomy

There are several existing scales depicting levels of autonomy [23,30,31] that vary from very basic teleoperation to fully intelligent systems for UUVs and other robots. An example is presented below for the US Army Future Combat System [31] that uses a scale with a reasonable amount of fidelity:

1. Remote control
2. Remote control with vehicle state knowledge
3. External pre-planned mission
4. Knowledge of local and planned path environment
5. Hazard avoidance or negotiation
6. Object detection, recognition, avoidance, or negotiation
7. Fusion of local sensors and data
8. Cooperative operations
9. Collaborative operations
10. Full autonomy

For UUVs, the Standard Guide on UUV Autonomy and Control [23] considers three components in assessing autonomy needs for missions: (1) situation awareness; (2) decision-making, planning, and control; and (3) external interaction. These three components are orthogonal and interrelated so they span the UUV "autonomy space". In specifying levels of autonomy for each component, a point is defined in the autonomy three space. The guide does not describe a particular UUV system, subsystem, sensor, architecture, or control system to implement the desired levels of autonomy. The guide addresses only the levels of autonomy and the capabilities within those levels. These are briefly summarized below; please refer to [23,31] for more details.

UUV situational awareness levels of autonomy

0	Raw	Unprocessed data collected from sensors
1	Semi-processed	
2	Feature	Data filtered, normalized, and characterized
3	Aggregate	Data aggregated over time, space, and/or feature characteristics (target detection)
4	Interpreted	Compared against some knowledge database
5	Inferred	Correlated to a knowledge base of situational or behavioural characteristics (further compared against the seabed)
6	Intent	Correlated to a knowledge base of known patterns of operation including recognizing plans and objectives and recognizing and/or predicting intent

UUV decision-making, planning, and control levels of autonomy

0	Direct control	Execution of externally defined commands, no decision-making
1	Sequenced execution	Execution of externally defined sequence of actions—limited decision-making
2	Adaptive execution	Execution of externally defined actions; closes loop on the actions; limited decision making
3	Objective achievement	Combine actions to achieve a single objective at a time
4	Multiple objective	Simplistic combining of multiple objectives
5	Joint objective achievement	Balance multiple objectives

UUV external interaction levels of autonomy

0	Teleoperation	Operator uses continuous video or sensor feedback to directly control UUV
1	Remote control	Operator provides frequent control, but UUV can autonomously execute some actions such as waypoint following and station keeping/idling; only initiative from UUV is abort of actions pending appropriate circumstances
2	Semi-autonomous control	UUV chooses from a priori list of actions to execute to achieve overall operator goal; capable of some operations during communications blackouts; operator can still override and redirect actions
3	Fully autonomous	UUV functions without operator intervention from launch until recovery; if within range, underway communications and redirection is still possible from the operator

The leadership for these standards has come from the National Institute of Standards and Technology's (NIST) autonomy levels for unmanned systems (ALFUS) Working Group in collaboration with other parties. This book addresses examples of the higher levels of autonomy for all three autonomy components outlined above. The parts of an architecture that tends towards the higher levels of autonomy described in this section are discussed next.

3 Parts of an Intelligent Autonomy Architecture

Parts of an intelligent autonomy architecture can include [32] Planning & Decision, Sensing & Perception, Monitoring & Diagnosis, Learning & Adaptation, and Networking & Collaboration. The parts of the architecture are briefly described next.

1. *Planning & Decision* develops actions to fulfil the UUV mission given its understanding of the world through Sensing & Perception's world model. Planning & Decision tempers its recommended actions against what is possible given the UUV system and subsystem health as known from Monitoring & Diagnosis. This capability allows the mission to be adapted in response to unanticipated events in the UUV or its environment. It draws on path planning (Paull et al., Chap. 4), decision-making (Benjamin et al., Chap. 2; Rajan et al., Chap. 3), learning (Carreras et al., Chap. 7), and cooperation (Redfield et al., Chap. 6) which are all areas of active research. NASA's Remote Agent Experiment is an example of a well-known advanced mission planner that maintains a given vehicle state and fulfils its mission [15].
2. *Sensing & Perception* develops a picture of the world or a world model as relevant to the UUV's mission. This is achieved through assembling and analysing data from its sensors—or other robots and operators through networking and collaboration—processing the sensor data, and interpreting as required. This picture is provided to the Planning & Decision part. Sub-capabilities here include obstacle avoidance, navigation, and collecting information to increase situational awareness. The mine countermeasures' (MCM) automated target recognition (ATR) example illustrates how the Sensing & Perception and Planning & Decision parts can interact. UUV systems with these capabilities can autonomously detect (Sensing & Perception), classify, identify and geo-locate (Planning & Decision) targets such as mines, and then plan a mission (Planning & Decision) to re-acquire the targets for optical images (Sensing & Perception) and further decisions for follow-on actions (Planning & Decision) [33].
3. *Monitoring & Diagnosis* tracks and monitors UUV systems through on-board sensors and subsystems. This part of the architecture performs fault and failure detection as well as monitors trends for the UUV systems and subsystems. When possible it reconfigures the vehicle sensors and/or systems to ensure vehicle survivability (so vehicle is not lost) and mission success. Generally this

is for monitoring and maintenance support. Some systems have the ability to model sensors and subsystems to provide estimates and performance measures to Planning & Decision. This part of the architecture deals with autonomy for the robot's robustness [34,35] as opposed to the autonomy for the mission. Some systems will send results from their Monitoring & Diagnosis part to help with Planning & Decision for mission planning. An example of state of the art in this area is given in Chap. 5 (Lane et al.).

4. *Learning & Adaptation* is the capability to adapt in situ to unanticipated events with no a priori information or knowledge. This can be used by all parts of the intelligent autonomy architecture towards greater mission effectiveness and UUV survivability. For example, the autonomy could have the UUV learn to recognize an underwater target based on its shape in a variety of orientations and sea bottom types. There are three relevant learning methodologies: model approximation, supervised learning, and reinforcement learning. Model approximation and supervised learning are fairly mature. Reinforcement learning is at a relatively earlier stage. However, unlike supervised learning reinforcement, learning does not need a database of examples to draw from. Rather, it learns from interaction with the environment against a reward value from a reinforcement function (generated by an operator). Reinforcement learning is good for adapting unexpectedly to changes in the environment. An update on reinforcement learning developments for applications to UUVs is covered in Chap. 7 (Carreras et al.).
5. *Networking & Cooperation.* Cooperation is defined here to mean robots working together. The degree of "working together" varies from collaboration, where there is tight cross-vehicle timing constraints to achieve a goal, to coordination, where the vehicles work relatively independent of one another. The inherent difficulty of the underwater environment (Sect. 6.1) impedes UUV cooperation making it more difficult than robots in other environments. The limited available bandwidth in an underwater communications channel usually means only one member of a cooperative can transmit at a time, which constrains the information being shared. This part of the architecture oversees how multiple autonomous systems (homogeneous or heterogeneous) work together and communicate the necessary information to fulfill mission goals. Cooperation does not exclude operators from being part of the network. Sub-capabilities include collision avoidance, navigation and communications as a cooperative member, decision-making, and planning, taking into account cooperative members may have dissimilar sensors, capabilities, etc. This is well covered in a later chapter (Redfield, Chap. 6).

The first four parts, and occasionally the fifth, are present in a UUV intelligent autonomy architecture. Their relative presence depends on the UUV's desired capabilities and tasking. To delve into more detail with respect to the intelligent autonomy for each part requires knowledge of the UUV's tasking and then interpreting its autonomy requirements. This will be discussed next for the two largest UUV users—the military and scientific communities.

4 Military Mission Requirements: Tools for Dull, Dirty, and Dangerous

> The nation is faced, currently and for the foreseeable future, with a multitude of military challenges that are unlike any seen in recent history. The enemy is diverse, not easily recognizable, and operates in atypical ways. These asymmetric threats have the ability to do great harm to our maritime forces and infrastructure, and the Navy must have the ability to address and defeat them in support of national Defense objectives, while continuing to execute its traditional roles.
>
> Unmanned systems have the potential, and in some cases the demonstrated ability, to reduce risk to manned forces, to provide the force multiplication necessary to accomplish our missions, to perform tasks which manned vehicles cannot, and to do so in a way that is affordable to the nation.
>
> -Program Executive Officer for Littoral and Mine Warfare [36]

When autonomous systems entered the battle space they changed the planning, organization, and execution of some military missions. Autonomous systems are valued for their ability to collect tactical, operational, and strategic intelligence data and implement solutions to a commander's objective. As a warfare platform their relatively small footprint and signatures, ease of clandestine deployment, and autonomous operation have provided definite advantages. They have been under consideration, and in some cases integrated, for a variety of missions.

For UUVs in particular, the US Navy has identified 40 distinct missions in mostly littoral waters [37,38]. The top eight (prioritized) missions are:

1. Intelligence, surveillance, and reconnaissance
2. Mine countermeasures
3. Antisubmarine warfare
4. Oceanographic data collection (battle space characterization)
5. Inspection/identification
6. Communications/navigation network node
7. Payload delivery
8. Information operations

Operational needs vary across such far-ranging missions. While UUVs could do these missions (some to a greater extent than others) it does not mean they necessarily should. Additional considerations include need, risk, alternatives, cost, and impact on current navy training and practices. The MCM mission using side-scan sonars on UUVs [32,39] has to date seen a fair amount of development and will be used as an example.

Initially, the MCM mission involves performing a route survey and a minehunting survey. With the route survey, side-scan sonar images of the sea bottom are collected when the sea bottom is known to be clear of mines. This provides a reference against which the subsequent minehunting survey can be compared and contrasted. In both cases, ATR tools are used to detect, identify, and classify targets as either mine-like or not. If a mine-like object has been detected in this manner there may be a requirement to return to that location to re-acquire the object with a higher-resolution sonar or optical camera to confirm its identity. If it is identified as a mine then the explosive ordinance disposal (EOD) process may follow. A general argument for using a robot is if the task is 'dull, dirty, or dangerous'—which fits the MCM mission.

Prior to UUVs, the minehunting survey was performed with manned ships towing side-scan sonars and the EOD with divers, so it was *dangerous* and *dirty* for obvious reasons. In this way, the sonar imagery was available to the operator in near real time and searching for mines involved the operator analysing endless side-scan imagery for a target that may or may not be there (*dull*). While autonomous systems can replace manned ships in minehunting surveys, perform EOD, or use ATR to assist with mine detection, the three *d*'s alone may not be sufficient justification to employ autonomous systems, as militaries are trained to work with dull, dirty, and dangerous.

Additionally, UUVs also provide scalable force multiplication for the MCM mission. In some cases, they may be economical compared to previous methods. As an off-board resource for a naval vessel, UUVs significantly extend the naval vessel's persistence and reach. Unlike a diver tasked to MCM operations, the UUV is more persistent (greater endurance) and will not become fatigued, hungry, bored, cold, disoriented, or as limited by sea state or poor visibility. The UUV increases the ship's reach beyond what it would otherwise achieve. These additional arguments are also consideration for integrating autonomous systems in military missions. Discussed next is the REMUS 100 UUV, as a specific example, which was the first wartime development of a UUV.

The REMUS (remote environmental monitoring units) 100 UUV has been under development since the late 1990s by the Woods Hole Oceanographic Institution. The man-portable REMUS 100 weighs 36 kg in air and has a depth rating of 100 m. It has been developed for a variety of missions including MCM [40–43]. REMUS 100 was used extensively in Iraq's port city of Umm Qasr in an MCM capacity after the 2003 invasion [44]. Its MCM performance was commendable and led to technological developments for the REMUS 600 [45], 1,000, and 6,000 UUVs. The REMUS class UUVs have also been platforms for UUV autonomy development [34,35], navigation [46,47], and scientific research [48–50] and they have now transitioned to the US Navy as operational systems [51].

Autonomy issues underlie UUV capabilities whether military or science-based. A mission draws on a variety of capabilities to fulfil its goals. The next sections looks at the extent to which intelligent autonomy enables some autonomy capabilities and where these capabilities are limited by factors beyond intelligent autonomy. This will start with an intelligent autonomy example, simultaneous localization and mapping (SLAM), which has applicability to the MCM mission.

4.1 *Intelligent Autonomy Example: Simultaneous Localization and Mapping*

Terrain-based navigation [52,53] is an autonomous navigation technique requiring intelligent autonomy. It takes bathymetric sonar, DVL, echo sounder, or altimeter measurements of the terrain and correlates it against features in a pre-existing digital terrain map to localize the UUV within that digital terrain map. These methods merge inertial navigation system (INS) input with a bathymetric sensor and a digital terrain map to estimate a global position measurement that integrates back into the INS. The drawback is that terrain-based navigation requires a pre-existing digital terrain map. A further development from terrain-based navigation is concurrent mapping and localization (CML) whereby the digital terrain map is constructed by the UUV in situ and concurrently used to localize the UUV within it. This is also referred to as SLAM. SLAM shows potential towards keeping the UUV submerged for long or covert missions.

This is discussed in more detail in Chap. 8 (Fallon et al.) which presents examples of underwater SLAM integrated into the decision process for intelligent autonomy.

4.2 *Military Autonomy Requirements*

Military missions are characterized by short-term, time-critical survey or exploration applications of which the MCM mission is an example [36]. Generally speaking, such missions require the UUV to have the capability to:

- *Reach or transit long distances*
  Other than greater mission endurance this also allows the UUV to be deployed covertly from further away and to operate at a greater stand-off from the ship and thus extend the ship's reach or influence.
- *Persistence*
  Long sampling times facilitate good situational awareness as well as the ability to work for the long time intervals needed for time-critical missions.
  Being at considerable range for extended periods from an operator places greater reliance on the initial three to four parts of the autonomy architecture (Sect. 3, Planning & Decision, Sensing & Perception, Monitoring & Diagnosis, and Learning & Adaptation). Limitations to these capabilities include access to high energy density power sources.
- *Collect data independent of human operator*
  If the desired reach and persistence are achievable, it likely brings the UUV beyond the range of standard underwater acoustic communications. The ability to make decisions, re-plan missions, sense and perceive, process data into

information, and mitigate UUV system issues supports the reach and persistence capabilities. This capability is limited by the sophistication of the initial three (and perhaps fourth) parts of the autonomy architecture.

- *Operate and coordinate multiple vehicles for force multiplication*
  Multiple UUVs allow the MCM mission, for example, to potentially achieve a higher coverage rate for the minehunting survey. Cooperation of multiple UUVs requires autonomy for UUV fleets to coordinate, reorganize, and re-plan amongst themselves, given limited operator connectivity, to respond to unexpected changes in members of the fleet, the environment, or mission objectives. This capability relies on all five parts of the autonomy architecture. This capability is limited by the difficult underwater communications channel (high attenuation and dispersion, limited bandwidth) which constrains the information flow and hence the potential level of cooperation possible.
- *Send back high-resolution information given poor bandwidth and high losses*
  This refers to the need for timely human imperative decision-making that may be required by some operational military procedures. On-board ATR is an asset as it reduces the large raw sonar image data files into more compact information. The requirement is to find ways to send the information despite the environment and, in some cases, maintain the covertness. The challenge here is the difficult underwater communications channel.
- *Track moving objects, for example, ships, submarines, and other UUVs*
  Like the SLAM mentioned earlier, this task is another that utilizes autonomy capabilities. In some instances, this is performed with the ship at large stand-offs from the UUV. There may also be a requirement to send the information back to the ship. These are particularly challenging autonomy capabilities to implement. This capability is limited by the sophistication of all five parts of the autonomy architecture, access to high energy density power sources, and the difficult underwater communications channel.
- *Detect, evaluate, and avoid potential threats*
  This requires sensors to collect data and process it in situ into information in order to detect the presence of threats such as mine-like objects. Mission re-planning may be necessary to confirm the identity of the threat by scanning it from a different perspective. Given capable sensors, this is limited by the sophistication of the first two parts of the autonomy architecture.
- *Provide covert and non-covert communications*
  Covert communications need terse (and possibly cryptic, using spread spectrum techniques [54]) communications at low power levels. The objective is to not reveal the transmitter or receiver of the communications. For non-covert communications, the UUV status may need to be relayed back to the ship on a

schedule particularly if the UUV has a problem it absolutely cannot rectify (e.g., cannot reconfigure subsystems to complete mission as in the case of a mission critical sensor failure). This is made more challenging by the difficult underwater communications channel.

- *Navigate given denied and/or degraded GPS*
  This refers not only to submerged condition operations but instances where the UUV cannot obtain a GPS calibration ("fix") when it surfaces due to deliberate jamming (denial). This also applies in the absence of jamming when it takes a long time (e.g., in excess of 15 min) for the UUV to obtain a calibration fix due to degraded conditions such as extreme weather or lack of satellites. The UUV is at risk of damage from surface traffic if it loiters on the surface too long.

In summary, autonomy (intelligent autonomy architecture, reasonable processor, and bus speeds) alone cannot provide all the autonomy capabilities desired. There are other limiting considerations such as the difficulty of the underwater communications channel to work in, underwater navigation, and the availability of high energy density power sources. These limitations will be discussed in Sect. 6 after the oceanographic research autonomy requirements of Sect. 5.

5 Oceanographic Research Requirements: Ocean Observation Tools

Most of the previous century could be called a century of under-sampling
–Walter Munk (Secretary of the Navy Research Chair in Oceanography [55])

Section 5 starts by looking at ocean observation problems and capabilities that marine robots can apply towards them and then discusses intelligent autonomy to support ocean observation for scientific research. There are, of course, military applications of ocean observation for the collection of hydrographic, oceanographic, and meteorological data in ocean environments to support real-time operations. In these cases, oceanographic data, information, and knowledge are needed in near real time for tactical support and rapid turnaround for battle space preparation [56].

5.1 Ocean Observation

Scientific research missions occur at a different pace from military ones. They also differ in scope since they are long-term, non-critical monitoring applications such as oceanographic data collection, pollution monitoring and detection, and observation of habitats. As well, scientific missions will consider using opportunistic measurements (Kanna et al., Chap. 3). For example, if the UUV is observing a marine habitat and happens upon a rare but interesting upwelling, it may be a good opportunity to divert from the main mission and observe the upwelling [57]. With intelligent autonomy for opportunistic measurements this is possible. Opportunistic measurements are not typically used for military missions, which are more time critical. Military operations, where possible, require products and procedures that yield data, information, and interpretation in a timely manner.

At a more fundamental level, engineering missions to develop new intelligent autonomy capabilities for research requirements have the possibility the mission may not execute as planned. This speaks to the need for sophisticated Planning and Decision as well as the Monitoring and Diagnosis parts in the intelligent autonomy architecture for test bed fail safes to validate experimental capabilities.

Adequately sampling a global, turbulent fluid with concurrent physical, biological, geological, and chemical processes is a fundamental ocean observation problem. The temporal and spatial attributes of these processes can be unpredictable. These processes manifest as forces from the atmosphere, ocean circulation, and land-sea interactions that impact the ocean environment. These forces fluctuate over a wide range of temporal and spatial scales, from global variability to local small-scale episodic events [58]. This mesoscale variability in the oceans was discovered in the 1960s. Ninety-nine percent of the ocean's kinetic energy is associated with mesoscale variability [59] and only a small percentage with the mean currents that were assumed up to that point. Mesoscale variability spans spatial scales on the order of 50–500 km and timescales on the order of 10–100 days. The awareness of mesoscale variability brought about the realization that some ocean measurements may be under-sampled. To be under-sampled is to sample at less than twice the highest frequency expected in the phenomena being observed. This is the Nyquist criterion which is observed when sampling or digitizing data. Under-sampling can result in aliasing, which makes high-frequency components indistinguishable from genuinely lower frequencies. To avoid this, the practice is to filter out frequencies that are above what the sampling system is capable of prior to sampling. However, this means a priori information has to exist on what the acceptable Nyquist rate should be in time and space. This can be a problem for episodic coastal events. To appreciate the under-sampling problem traditional ocean observation techniques are discussed next.

Traditional ocean observations used ships, fixed observatories, and satellites. Ship surveys and fixed observatories, such as moored arrays, yield observations of the subsurface ocean structure and variability. Moored arrays can sample for several months at high resolutions. However, these operations are expensive and it is not feasible to deploy enough moored arrays to resolve the short spatial scales of the coastal ocean, for example. Ship surveys, on the other hand, provide better spatial coverage and resolution but usually last only weeks because of expense and weather limitations. Remote observations through satellites, airborne LIDAR (laser imaging detection and ranging), and shore-based high-frequency radar yield surface characteristics only (since EM waves do not penetrate far into the water, Sect. 6.1). It is more difficult to collect data on the subsurface structure which is not as well understood. In the end, correlated data in space and time, at the required resolutions, above and below the water surface are required.

Therefore, an ocean observation capability that merges the high spatial resolution of ship surveys with the endurance and temporal resolution of moored arrays is desired. Some of these capabilities exist in marine robots. In some cases, UUVs are economically competitive against traditional ocean observation systems such as ships. The initial cost of an underwater glider is comparable to a few days at sea on a scientific ship [60]. As with the military motivation for employing UUVs, other advantages that marine robots bring to ocean observation have to be considered as well.

5.2 *Marine Robots Applied to Ocean Observation*

Marine robots can scale to the phenomena of interest and be deployed to remote locations (e.g., under ice). They can support scientific studies on climate change, which requires long-term sustained observations on a global scale. At the other extreme, they can support short-term observations in a local area for transient and rapidly evolving phenomena such as the evolution of harmful algae blooms in coastal ocean environments [61]. Used as networks [62], marine robots provide spatial and temporal coverage to complement data obtained from the more traditional ocean observation systems mentioned earlier. For example, fleets of gliders can sample over *large areas* for *long times*. To complement this, UUVs can sample smaller areas with a larger suite of instrumentation. The data obtained by both could be used to complement satellite surface measurements of ocean properties [63].

Over the last decade marine robots have transformed ocean observation [64,65]—especially in physical oceanography [66]. They can support diverse suites of advanced sensors to resolve interacting physical, chemical, biological, and geological (sedimentary) phenomena. Examples of such data collection include conductivity, temperature, density, salinity, methane content, nitrite content, sound speed, current velocity, oxygen content, images of seafloor bathymetry, mass spectrometry of ocean water constituents, and particle (suspended sediment) concentration.

Two oceanographic disciplines where marine robots can support scientific research are briefly discussed next to illustrate autonomy requirements. These two disciplines differ in their temporal and spatial scales.

5.2.1 Observations for Ocean Role in Climate

The ocean is vital in atmospheric and Earth habitability and is an important source of accessible carbon [65]. Temperature gradients across the lower atmosphere, sea surface, and subsurface have ramifications on global heat and moisture exchanges. Marine robots can provide persistent and systematic measurements of circulation and properties' distributions. UUVs such as gliders are well suited to cover the required 200 km (and larger) spatial scales and seven days (and longer) timescales.

UUVs can also observe deep and intermediate water masses that form during high winds and strong heat flux transfers from the ocean to the lower atmosphere. They can persistently observe circulation and property distributions of the remote oceans as a function of depth during all weather conditions to determine their spatial scales. UUVs can be equipped to determine the along-current extent of circulation patterns that are not readily apparent from ship-collected data. This longitudinal span of zonal current systems is not monitored with ship studies because of the mesoscale variability. As well, these current systems can be sometimes too remote for ship monitoring.

With ocean research into climate, marine robots measure the vertical structure of the ocean, complement satellite measurements, and help validate of models that infer sea-surface temperature from satellite altimeters.

5.2.2 Coastal Ocean Observation

The coastal ocean environment has complex spatial patterns varying over a range of timescales [65]. Marine robots can observe coastal ocean processes that combine to create short spatial (approximately 1 km) and timescale (less than a day) variability, as well as chemical and biological processes that add to the complexity. These processes include forcing by wind and buoyancy (i.e., rivers), variations in bathymetry, mixing from large tides, and the presence of weather fronts. Capturing this variability is vital.

Insight into along-the-continental shelf transport and exchange processes between coastal and deep ocean waters is needed for search and rescue operations, addressing environmental hazards like spills, and monitoring coastal ecosystems that require transport pathway knowledge. It is challenging to measure the correct spatial and temporal variability and consequently this has limited progress in the coastal ocean using traditional methods. Marine robots have advanced the understanding of coastal ocean transport and exchange processes. They yield more data per unit cost than ships or moored systems at the required spatial and temporal resolution. At the same time they are also less affected by weather or ship availability.

Based on the wide-ranging requirements for ocean observation, marine robots can well complement existing technologies with their ability to perform adaptive sampling.

5.3 Intelligent Autonomy Example: Adaptive Sampling

Marine robots use adaptive sampling techniques to distinguish between spatial and temporal variability. Such adaptive sampling capabilities use intelligent autonomy to implement.

Adaptive sampling is the ability to autonomously determine optimal routes, locations, and times (or frequencies) for oceanographic data collection based on in situ sensor measurements and analysis of the ocean process of interest [67–71]. Various measures can be used to quantify how well this is being achieved based on the spatial and temporal scales involved. As an example, a broad-area survey may collect data to minimize estimation errors of the process, whereas a small-scale coastal phenomenon may collect data in and around features of interest at positions with the largest gradients. Some strategies also optimize the adaptive sampling with consideration for the marine robot limitations (speed, endurance, depth, sensor resolution, etc.) as well. The concept is to devise a strategy to direct the robot to locations that most likely yield information about the sensed quantity. Examples of quantities sensed in this manner include currents, temperature [72,73], salinity, dissolved oxygen, fronts [74], chemical plumes [75], algae blooms [76], hydrocarbons [77,78], carp incursions [79], marine habitats [80], and marine mammals [81]. These strategies are generally deliberative in nature. Adaptive sampling has also found applications in military missions for rapid environmental assessment [82].

In the time it takes to perform a UUV survey the ocean quantity being measured may have changed. Consequently, it is natural to consider deploying multiple UUVs [83–86] as an option to sample the sensed quantity to the required temporal and spatial resolutions. This is covered in a subsequent chapter on cooperation between marine robots (Redfield, Chap. 6).

Given the background and example, the reader can better appreciate autonomy for oceanographic research applications. As in the military example developed earlier, we will next explore the extent to which intelligent autonomy enables select autonomy capabilities and where these capabilities are limited by factors beyond intelligent autonomy.

5.4 Autonomy for Oceanographic Research

There is a scientific need for measurements to correlate in space and time as well as above and below the water surface (i.e., across multiple environments). The value of a deployment is improved if the intelligent autonomy directs the UUV—based on in situ measured data and processed information—to the strategic area to collect its data. Below is an evaluation against the autonomy capabilities from the military list discussed earlier:

- *Reach and transit long distances*
  Unlike the military case, there is no requirement for covert deployments. There is certainly a preference to deploy UUVs from shore stations for coastal ocean measurements to minimize ship use. Similar to the military mission, it is an advantage if the ship were at a greater range from the UUV. Here, the reason is so the ship (with its radiated acoustic and magnetic noise) does not interfere

with the phenomenon being studied. Greater mission endurance is necessary to capture even a fraction of the mesoscale spatial variability (on the order of 50–100 km).

- *Persistence*
  This is desired to achieve long sampling times to adequately capture the environment. The mesoscale variability for global ocean phenomenon is for timescales that are on the order of 10–100 days which is longer than the typical MCM missions.
  As in the military case, reach and persistence have greater reliance on the initial three to four parts of the autonomy architecture. As in the military case this is limited by access to high energy density power sources. The high energy density needs are quite daunting given the mesoscale variations.

- *Collect data independent of human operator*
  As in the military case, if the desired reach and persistence are achievable, it likely brings the UUV beyond the range of standard underwater acoustic communications. The ability to make decisions, re-plan missions, sense and perceive, process data into information, and mitigate UUV system issues supports the reach and persistence capabilities. Given the mesoscale spatial requirements and generally less than optimal propagation, the UUV will easily go beyond the range for acoustic communications. As in the military case, this is limited by the sophistication of the initial three (and perhaps the fourth) parts of the autonomy architecture.

- *Operate and coordinate multiple vehicles for force multiplication*
  Multiple UUVs may allow mesoscale phenomenon to be sampled at the required resolution. Cooperation of multiple UUVs requires autonomy for fleets of UUVs to coordinate, reorganize, and re-plan amongst themselves, given intermittent connectivity with an operator, in response to unexpected changes in members of the fleet, the environment, the phenomenon, or mission objectives. As an example, the deployment of gliders to sample spatially large phenomena [87]. As in the military case, this capability relies on all five parts of the autonomy architecture and is limited by the difficult underwater communication channel which constrains the information flow rates and hence the level of cooperation possible.

- *Send back high-resolution information given poor bandwidth and high losses*
  This refers to the need for human imperative decision-making which may be useful for some scientific work. There is no requirement for covertness, but as in the military case the difficult underwater communications channel is the most significant limitation.

- *Track moving objects, for example, tiger sharks, baleen whales, chemical plumes, and thermoclines*
  This does not have to be performed at large stand-offs from the support ship but often is so that ship noise does not interfere with the object being tracked. As in the military case this is limited by the sophistication of all five parts of the autonomy architecture, access to high energy density power sources, and the difficult underwater communications channel.
- *Detect, evaluate, and avoid potential threats*
  This is not a similar requirement to the military case. Obstacle avoidance of submerged objects (e.g., shoals) when submerged, and ships while transmitting on the surface, is the most pressing requirement here. This is limited by the sophistication of the first two parts of the autonomy architecture and potentially the difficult underwater communications if high-bandwidth information must be relayed back for timely follow-on actions.
- *Provide covert and non-covert communications*
  Covert communications is not a requirement as in the military case. However, the sensitivity of marine mammals to potential UUV transmissions is a consideration. As in the military case, the UUV's status may have to be relayed back to the ship if it cannot reconfigure itself to cope with a problem (e.g., broken mission critical sensor). This is limited by, as in the military case, the difficult underwater communications channel.
- *Operate given denied and/or degraded GPS*
  GPS will generally be accessible on the surface for research purposes (i.e., not denied). As in the military case, it can take time (occasionally, up to 15 min) for the UUV to obtain a GPS calibration as conditions can be degraded due to lack of satellites or weather.

In summary, oceanographic research does not have strong requirements for:

- Covert deployment
- Covert communications
- Detecting, evaluating, and avoiding potential threats (beyond obstacle avoidance)
- Expected operation with denied GPS capabilities

The oceanographic research requirements for autonomy also include:

- *Opportunistic measurements*
  These capabilities require advanced deliberation in their planning and execution. The research performed at the Monterrey Bay Aquarium Research Institute (Rajan et al., Chap. 3) and the NASA Jet Propulsion Laboratory [57] are good examples.

- *Intelligent tasking of scientific sensors* (e.g., mass spectrometers, magnetometers) The Woods Hole Oceanographic Institute has worked extensively on integrating mass spectrometers on-board UUVs for oceanographic purposes, for example, to determine chemical constituents and to track chemical plumes. The ability to task these sensors to track plumes or specific tracers to their source requires intelligent autonomy.
- *Robust monitoring & diagnosis and planning & decision parts in the architecture* This is to provide fail safes within the intelligent autonomy architecture when experimental autonomy capabilities are being prototyped.
- *Capturing mesoscale variability* If only UUVs are used, this puts a significant requirement on high energy density power sources.

There is still a real need for research ships as platforms for oceanographic research. The emergence of unmanned systems does not entirely displace ships or fixed observation stations. Research ship designs now include support for marine robots [64]. Until there are transformational changes in the energy density of on-board power sources, surface ships are needed to augment marine robots for ocean observation. To date, the investigation of the ocean by UUVs is confined only to the upper ocean.

There is commonality in the autonomy desired by military and scientific users albeit for different reasons. An intelligent autonomy architecture implemented with good processor and bus speeds will not provide all the autonomy desired for military or scientific applications. This is because an intelligent autonomy architecture cannot change the physics of the underwater communications channel or the energy to support long missions—the limitations to marine autonomy. At best, the intelligent autonomy architecture provides partial solutions to deal with these limitations.

6 Limitations to Marine Autonomy

Limitations to marine autonomy arise from the properties of the underwater communications channel, underwater navigation, and access to high energy density sources. The first two are a consequence of the UUV environment and will be discussed next.

The amount of autonomy in a robot varies inversely with the communications bandwidth needed—with lower levels (e.g., teleoperation) requiring the most and higher levels (e.g., complete autonomous control) requiring the least. A compelling argument for marine robot autonomy comes from the underwater environment where high bandwidth and data rates as well as continuous communications with an operator is generally not possible.

The differences between the environments that unmanned vehicles operate in become apparent when the focus is on communications. For example, in-air communications between an unmanned aerial vehicle and a ground station (with eight-inch-diameter directional antennas) can be 10 Gbps with 1 W of radiated power. The most capable underwater communication system offers about 10 Kbps using acoustic communications with a high signature for detection [32].

6.1 Underwater Communications

Communication is needed between the UUV and support platform (e.g., ship) for transmission of control, data, information, and, occasionally, video. Control can include navigation, status, and commands for the UUV. Transmitted data is collected from, for example, hydrophones, current meters, and seismometers as well as sonars. The available bandwidth, range (between source and receiver), data rate, deployed depth, network infrastructure, and detectability are parameters that affect the communication channel's effectiveness. Detectability becomes a consideration when communications is desired without revealing the sender or receiver to possible interception. Additional communication challenges are associated with multiple UUVs operating cooperatively.

Underwater communication involves the transmission of control, data, information, etc. in the form of electromagnetic (EM), free-space optical (FSO), or acoustic (pressure) waves.

6.1.1 Electromagnetic Propagation Underwater

Electromagnetic (EM) waves in the radio frequency part of the spectrum (tens to hundreds of MHz) do not propagate far underwater because of their high absorption, $a = \sqrt{\pi f \mu \sigma}$ (f = frequency, μ = magnetic permeability, and σ = electrical conductivity), in water. The high electrical conductivity of sea water from dissolved salts makes its absorption two orders of magnitude higher than in freshwater. The propagation speed of EM waves in sea water is given as [88]

$$c \approx \sqrt{\frac{4\pi f}{\mu\sigma}}.$$

In highly conductive media like sea water the absorption and propagation speed of EM waves are proportional to the square root of frequency; consequently, lower frequencies travel further.

FSO waves for underwater communications are limited to short ranges as they also suffer high absorption in water and, additionally, are susceptibility to backscatter from suspensions [89]. Water free of suspensions attenuates optical frequencies 1,000 times more than air with no suspensions. However, underwater, blue-green wavelength FSO waves offer options for high-bandwidth communication (10–150 Mbps) over moderate ranges (10–100 m). FSO waves are less effective for communications in the higher levels of ambient light at shallow water depths.

For the ranges where EM and FSO waves propagate well, they have the advantage of high propagation speeds. The speed of light in water (3×10^8 m/s) is five orders of magnitude faster than acoustic waves (1.5×10^3 m/s). Additionally, the relatively higher frequencies of EM/FSO waves (compared to say acoustic frequencies) mean they have more available bandwidth. By definition, the bandwidth cannot exceed the highest transmitted frequency and usually, the bandwidth is a fraction of the highest frequency. Therefore, bandwidth increases with frequency. Consequently, EM and FSO waves are good options for communications between closely spaced nodes in a network.

Of the three (EM, FSO, and acoustic), acoustic communication is the best for underwater applications due to its relatively low absorption in water (particularly in thermally stable, deep water). Despite this, underwater acoustic propagation still has significant challenges to efficient communications. The highly variable and unpredictable spatial and temporal water properties means there is no universal propagation model for the underwater communications channel that captures performance across a wide range of underwater environments. For example, a system designed for shallow water does not work well in deep water.

Losses, multiple paths, high ambient noise, and surface scattering limit the underwater propagation of acoustic signals. Loss mechanisms include geometric spreading, absorption, and scattering. Multi-path is created by the presence of the upper and lower boundaries (sea floor and water surface, respectively) and the variable sound speed in water. The acoustic ambient comes from underwater mammals, noise from natural water movement, shipping, and industrial activities. Surface scattering adds to the multi-path problem.

6.1.2 Difficult Environment for Acoustic Propagation

As far as acoustic signal propagation is concerned, the ocean can be modelled as a waveguide with a sea surface and an ocean bottom. This waveguide, or channel, is characterized by a channel coherence time (a measure of how static the channel is) over which the channel impulse response is assumed constant. However, temporal changes in the channel occur because of the environment and motion of the transmitting or receiving platforms (like UUVs). Environmental variation can give rise to rapid temporal fluctuations in the channel making it a challenge to estimate the parameters of a fluctuating system (i.e., the channel impulse response) with a large number of independent parameters [90].

The underwater sound channel performance is characterized by low sound speed and loss mechanisms such as spreading, absorption, scattering, and multi-path.

1. *Low Sound Speed*

As mentioned earlier, the sound speed in water is low compared to the speed of light—five orders of magnitude less. The relatively slow sound speed in sea water, compared to EM waves, has important ramifications on underwater communications. First, underwater channel coherence times are on the order of 40 ms, so the link quality can change significantly in a short period. Thus, for source to receiver separations of more than about 100 m in such dynamic environments, channel state information from a receiver to a transmitter may be irrelevant before it is received and implemented at the transmitter. Another impact of this high channel latency is the penalty from message acknowledgments that require significant handshaking between source and receiver. Finally, the relatively slow sound speed creates noticeable Doppler shifts of received signals due to UUV motion or scattering off the moving sea surface. The Doppler shift, f_d, of a received signal is given by $f_d = f_o U/c$ where f_o is the original frequency of the signal and U is the rate of change of the propagation path length (i.e. UUV velocity). Thus, even at typical UUV speeds of $U = 2$ m/s and $f_o = 25$ kHz (typical acoustic carrier frequency), the Doppler shift is roughly 33 Hz [91]. These Doppler shifts result in a reduction in the coherence time or apparent increase in the rate of channel fluctuation. This complicates the problem of channel tracking at a receiver and further aggravates the problem discussed earlier regarding feedback of channel state information from a receiver to a transmitter.

2. *Losses*

Geometric spreading losses occur when the originally transmitted acoustic energy spreads out over a larger surface area as it propagates away from its source. At relatively short range, r, this expanding surface area is spherically shaped, so the acoustic energy decay varies as $1/r^2$. At larger ranges from the acoustic source, the energy transitions to cylindrical spreading due to the waveguide ocean model that reflects energy from the sea surface and ocean bottom. This transition occurs when the acoustic energy is at ranges greater than the local water depth from the source. Cylindrical spreading losses vary as $1/r$.

Absorption of underwater acoustic signals occurs when the acoustic energy is converted into heat due to the inelasticity of water. In sea water, the absorption loss of acoustic signals varies as $e^{-\alpha(f)R}$ such that $\alpha(f)$ is the absorption coefficient at frequency f and range R [92]. The absorption loss for pure water is $a(f) = A_3P_3f^2$ such that A_3 is a constant and P_3 is the pressure dependencies. At short range, spherical spreading is more significant than absorption losses. However, even at short ranges (e.g., 400 m) the absorption loss at 100 kHz exceeds that at 25 kHz by almost 15 dB. The frequency dependence of absorption loss means the underwater channel is band-limited and available bandwidth is a decreasing function of range (as well as transmission frequency). Consequently, UUVs operating at greater ranges from their support ship must have higher levels of autonomy. This characteristic impacts the choice of modulation and multi-access techniques as well as the problem of optimizing network topology.

Scattering is where non-uniformities in the water, such as suspensions and bubbles, cause acoustic waves to deviate from their normal trajectory. It also includes deviation of reflected radiation from the angle predicted by reflection under typical conditions. This is especially relevant to underwater channels. When the wind speed increases, the sea surface roughens and the effect of surface scattering becomes evident [93]. Surface scattering introduces not only power loss but also spreading in the delay of each surface bounce path.

The combined loss effects due to spreading, absorption, and scattering can be assembled into a total loss term [89] for the underwater acoustic channel. If the channel has P paths, and ξ_p denotes scattering loss, R_p propagated range, and τ_p the propagation delay of the pth path, then the loss along the pth path is$R_\beta^p \times e^{a(f)R_p} \times \xi_p$. For the channel coherence time interval, its transfer function at frequency f can be described as

$$H(f) = \sum_{P=0}^{P-1} \frac{1}{\sqrt{R_p \exp^{a(f)R_p \xi_p}}} e^{-j2\pi f \tau p}$$

such that $1 < \beta < 2$ captures how spherical or cylindrical the geometric spreading is. The channel attenuation depends on range and frequency. Since $\alpha(f)$ increases as f increases, high-frequency waves become attenuated at short range, while low-frequency acoustic waves propagate further. As a result, the use of lower frequencies for long range limits the available bandwidth while for short-range applications several tens of kHz of bandwidth is possible [94].

Therefore, underwater communications with acoustics is usually between 10 Hz and 1 MHz, with the usable frequency band depending on the transmission distance. Very high frequencies (>50 kHz) propagate only short ranges. For moderate ranges, 20–50 kHz is often used. Low-frequency waves (<10 kHz) are effective for very long-range communication on the order of tens of kilometers. The bandwidth for acoustic communications is typically kHz to tens of kHz, which is much less than that of radio frequency communications. How to utilize the limited bandwidth efficiently is a challenge.

3. *Multi-path*

As mentioned earlier, the ocean waveguide reflects acoustic waves from the sea surface and ocean bottom and, additionally, *refracts* acoustic waves due to its spatially varying sound speed. This creates multiple propagation paths (multi-path) for acoustic waves between the source and receiver. Such multi-path creates a delay spread (time between first and last arrival of a given signal) in the time-varying impulse response of the communications channel. Typically, delay spreads are 10 ms but, significantly, can be as high as 80–100 ms [95]. Large delay spreads cause time dispersion of the acoustic signal which leads to inter-symbol interference [90]. For example, a channel with a bandwidth of 4 kHz has a symbol interval of *1/B* or 0.25 ms. With a delay spread of 10 ms (from a 15 m difference in path lengths), each symbol affects the subsequent 10/0.25 = 40 symbols due to delay spread/time dispersion. For a delay spread of 100 ms, the subsequent 400 symbols suffer inter-symbol interference.

4. *Ambient Noise*

Underwater ambient noise is due to unidentified multiple sources typically, with no one source dominating [96]. Common sea-surface noise sources include the ship-radiated noise (on-board machinery, flow-induced noise, propeller cavitation), blasting, breaking waves along with consequent bubble production and collapse, wind, and marine animals (e.g., shrimp and whales). Moreover, surface ships that cross ocean basins can produce a general low-frequency background traffic noise that may not appear like shipping noise. The ambient noise can vary greatly depending on time of day, location, and water depth.

For short-range acoustic communication as might be seen in some military missions, the level of ambient noise may be well below the desired signal. For long-range or covert acoustic communication, the noise level may be a limiting factor for communication performance.

5. *Surface Scattering*

The scattering of acoustic energy by the moving sea surface presents a complexity to the underwater channel. The rough sea surface gives rise to channel delay spreads for each surface bounce path, reducing the spatial correlation of scattered signals, and potentially resulting in high intensity and rapidly fluctuating arrivals in the channel impulse response. For a flat calm surface, each surface scattered path results in an arrival in the impulse response that is fairly stable and localized in delay. In such cases, the impulse response of the channel is sparse. As the sea surface roughens, the arrivals not only fluctuate in time but also become spread in delay. This results in the need to track a more rapidly varying and less sparse impulse response for the channel.

The objective of underwater acoustic communication is to overcome the performance limitations induced by the high loss and dispersive underwater channel while at the same time improve the bandwidth efficiency. Research in underwater communications has expanded over the last decade from point-to-point communications to include underwater networks as well [92]. However, improvements in underwater acoustic communications are unlikely (in the short term) to lead to transformational changes in the achievable range and bandwidth.

Autonomy does not imply that communications with an operator is not necessary or undesirable. For uploading data and human imperative decisions there is a requirement to communicate. Rather, autonomy looks at how the vehicle should cope when it needs to communicate but cannot. Techniques such as inference are considered to deal with communications drop-outs [97]. The difficult underwater communications channel puts emphasis on the autonomy architecture to adequately manage in order to fulfil a mission.

6.2 Underwater Navigation

This section briefly reviews some basic concepts behind current underwater navigation practices that are referenced in the subsequent chapters. In Table 1.1, typical navigation and payload sensors used on-board UUVs are summarized. The distinction between the two is sometimes blurred. For example, a forward-looking or bathymetric sonar sensor may be the primary mission sensor but could also be used for navigation. It could therefore be used to both sense and adapt, which is an intelligent autonomy function.

As an example to keep in mind, a UUV mission that requires accurate and precise navigation and positioning would be a large-area survey (or search) to acquire a map of the seabed and to revisit discovered targets. For such a mission the UUV has to navigate an unfamiliar area, create a map, and then be able to navigate back to previously discovered targets.

Navigation accuracy is the precision with which the UUV guides itself from one geo-referenced point to another. Positional accuracy is the error of how well the UUV localizes itself within a geo-referenced map. Accuracy is how well a measurement maps the actual value of the quantity (as affected by random noise) and precision refers to repeatability of a measurement (as affected by biases and other forces). Global navigation satellite systems such as the global positioning system (GPS) provide accurate and precise positioning (as well as time, altitude, and speed) for mobile water surface, terrestrial, and aerial robotic vehicles to navigate. With GPS augmentation, sub-metre positioning accuracy is theoretically available. GPS is accessed through receiving electromagnetic signals that are greater than 1 GHz which, as mentioned earlier, do not propagate far underwater. However, there is no analogous universal positioning system for underwater robots such as UUVs.

Consequently, underwater navigation relies on communications between UUVs and external references that know their geo-referenced positions (e.g., transponders, ships). Once range and bearing between the UUV and such a system is measured (through other instruments), the geo-referenced position of the UUV can be calculated. Despite the underwater acoustic propagation limitations, acoustic signals still propagate significantly better than EM or optical ones, so navigation sensors are built around acoustic transducers. The challenge addressed by these acoustic transducers is the accuracy of the UUV position over the course of a long mission.

Generally, there are three main underwater navigation methods: (1) inertial, (2) acoustic which includes transponders and modems, and (3) geophysical (SLAM/terrain-based navigation).

6.2.1 Dead Reckoning

Traditional dead-reckoned position estimates are based on the sum of the previous position and the current UUV *heading, velocity*, and time elapsed. Consequently, position errors are cumulative and grow unbounded with distance travelled.

Table 1.1 Typical sensors on-board UUVs

Device		Functionality	Sample applications	Strength	Weakness
Non-acoustic	Inertial measurement unit (IMU)	Measures accelerations in three orthogonal directions and three rotation rates in inertial frame	Dead reckon UUV position	Precise dead reckoning for short time periods	Unbounded growth in position error
	Magnetic compass	Digital 3-axis compass provides a global reference for orientation relative to Earth's magnetic field	Good heading reference	Uses only Earth's magnetic field and independent of depth or water conditions	Compass bias, proximity to ship UUV magnetic signature affects it
Sonars	Doppler velocity log	Determines UUV speed over seabed and altitude	UUV navigation	Good complement as an aid to IMU	
	Pencil beam scanning sonar/echo sounder	A narrow single sonar beam scans across the surface	Profiling; point measurements of depths	Accurately describes cross section of target area	Accuracy depends on the UUV attitude
	Multi-beam/bathymetric	Multiple beams that measure time of flight of returns (depths)	Create 3D images of the seabed for classification or mapping; underwater SLAM	Gathers echo sounding data more efficiently	Resolution inversely proportional to frequency; may not always capture the first ping
	Acoustic Doppler current profiler	Measures the current profile	Aiding INSs with UUV navigation; measuring current profile	Can produce an entire profile for the current	Resolution inversely proportional to frequency; may not always capture the correct return;

(continued)

Table 1.1 (continued)

Device	Functionality	Sample applications	Strength	Weakness
Sub-bottom profiler	Low-frequency echo sounders that penetrate into the seabed	Penetrate seabed to determine contents; look for buried targets		
Forward-looking sonar	Similar in principle to a multi-beam sonar only forward looking; beams may be mechanically steered to scan	Underwater navigation, target detection, obstacle avoidance		
Side scan	Multiple beams that measure intensity of returns to create a 2D image	Large-area 2D scans to search and detect targets within an area	Generally works at relatively high speeds (10kt) to give high area coverage	Resolution inversely proportional to range 1.8 MHz produces a 40 m range
Synthetic aperture sonar	Coherent processing of consecutive displaced returns to create a virtual array/antenna	Large-area 2D scans to search and detect targets within an area, etc.	Very high resolution (cms) and large swaths (100 s of metres); good target classification; wide dynamic range and high SNR allows detection of even weak targets; range resolution; multi-aspect images ($>30^{\circ}$)	Good only under tightly prescribed speed and path; works only at low speeds (3 knots)—function of pulse rate and sensor size; sensor position known to within $\lambda/8$; in shallow water multi-path and direct surface returns

To combat this UUVs dead reckon when submerged and surface for a GPS update to recalibrate their position when the position error (usually the cross-track one) grows beyond an acceptable threshold. Generally, the desire is to minimize such surfacing as it makes the mission inefficient. As well, autonomous operation in deep water, under ice, or for covert purposes requires the UUV to be submerged for as long as possible.

Current UUV navigation is based on combining three complementary methods (sensor suites): dead reckoning (DVL and INS) with absolute acoustic positioning (long baseline or ultrashort baseline) and a heading reference (fibre optic gyroscope, ring laser gyro, magnetic compass, etc.). Dead reckoning with a DVL aided by an INS and a heading reference gives precise, *relative* position estimates but whose uncertainty grows unbounded with distance travelled. Acoustic positioning alone provides a geo-referenced *absolute* position but is not precise and subject to outliers and dropouts due to the environment. In combining these complementary measurement types, the overall solution provides more optimal long-term geo-referenced accuracy [98,99]. Kalman and particle filters [100] are two state estimation techniques that aid the INS in reducing the position uncertainty and therefore increase the time between UUV surfacing for position fixes.

6.2.2 Aided Dead Reckoning

Aided dead reckoning with UUVs improves upon traditionally dead reckoning with a DVL, at the very least, and often includes an INS and maybe an acoustic Doppler current profiler (ADCP) as well.

1. *Doppler Velocity Log*

The DVL *uses acoustic measur*ements to do sea bottom (or a reference water layer) tracking in order to determine the UUV speed over ground. It determines the UUV surge, sway, and heave velocities by transmitting acoustic pulses and measuring their Doppler-shifted returns off the seabed. Therefore, DVLs must use the local sound speed in their calculation of the UUV velocity. Some DVLs have integral sensors (e.g., temperature and depth; temperature, depth, and salinity; conductivity, temperature, and density) to determine local sound speed while others use an externally provided reference sound speed (e.g., 1,500 m/s) and correct the UUV velocities in post-processing. These updated velocity vectors are integrated with an on-board heading reference (magnetic compass, fibre optic gyro, etc.) to provide estimates of UUV position.

DVLs can work down to 6,000 m depth. Their accuracy depends on the operating frequency with higher frequencies yielding better resolution (though at decreased range as explained earlier). DVL frequencies vary from 100 s of kHz (range hundreds of metres) to 1,000 s of kHz (range tens of metres). A DVL may also have an integral altimeter to measure UUV altitude above the seabed. Altimeters generally operate by measuring time of flight between its transmitted acoustic pulse and the first received echo (i.e., first return) which exceeds a specified signal threshold. This is accurate when the ambient is low and the water is free from bubbles or suspensions (not often the case). When the altimeter cannot reach the bottom due to large UUV altitudes (or attitudes, i.e., rolled or pitched to high angles) it returns a zero reading.

Some DVLs also have an integral ADCP. This is a sonar that measures water current velocities for a range of depths to produce a vertical profile of the current velocities. Suspensions in the water currents create returns from the ADCP transmission that are detected by the ADCP receiver. Later arriving returns are from particles at greater depths. The ADCP simultaneously senses in several directions to calculate Doppler-shifted returns to distinguish between receding and approaching particles. ADCP ranges vary from fractions of a metre to a thousand metres and inversely so with frequency. Whether a DVL has an integral CTD (conductivity, temperature, and density) sensor, altimeter, and ADCP depends on its cost. If the UUVs possess all three and they are mounted to the UUV, the DVL can determine water current, UUV velocity, and altitude above the seabed—all of which contribute to the on-board UUV navigation. The DVL is often aided by an INS.

2. *Inertial Navigation Systems*

An INS calculates position, velocity, and attitude using high-frequency measurements from an inertial measurement unit (IMU). An IMU usually consists of three accelerometers to measure accelerations due to surge (back/forth), sway (left/right), and heave (up/down) and three gyros to measure angular rate relative to inertial space. Position and velocity result from integrating the accelerations against a heading reference. In this manner, the UUV continuously dead reckons its position, velocity vector, attitude, and true heading without external references. However, the INS error increases unbounded with elapsed time (or distance travelled) due to the inherent bias errors (e.g., drift) of the accelerometers and gyros in the IMU. This error growth leads to large UUV position errors. The accuracy of an INS is quantified by the standard deviation of its positional error growth in free inertial (unaided) performance mode. The standard deviation for navigational grade INSs is typically about 1 nautical mile per hour. This accuracy strongly depends on whether the INS uses fibre optic gyros, ring laser gyros, inclinometers, etc. A free unaided INS will, after a short period of time, have unacceptable position errors and require surfacing for a position fix. State estimation techniques, such as Kalman filters, are used to provide better navigation solutions for UUVs that use INS and allow them to stay submerged longer. These filters leverage the complementary nature of dead reckoning (with a heading reference) and acoustic positioning. The filter can use absolute positioning from the acoustic positioning to constrain the unbounded uncertainty of dead reckoning. By using information from both navigation methods the filtered solution combines the strengths of all the methods.

3. *State Estimation Techniques*

The objective of state estimation is to estimate the states of a system, x, given inputs, u, and measurements, z. The Bayes filter is the most general state estimation algorithm (Algorithm 1). The input is the belief distribution of states x at time t-

1, bel(x_{t-1}). The algorithm consists of a predict then update cycle. The prediction state (step 2 of Algorithm 1) estimates the probability distribution of the current state based on a prediction model $p(x_t|u_t, x_{t-1})$. The update is performed in step 3 with update model $p(z_t| x_t)$ where ν is a normalization constant. The Bayes filter is optimal; however, step 2 is intractable. The Kalman filter and its variants approximate the Bayes algorithm.

Algorithm 1: Bayes filter **Require**: (bel(x_{t-1}), u_t, z_t)

1: **for** states in x_{t-1} **do**
2: $\overline{\mathbf{bel}}(x_t) = \int p(x_t \mid u_t, x_{t-1})\mathbf{bel}(x_{t-1})dx_{t-1}$
3: $\mathbf{bel}(x_t) = \nu p(z_t \mid x_t)\overline{\mathbf{bel}}(x_t)$
4: **end for**
5: **return** (bel(x_t))

The Kalman filter is the optimal Bayesian estimator of the state given the system is linear, error distributions are additive zero mean white and Gaussian, and the system satisfies the Markov assumption (current state depends only on previous state). Each state distribution bel(x) is assumed Gaussian and therefore fully parameterized by its mean, μ, and its covariance matrix, Σ. Consider process (μ) and observation (z) models of a linear system given by

$$\begin{aligned} \mu_t &= A\mu_{t-1} + Bu_t + w_t \\ z_t &= C\mu_k + \upsilon_k \end{aligned},$$

where w_t $N(0,Q_t)$ and υ_t $N(0, R_t)$ are the normal process and measurement noise with covariances Qt and Rt, respectively. The Kalman filter algorithm is outlined in Algorithm 2.

Algorithm 2: Kalman filter **Require**: (μ_{t-1}, Σ_{t-1}, u_t, z_t)

$$\overline{\mu}_t = A_t\mu_{t-1} + B_t u_t$$

$$\overline{\Sigma}_t = A_t\Sigma_{t-1}A_t^T + Q_t$$

$$K_t = \overline{\Sigma}_t C_t^T (C_t\overline{\Sigma}C_t^T + R_t)^{-1}$$

$$\mu_t = \overline{\mu}_t + K_t(y_t - C_t\overline{\mu}_t)$$

$$\Sigma_t = (I - K_tC_t)\overline{\Sigma}_t$$

return(μ_t, Σ_t)

Where the system is nonlinear the extended Kalman filter (EKF) linearizes the system about the mean with a first-order Taylor approximation. Consider process and observation models *f* and *h*, respectively:

$$\mu_t = f(\mu_{t-1}, u_t, w_t)$$

$$z_k = h(\mu_k, \upsilon_k)$$

such that w_t $N(0, Q_t)$ and υ_t $N(0, R_t)$ are normal process and measurement noise with covariances Q_t and R_t, respectively. The EKF algorithm is shown in Algorithm 3.

Algorithm 3: Extended Kalman filter **Require**: $(\mu_{t-1}, \Sigma_{t-1}, u_t, z_t)$

$$\overline{\mu}_t = f(\mu_{t-1}, u_t, 0)$$

$$\overline{\Sigma}_t = A_t \Sigma_{t-1} A_t^T + W_t Q_{t-1} W_t^T$$

$$K_t = \overline{\Sigma}_t H_t^T (H_t \overline{\Sigma}_t H_t^T + V_t R_t V_t^T)^{-1}$$

$$\mu_t = \overline{\mu}_t + K_t (z_t - h(\overline{\mu}_t, 0))$$

$$\Sigma_t = (I - K_t H_t) \overline{\Sigma}_t$$

return(μ_t, Σ_t)
such that

$$A_t = \frac{\partial f(\mu_{t-1}, u_t, 0)}{\partial \mu_t}$$

$$W_t = \frac{\partial f(\mu_t, u_t, 0)}{\partial \mu_t}$$

$$H_k = \frac{\partial h(\overline{\mu}_t, 0)}{\partial \overline{\mu}_t}$$

$$V_k = \frac{\partial h(\overline{\mu}_t, 0)}{\partial \upsilon_k}$$

In short, the EKF uses measurements that are observed over time and contain random noise and other inaccuracies (e.g., IMU drift and biases) to yield values that are closer to the true measurement values and quantities derived from these measurement values. It achieves this by producing *estimates* of the true measurement values and quantities derived from these values. The EKF makes a prediction, estimates its uncertainty, and computes a weighted average of the predicted value and the measured value. The solution is weighted to favour the value with the least uncertainty. In this way, the estimates are closer to the true values than the measurements since the weighted average has a better estimated uncertainty than either of the values that went into the weighted average [101–103].

The EKF based on an error-state model provides better navigation performance than independent navigation sensors. As a result, the growth of the UUV position error is much slower, so the UUV can remain submerged for longer periods and surface less often for a GPS calibration.

6.2.3 Acoustic Navigation

1. *Positioning Systems*

At the heart of an acoustic positioning system is the ability to precisely measure the distance between two acoustic nodes. An underwater acoustic positioning system provides the UUV with the acoustic range and bearing to acoustically triangulate its position relative to a ship or beacon that knows its geo-referenced position. This can be in the form of a long baseline (LBL), short baseline (SBL), or ultra-short baseline (USBL) (also referred to as super short baseline, SSBL) [104,105].

With LBL navigation the UUV(s) operate in a network of seabed-deployed baseline transponders as widely spaced known reference points. LBL yields high position accuracy and position stability independent of water depth. The position is determined by measuring the slant ranges from three, or more, widely spaced transponders. These systems operate at frequencies around 10 kHz with ranges on the order of 5–10 km and resolutions that vary from a fraction of a metre to several metres. The round trip or two-way travel time (TWTT) is used to determine the UUV range from the deployed transponders at known locations. This has limitations including the change in the UUV in the time it takes for the acoustic signal to make a round trip. Recently, there has been interest in one-way travel time (OWTT) implementations of LBL [106]. Other LBL limitations include its cost and the time associated with setting up a network of transponders.

USBL or SBL is similar in principle to LBL except that the transponders are much closer together, often mounted on a ship or platform. When a surface reference, such as a ship, is available, USBL or SBL positioning can be used to calculate the UUV position relative to the ship's known GPS position from measured acoustic ranges and bearings. With USBL the UUV transceiver transmits an acoustic pulse to the shipboard transponder. The ship replies with its acoustic pulse which is detected by the UUV transceiver. The time from the transmission of the UUV's initial acoustic pulse to the detected reply is measured and used to determine a range (TWTT). The relative phases of the acoustic signals received are measured by closely spaced elements in a transducer (hydrophone). The angle between the UUV and the ship is measured through a transducer array on the UUV. Update rates are typically 0.5–10 s for ranges of 500–2,000 m or better with accuracies to as good as fractions of a metre in range and a few degrees in bearing.

SBL offers improved positioning over USBL through a larger baseline, achieved with larger transducer separations made possible by a larger ship. The position is determined through the relative arrival times at three (or more) UUV-mounted hydrophones.

Table 1.2 Comparison of common acoustic positioning systems to aid UUV navigation [107]

Acoustic positioning method	Advantages	Disadvantages
Long baseline (LBL)	■Highest potential accuracy ■Accuracy preserved over wider area	■Requires multiple subsea transponder ■Requires two-way ranging ■Long update intervals ■Acoustic ray bending
Short baseline (SBL)	■Good potential accuracy ■Requires single subsea transponder	■Accuracy depends on shipboard roll, pitch, and heave
Ultrashort baseline (USBL)	■Requires single subsea transponder	■High noise susceptibility ■Accuracy depends on ship roll, pitch, and heave

Acoustic positioning systems, unlike INS, have no cumulative error with time, but their update rates are low. As well they are susceptible to random noise, outliers, and dropouts. A comparison of the relative merits of the three methods is shown in Table **2** [107].

2. *Underwater Modems*

Progress in underwater acoustic communications has had impact on underwater navigation capabilities. The underwater modem allows simultaneous communication of small packets and ranging based on time of flight. If the transmitted position is included in the information communicated, then the receiver can bound its position to a sphere centred on the transmitter. This capability opens up many possibilities. This removes the need for the beacons to be fixed or localized prior to the mission. In addition, it allows for inter-UUV communication, which permits teams of UUVs to cooperatively navigate.

An acoustic modem used for navigation is the WHOI Micromodem for which there are two models. Micromodem-1 [108] uses frequency shift keying with frequency hopping (FH-FSK), or variable rate phase-coherent keying (PSK). Micromodem-2 [109] adds a number of innovations, one of which is a precise pulse-per-second clock. This addition importantly allows ranges to be calculated based on OWTT, which is faster and more accurate than TWTT ranging. Typically, due to the limited bandwidth underwater, the communication channel is shared using a time division multiple access (TDMA) scheme. In TDMA, each member in the group is allotted a time slot within a communication cycle with which to broadcast information. Unfortunately, the communication cycle increases with group size. The Micromodem is a viable option for deep oceanographic surveys, and has achieved 200 bits per second at depths of 11 km [110]. Comprehensive overviews of the WHOI Micromodem capabilities exist [111,112].

6.2.4 Geophysical Navigation

These methods, also referred to as terrain-based navigation or SLAM, use physical features within the environment to produce an estimate of the UUV location. This can be done either by giving the UUV a map or by building the map over the course of the mission. This has typically been performed with sonars (imaging or side scan) although there has been work using magnetic and optical sensors. The features can be existing or deployed specifically for navigation purposes. This method was briefly discussed earlier and will be covered in detail in Chap. 8.

Examples of UUV systems that navigate well with aided dead reckoning include [113]. Some use a fusion of several sensors and techniques mentioned above [114]. Several good review papers exist that cover developments in underwater navigation [115–117] as well as in the following chapters. The challenge of intelligent autonomy is to give the best combination of navigation techniques (based on affordability and availability) to achieve the required UUV positioning accuracy and precision, keep the UUV submerged for long intervals with much less reliance on external acoustic positioning systems, re-plan missions when the UUV discovers it is off course, re-plan a mission to compensate for detected currents, etc. If the underwater communications and navigation were not challenges, then the on-board energy would limit what intelligent autonomy can achieve.

6.3 Energy

UUV endurance can range from hours to months depending on the energy source and vehicle form, speed, and on-board equipment. Small UUVs are incapable of long endurance missions and thus cannot be far from an operator and/or ship. They are also less likely to be considered for jetty-launched missions for long periods. Consequently, they do not need much autonomy (unless transformational high energy density batteries appear). However, small UUVs are good test beds for developing intelligent autonomy for larger UUVs. Very long endurance missions on the order of hundreds of kilometres (many days), such as those in the Canadian UNCLOS mission [118], can only be achieved with large UUVs. Larger UUVs have extended endurance, so they are generally the ones with a requirement for intelligent autonomy.

On-board power is used for UUV propulsion, vehicle equipment, mission or payload sensors, and on-board processing. Vehicle equipment can include control planes, buoyancy compensation systems, navigation equipment, modems and other communications equipment, fault monitoring sensors, vehicle computer, and payload computers. Mission sensors, for which the UUV is the platform, can include side-scan or bathymetric sonars, and magnetometers. (see Table **1**). Mission-related on-board processing, the control architecture (planning, decision-making, fault tolerance, middleware, etc.), and the ATR reside on the payload computer (Sect. 1). To estimate the total power required, the power for propulsion, vehicle equipment, mission sensors, and payload computer must be included [119]. The following is a description of the energy to complete a mission for a survey of range, R, and speed, U:

$$E = \frac{(p_p + p_v + p_{ms} + p_{pc}) \times R}{3600 \times U} \ (kWh)$$

such that $E =$ on-board energy (kWh)

$p_p =$ propulsion power (Watts)
$p_v =$ vehicle equipment power (W)
$p_{ms} =$ mission sensor power (W)
$p_{pc} =$ payload computer (W)
$R =$ survey range (km)
$U =$ velocity (metres/second)

The propulsion power can be expressed as $p_p = F_D \times U$ such that drag force $F_D = \frac{1}{2}\rho U^2 S \times C_D$ (implies p_p U^3), C_D is the drag coefficient for the UUV form and S is the UUV surface area.

On long missions, most of the energy consumed is for propulsion. The on-board space allotted for energy, in the form of batteries, is finite, so the objective is to carry batteries of the highest *specific energy density* possible. Given a constant energy density source, a larger UUV will have more on-board energy than a smaller one. Current UUVs with endurances greater than 24 h have a dry weight of at least 1 Ton. Man-portable UUVs have endurances on the order of hours. This limitation could be overcome again, with transformational high density energy sources. *Generally, the size of the UUV correlates directly with the speed and endurance possible. High-endurance and high-speed UUVs have to be larger.* That is why intelligent autonomy is more important for larger UUVs.

The term p_{pc} for the payload computer is tied to the autonomy requirement. The executions of the control architectures and agents used in the multi-objective optimizers, planners, executives, and on-board processing of sensor data all demand a high duty cycle from the payload computer, which can impact the UUV endurance.

While smaller man-portable UUVs have payload computers, the energy consumption on smaller UUVs for the payload computer is disproportionately greater than on larger UUVs. This is not to say that small UUVs do not have payload autonomy; gliders are examples of UUVs that do [120]. However, gliders use variable buoyancy engines to achieve their propulsion [121], so all on-board energy can be dedicated to p_v, p_{ms}, and p_{pc} terms. For gliders, the p_p term is the smallest in the energy equation.

Effort has gone into investigating energy sources for UUVs [122–125], ranging from the common lead acid and lithium-based batteries to modern fuel cells [124,125]. Fuel cells are an emerging technology that will affect how UUVs are used for long missions and for accommodating intense on-board autonomy computations. See [37] for how UUV power and energy requirements can be met with electrochemical (e.g., lithium-based batteries [126]) or air-independent energy

Table 1.3 Comparison of energy sources for UUVs [127]

	Type	Energy density (Wh/dm^3)	Endurance(h)	Safety	Cost	Logistic/maintenance
Lead acid	Rechargeable	10–20	4–8	High	Low	Low
NiCd/NiMH	Rechargeable	10–30	4–12	High	Low	Low
Alkaline batteries	Primary	10–30	4–12	High	Low/high	Low
Ag Zn	Rechargeable	30–50	12–20	Medium	High	Medium
Lithium ion(D-cells)	Rechargeable	40–70	16–28	Medium	Medium	Low
Lithium polymer (poach)	Rechargeable	50–75	23–30	Medium	Medium	Low
Al O_2	Semi-fuel cell	80–90	32–36	Medium	Medium	High
Hydrogen-oxygen	Fuel cell	100+	40+	Low	Medium	High
Lithium batteries	Primary	100–150	40–60	Low	Low	Low

sources (fuel cells) [124,125]). Table **3** summarizes some common energy sources used on-board UUVs.

Longer-range and more autonomous missions will require higher energy density batteries to be developed. This autonomy requirement is directly tied to higher energy density batteries. Autonomy could also consider mission plans that are more energetically favourable (e.g., to re-acquire targets [5]) because the UUV is unexpectedly low in energy.

High energy density power sources could multiply the benefits of UUVs by conferring extended persistence and reach through superior endurance. UUVs would then be able to operate continuously over many days or weeks with powerful sensors and, at the same time, be just as capable with a smaller footprint. This would facilitate operations from vessels of opportunity. Thus on-board high energy density power sources are a critical enabler towards achieving autonomy on-board UUVs.

References

1. Roberts GN, Sutton R (eds) (2006) Advances in unmanned marine vehicles. Institution of Electrical Engineers, Michael Faraday House, Six Hills Way, Stevenage, UK
2. Brooks R, (1986) A robust layered control system for a mobile robot IEEE J Robotics and Automation, RA-2:14–23
3. Herman M, Hong T, Swetz S, Oskard D (1988) Planning and world modelling for autonomous undersea vehicles. In: Proceedings of 3rd international symposium on intelligent control, Arlington, Virginia
4. Seto ML (2010) An agent to optimally re-distribute control in an underactuated AUV. Int J Intell Defence Support Syst 4(1):3–19
5. Seto ML (2011) Autonomous mission-planning with energy constraints for AUVs with side scan sonars. In: Proceedings of IEEE international conference machine learning and applications
6. Hudson J (2011) Three-dimensional path-planning for a communications and navigations aid working cooperatively with autonomous underwater vehicles. In: Proceedings of autonomous intelligent systems conference, Povoa de Varzim, Portugal, p 12
7. Li H, Popa A, Thibault C, Seto M (2010) A software framework for multi-agent control of multiple AUVs for underwater MCM. In: Proceedings of IEEE autonomous intelligent systems conference, Povoa de Varzim, Portugal
8. Carlesi N, Fabien M, Jouvencel B, Ferber J (2011) Generic architecture for multi-AUV cooperation based on a multi-agent reactive organizational approach. In: Proceedings of 2011 IEE/RSJ international conference intelligent robots and systems, p 7
9. Benjamin M, Schmidt H, Newman PM, Leonard JJ (2010) Nested autonomy for unmanned marine vehicles with MOOs-IvP. J Field Robotics 27(6):834–875
10. Patron P, Miguelanez E, Cartwright J, Petillot YR (2008) Semantic knowledge-based representation for improving situational awareness in service oriented agents of autonomous underwater vehicles. In: Proceedings of 2008 MTS/IEEE oceans 2008 Conference, Quebec City, Canada
11. McGann C, Py F, Rajan K et al. (2008) A deliberative architecture for AUV control. In: Proceedings of International Conference on Robotics and Automation, CA, USA
12. Curcio J, Schneider T, Benjamin M, Patrikalakis A (2008) Autonomous surface craft provide flexibility to remote adaptive oceanographic sampling and modeling. In: Proceedings of MTS/IEEE Oceans 2008 Conference, Quebec City, Canada

13. Zaychik-Moffit V, Franke JL et al (2006) Mixed-initiative adjustable autonomy in multi-vehicle operations. In: Proceedings of 2006 AUVSI unmanned systems North America, US, Orlando
14. Havelund K, Lowry M, Penix J (1998) Formal analysis of a space craft controller using SPIN. In: Proceedings of 4th international SPIN workshop, Paris, France
15. Bernard DE, Gamble EB et al. (2000) Deep space 1 technology validation report – remote agent experiment. In: Technical Report NASA Ames
16. Jonsson A, Morris P, Muscettola N, Rajan K, Smith B (2000) Planning in interplanetary space: theory and practice. In: Proceedings of Conference on Artificial Intelligence Planning Systems 2000, pp. 177–186
17. Bresina J, Golden K, Smith DE, Washington R (1999) Increased flexibility and robustness of mars rover. In: Proceedings of 5th international symposium on ai, robotics, and automation in space, ESTEC, Noordwijk, Netherlands
18. Bresina J, Washington R (2001) Robustness via run-time adaptation of contingent plans. In: Proceedings of AAAI-2001 spring symposium: robust autonomy, Stanford, USA
19. Matthews W (2008) Computer takes over to land stricken aircraft. C4ISR – Magazine of Net-Centric Warfare
20. Blackburn M (2002) Autonomy and intelligence – a question of definitions. In: Proceedings of ONR-UCLA autonomous intelligent networks and systems symposium, Los Angeles, USA.
21. Barber KS, Martin CE (2000) Autonomy as decision-making control. In: Proceedings of 7th international workshop on agent theories, architectures, and languages, USA, Boston
22. Lin P, Bekey G Abney K (2008) Autonomous military robotics: risk, ethics, and design. Department of the Navy, Office of Naval Research (#N00014–07–1–1152 and N0001408–1–1209)
23. Standard Guide for Unmanned Undersea Vehicles (UUV) Autonomy and Control. ASTM F2541–06.
24. Carlesi N, Michel F, Jouvencel B, Ferber J (2011) Generic architecture for multi-auv cooperation based on a multi-agent reactive organizational approach. In: Proceedings international conference on intelligent robots and systems, pp 5041–5047
25. McGann C, Py F, Rajan K, Thomas H, Henthorn R, McEwen R (2008) A deliberative architecture for AUV control. In: Proceedings of 2008 IEEE international conference robotics and automation, pp 1049–1054
26. Tangirala S, Kumar R, Bhattacharyya S, O'Connor M, Holloway LE (2005) Hybrid-model based hierarchical mission control architecture for autonomous underwater vehicles. In: Proceedings of 2005 American control conference, p 6
27. Teck TY, Chitre M, Vadakkepat P (2010) Hierarchical agent-based command and control system for autonomous underwater vehicles. In: Proceedings of 2010 international conference autonomous and intelligent systems, p 6
28. Goldberg D (2011) Huxley: a flexible robot control architecture for autonomous underwater vehicles. In: Proceedings of IEEE oceans 2011-Spain Conference, p 10
29. Eickstedt DP, Sideleau SR (2010) The backseat control architecture for autonomous robotic vehicles: a case study with the IVER2 AUV. Mar Tech Soc J 44(4):42–54
30. ONR's Uninhabited Combat Air Vehicles Program (2000) Naval Studies Board, National Research Council, p 58
31. Huang H, Pavek K, Albus J, Messina E (2005) Autonomy levels for unmanned systems (ALFUS) framework: an update. In: 2005 SPIE Defense and Security Symposium, Florida, p 10
32. Naval Studies Board (2005) Autonomous Vehicles in Support of Naval Operations. The National Academies Press, Washington, DC, p 257
33. Krogstad TR, Wiig M, Cleophas P, Midtgaard O (2011) Autonomous object identification in mine countermeasure missions. In: Proc. 2011 oceans conference, p 9
34. Patron P Miguelanez E, Petillot YR, Lane DM, Salvi J (2008) Adaptive mission plan diagnosis and repair for fault recovery in autonomous underwater vehicles. In: Proceedings of oceans 2008 conference, p 9

35. Miguelanez E, Patron P, Brown KE, Petillot YR, Lane DM (2011) Semantic knowledge-based framework to improve the situation awareness of autonomous underwater vehicles. IEEE Trans Knowl Data Eng 23(5):759–773
36. The Navy unmanned surface vehicle (USV) master plan. Department of the Navy – USA, 2007
37. The Navy unmanned undersea vehicle (UUV) master plan. Department of the Navy – USA, 2004
38. Button RW Kamp J, Curtin TB, Dryden J (2009) A survey of missions for unmanned undersea vehicles. Sponsored by the US, RAND Corporation, Navy
39. Naval mine warfare: operational and technical challenges for naval forces. Committee for Mine Warfare Assessment, Naval Studies Board, National Research Council, 2001
40. von Alt C, Allen B, Austin T, Forrester N, Goldsborough R, Purcell M, Stokey R (2001) Hunting for mines with REMUS: a high performance, affordable, free swimming underwater robot. In: Proceedings of MTS/IEEE 2001 oceans conference and exhibition, vol 1, pp 117–122
41. Stokey R, Austin R, Allen B, Forrester N, Gifford E, Goldsborough R, Packard G, Purcell M, von Alt C (2001) Very shallow water mine countermeasures using the REMUS AUV: a practical approach yielding accurate results. In: Proceedings of MTS/IEEE 2001 oceans conference and exhibition, vol. 1, pp 149–156
42. Stokey R, Austin R, Allen B, Forrester N, Gifford E, Goldsborough R, Packard G, Purcell M, von Alt C (2001) Very shallow water mine countermeasures using the REMUS AUV: a practical approach yielding accurate results. In: Proceedings of MTS/IEEE 2001 oceans conference and exhibition, vol 1, pp 149–156
43. von Alt C (2003) REMUS 100 transportable mine countermeasure package. In: Proceedings of oceans 2003 conference, vol 4, pp 1925–1930
44. Giger G, Kandemir M, Dzielski J (2008) A graphical mission planning tool for use in mine counter measure operations. In: Proceedings of MTS/IEEE oceans 2008 conference, pp 1–11
45. Stokey RP, Roupe A, von Alt C, Allen B, Forrester N, Austin R, Goldsborough R, Purcell M, Jaffe M, Packard G, Kukulya A (2005) Developments of the REMUS 600 autonomous underwater vehicle. In: Proceedings of MTS/IEEE oceans 2005 conference, vol 2, pp 1301–1304
46. Fallon MF, Kaess M, Johannsson H, Leonard JJ (2011) Efficient AUV navigation fusing acoustic ranging and side-scan sonar. In: Proceedings of 2011 IEEE international conference robotics and automation, pp 2398–2405
47. Kukulya A, Plueddemann A, Austin R, Stokey R, Purcell M, Allen B, Littlefield R, Freitag L, Koski P, Gallimore E, Kemp J, Newhall K, Pietro J (2010) Under-ice operations with a REMUS-100 AUV in the Arctic. In: Proceedings of 2010 IEEE/OES autonomous under water vehicles, pp 1–8
48. Plueddemann AJ, Packard G, Lord J, Whelan S (2008) Observing Arctic coastal hydrography using the REMUS AUV. In: Proceedings of autonomous under water vehicle 2008 conference, p 4
49. Levine ER, Goodman L, Lubke C (2008) AUV based swell characterization. In: Proceedings of IEEE/OES autonomous under water vehicles conference, p 6
50. Goodman L, Levine ER (2010) Subsurface observations of surface waves from an autonomous underwater vehicle. IEEE J Oceanic Eng 35(4):779–784
51. McCarthy K (2003) REMUS - a role model for AUV technology transfer. Int Ocean Syst 7(6):22–23
52. Hagen P, Hegrenaes O, Jalving B, Midtgaard O, Wiig M, Hagen K (2008) Making AUVs truly autonomous. In: Inzartsev AV (ed) Underwater vehicles. I_Tech, Vienna, Austria, p 582
53. Carreno S, Wilson P, Ridao P, Petillot Y (2010) A survey on terrain based navigation. In: Proceedings of the MTS/IEEE oceans 2010 conference, p 7
54. Yang TC, Yang W (2008) Low probability of detection underwater acoustic communications for mobile platforms. In: Proceedings of oceans 2008 Conference, p 6
55. Munk W (2002) Oceanography, 15(1), pp. 135–141

56. Jones C, Creed E, Glenn S, Kerfoot J, Kohut J, Mugdal C, Schofield O (2005) Slocum glider – a component of operational oceanography. In: Symposium on unmanned untethered submersible technology, 21–24 Aug 2005, Durham, NH, USA
57. Huntsberger T, Woodward G (2011) Intelligent autonomy for unmanned surface and underwater vehicles. In: Proceedings of oceans 2011 conference, p 10
58. Robinson AR (ed) (1983) Eddies in marine science. Springer, Berlin
59. Munk WF, Worcester PF Ocean acoustic tomography: the sense of excitement and discovery remains,. Oceanography, pp 8–10
60. Schofield O, Kohut J, Aragon D, Creed L, Graver J, Haldeman C, Kerfoot J, Roarty H, Jones C, Webb D, Glenn SM (2007) Slocum gliders: robust and ready. J Field Robotics 24(6):1–14
61. Anderson D (2004) The growing problem of harmful algae. Oceanus 43(2):1–5
62. Cronin D, Landrum GW, Sharp K (2008) Application of autonomous underwater vehicle systems in distributed ocean observing networks. In: Proceedings of oceans 2008, p 3
63. Sandwell DT (1990) Geophysical applications of satellite altimetryReviews of Geophysics Supplement 132–137
64. Science at sea: meeting future oceanographic goals with a robust academic research fleet (2005) Committee on Evolution of the National Oceanographic Research Fleet, Ocean Studies Board, National Research Council of the National Academies, p 121
65. Rudnick DL, Perry MJ (eds) (2003) Autonomous and lagrangian platforms and sensors - workshop report, p 64. www.geo-prose.com/ALPS
66. Fifty Years of Ocean Discovery: National Science Foundation (1950:2000). Ocean Studies Board National Research Council, National Academy of Sciences, 2000
67. Yilmaz NK, Evangelinos C, Lermusiaux PFJ, Patrikalakis NM (2008) Path planning of autonomous underwater vehicles for adaptive sampling using mixed integer linear programming. IEEE J Oceanic Eng 33(4):522–537
68. Cannell CJ, Stilwell DJ, Austin JA (2004) A simulation tool to support the development of adaptive sampling algorithms for multiple autonomous underwater vehicles. In: Proceedings of 2004 IEEE/OES autonomous underwater vehicles conference, pp 127–133
69. Eickstedt DP, Benjamin MR, Wang D, Curcio J, Schmidt H (2007) Behavior based adaptive control for autonomous oceanographic sampling. In: Proceedings of 2007 IEEE international conference on robotics and automation, Rome, Italy, pp 4245–4250
70. Ho C, Mora A, Saripalli S (2012) An evaluation of sampling path strategies for an autonomous underwater vehicle. In: Proceedings of 2012 IEEE international conference on robotics and automation, p 6
71. Balasuriya A, Petillo S, Schmidt H, Benjamin M (2010) Behavior-based planning and prosecution architecture for autonomous underwater vehicles in ocean observatories. In: Proceedings of 2010 IEEE oceans 2010 conference, Sidney, Australia, p 5
72. Cruz NA, Matos AC (2010) Adaptive sampling of thermoclines with autonomous underwater vehicles. In: Proceedings of oceans 2010 Conference, p 6
73. Woithe HC, Ulrich K (2011) A lightweight scripting engine for the slocum glider. In: Proceedings of 2011 oceans Conference, p 7
74. Gottlieb J, Graham R, Maughan T, Py F, Elkaim G, Rajan K (2012) An experimental momentum-based front detection method for autonomous underwater vehicles. In: Proceedings of 2012 IEEE international conference robotics and automation, p 6
75. Farrell JA, Pang S, Li W (2005) Chemical plume tracing via an autonomous underwater vehicle. IEEE J Oceanic Eng 30(2):15
76. Zhang Y, McEwen RS, Ryan JP, Bellingham JG (2010) Design and tests of an adaptive triggering method for capturing peak samples in a thin phytoplankton layer by an autonomous underwater vehicle. IEEE J Oceanic Eng 35(4):12
77. Camilli R (2007) Characterizing marine hydrocarbons with in-situ mass spectrometry. In: Proceedings of oceans 2007 Conference, p 7

78. Jakuba MV, Steinberg D, Kinsey JC, Yoerger DR, Camilli R, Pizarro O, Williams SB (2011) Toward automatic classification of chemical sensor data from autonomous underwater vehicles. In: Proceedings of 2011 IEEE/RSJ international conference on intelligent robots and systems, p 6
79. Tokekar P, Bhadauria D, Studenski A, Isler V (2010) A robotic system for monitoring carp in Minnesota lakes. J Field Robotics, Special Issue on Maritime Autonomy Part 1 27(6): 779–789
80. Rigby P, Pizarro O, Williams SB (2010) Toward adaptive benthic habitat mapping using Gaussian process classification. J Field Robotics, Special Issue on Maritime Autonomy, Part. 1 27(6):741–758
81. Dassatti A, van der Schaar M, Guerrini P, Zaugg S, Houegnigan L, Maguer A, Andre M (2011) On-board underwater glider real-time acoustic environment sensing. In: Proceedings of oceans 2011, Spain, p 8
82. Wang D, Lermusiaux PFJ, Haley PJ, Eickstedt DP, Leslie WG, Schmidt H (2009) Acoustically focused adaptive sampling and on-board routing for marine rapid environmental assessment. J Marine Syst 78(S1):S393–S407
83. Heaney KD, Duda TF (2005) Non-linear optimization of multi-vehicle ocean sampling networks for cost-effective ocean prediction systems. In: Proceedings of oceans 2006, Asia-Pacific, p 5
84. Fiorelli E, Leonard NE, Bhatta P, Paley DA, Bachmayer R, Frantantoni DM (2006) Multi-AUV control and adaptive sampling. IEEE J Oceanic Eng, 31(4):14
85. Popa DO, Sanderson AC, Komerska RJ, Mupparapu SS, Blidberg DR, Chappel SG (2004) Adaptive sampling algorithms for multiple autonomous underwater vehicles. In Proceedings of 2004 IEEE/OES autonomous underwater vehicles conference, pp 108–118
86. Woithe HC, Kremer U (2011) Using slocum gliders for coordinated spatial sampling. In: Proceedings of 2011 oceans conference, Spain, p 6
87. Leonard NE, Paley DA, Davis RE, Fratantoni DM, Lekien F, Zhang F (2010) Coordinated control of an underwater glider fleet: An adaptive ocean sampling field experiment in Monterey Bay. J Field Robotics 27(6):718–740
88. Balanis CA (1973) Advanced engineering electromagnetics. John Wiley & Sons, New York
89. Liu L, Zhou S, Cui J (2008) Prospects and problems of wireless communication for underwater sensor networks. Wiley WCMC Special Issue on Underwater Sensor Networks, 8(8):877–994
90. Preisig J (2006) Acoustic propagation considerations for underwater acoustic communications network development. In: Proceedings of WUWnet 2006,p 5
91. Preisig J (2005) Performance analysis of adaptive equalization for coherent acoustic communications in the time-varying ocean environment. J Acoust Soc Am 118(1):263–278
92. Chitre M, Shahabudeen S, Stojanovic M (2008) Underwater acoustic communications and networking: recent advances and future challenges. J Marine Tech Soc, 42(1):103–116
93. Dahl PH (2004) Forward scattering from the sea surface and the van Cittert-Zernike theorem. J Acoust Soc Am 115(2):2067–2080
94. Stojanovic M (2006) On the relationship between capacity and distance in an underwater acoustic communication channel. In: Proceedings of first ACM international workshop on underwater networks/MobiCom conference, Los Angeles, CA
95. Kilfoyle D, Baggeroer AB (2000) The state of the art in underwater acoustic telemetry. IEEE J Oceanic Engineering 25(1): 4–27.
96. National Research Council (NRC) (2003) Ocean noise and marine mammals. National Academies Press, Washington, DC
97. Soltzing C, Lane DM (2010) Improving the coordination efficiency of limited communication multi-AUV operations using a multi-agent architecture. J Field Robotics 27(4):412–429
98. Nicosevici T, Garcia R, Carreras M, Villanueva M (2004) A review of sensor fusion techniques for underwater vehicle navigation. In: Proceedings of MTS/IEEE 2004 oceans conference, p 6

99. Whitcomb LL, Yoerger DR, Singh H (1999) Combined doppler/LBL based navigation of underwater vehicles. In: Proceedings of international symposium on unmanned untethered submersible technology, Durham, NH
100. Donovan GT (2011) Development and testing of a real-time terrain navigation method for AUVs. In: Proceedings of oceans 2011 conference, p 9
101. Thrun S, Burgard W, Fox D (2006) Probabilistic robotics. The MIT Press, Cambridge
102. Grewal MS, Weill LR, Andrews AP (2007) Global positioning systems, inertial navigation, and integration, Ch 8, 2nd edn. Wiley, New York
103. Brown RG, Hwang PYC (1996) Introduction to random signals and applied Kalman filtering with MATLAB exercises and solutions,Ch 5–7, 3rd edn. Wiley, New York
104. Philip DRC (2003) An evaluation of USBL and SBL acoustic systems and optimisation of methods of calibration - Part I. Hydrographic J 108:18–25
105. Tan HP, Diamant R, Seah WKG, Waldmeyer M (2011) A survey of techniques and challenges in underwater localization. J Ocean Eng 38(14–15):1663–1676
106. Crosbie B, Anderson M, Wolbrecht E, Canning J, Edwards D (2010) Synchronous navigation of AUVs using WHOI micro-modem 13-bit communications. In: Proceedings of oceans 2010 conference, p 5
107. Thomson D (2005) True redundant acoustic dynamic positioning sensors. In: Marine technology society dynamic positioning conference
108. Freitag L, Grund M, Singh S, Partan J, Koski P, Ball K (2005) The WHOI micro-modem: an acoustic communications and navigation system for multiple platforms. In: Proceedings of MTS/IEEE oceans 2005 conference, vol 2, pp 1086–1092
109. Gallimore E, Partan J, Vaughn I, Singh S, Shusta J, Freitag L (2010) The WHOI micromodem-2: a scalable system for acoustic communications and networking. In: Proceedings of oceans 2010 conference, pp 1–7
110. Singh S, Webster S, Freitag L, Whitcomb L, Ball K, Bailey J Taylor C (2009) Acoustic communication performance of the WHOI micro-modem in sea trials of the NEREUS vehicle to 11,000 m depth. In: Proceedings of MTS/IEEE oceans 2009 conference, Biloxi, USA, p 6
111. Singh S, Grund M, Bingham B, Eustice R, Singh H, Freitag L (2006) Underwater acoustic navigation with the WHOI micro-modem. In: Proceedings of oceans 2006 conference, p 4
112. Eustice R, Whitcomb L, Singh H, Grund H (2006) Recent advances in synchronous-clock one-way-travel-time acoustic navigation. In: Proceedings of oceans 2006 conference, p 6
113. Jalving B, Gade K, Hagen OK, Vestgard K (2003) A toolbox of aiding techniques for the HUGIN AUV integrated inertial navigation system. In: Proceedings of MTS/IEEE oceans 2003 conference, vol 2, pp 1146–11153
114. Miller PA, Farrell JA, Zhao Y, Djapic V (2010) Autonomous underwater vehicle navigation. J Oceanic Eng 35(2):663–678
115. Kinsey JC, Eustice RM, Whitcomb LL (2006) A survey of underwater vehicle navigation: recent advances and new challenges. In: Proceedings of IFAC conf. on Manoeuvering and control of marine craft, p 12
116. Stutters L, Liu H, Tiltman C, Brown DJ (2008) Navigation technologies for autonomous underwater vehicles. IEEE Trans Syst, Man and Cybernetics - Part C.: Appl Rev 38(4): 581–589
117. Bingham B (2008) Navigating autonomous underwater vehicles. In: Underwater Vehicles, Ed. by Alexander Inartzev, ISBN 978-953-7619-49-7, pp. 582, December 2008, I-Tech, Vienna, Austria
118. Kaminski C, Crees T, Ferguson J, Forrest A, Williams J, Hopkin D, Heard G (2010) Twelve days under-ice - an historic AUV deployment in the Canadian Arctic. In: Proceedings of IEEE/OES AUV 2012 conference, p 8
119. ISE - Build your Own AUV Design Info (2010)
120. Fernandez-Perdomo E, Cabrera-Gamez J, Hernandez-Sosa D, Isern-Gonz´alez J, Domınguez-Brito AC, Redondo A, Coca J, Ramos AG, Fanjul EA, Garcıa M (2010) Path planning for gliders using regional ocean models: application of Pinzon path planner with the ESEOAT model and the RU-27 trans-Atlantic flight data. In: Proceedings of oceans 2010, Sidney, Australia, p 10

121. Davis RE, Eriksen CC, and Jones CP (2002) Autonomous buoyancy-driven underwater gliders. In: Griffiths G (ed) Technology and applications of autonomous underwater vehicles. Taylor and Francis, London
122. Hawley JG, Reader GT (1992) A knowledge-based aid for the selection of autonomous underwater vehicle energy systems. In: Proceedings of 1992 symposium on autonomous underwater vehicle technology, pp 177–180
123. Abu Sharkh SM, Griffiths G, Webb AD (2003) Power sources for unmanned underwater vehicles. In: Griffiths G (ed) Technology and applications of autonomous underwater vehicle. Taylor & Francis, London, pp 19–35. ISBN 0–415–30154–8
124. Hasvold O, Johansen KH (2002) The alkaline aluminum hydrogen peroxide semi-fuel cell for the HUGIN 3000 autonomous underwater vehicle. In: Proceedings of 2002 workshop on autonomous underwater vehicles, pp 89–94
125. Yamamoto L, Aoki T, Tsukioka S, Yoshida H, Hyakudome T, Sawa T et al. (2004) Fuel cell system of AUV Urashima. In: Proceedings of MTS/IEEE oceans 2004 conference, pp 1732–1737
126. Tarascon JM (2010) Review - key challenges in future Li-Battery research. Phil Trans R Soc A 368:3226–3241
127. Storkersen N, Hasvold O (2004) Power sources for AUVs. In: Proceedings of science and defence conference, Brest, France, p 5

Chapter 2
Autonomy for Unmanned Marine Vehicles with MOOS-IvP

Michael R. Benjamin, Henrik Schmidt, Paul M. Newman, and John J. Leonard

Abstract This chapter describes the MOOS-IvP autonomy software for unmanned marine vehicles and its use in large-scale ocean sensing systems. MOOS-IvP is comprised of two open-source software projects. MOOS provides a core autonomy middleware capability and the MOOS project additionally provides a set of ubiquitous infrastructure utilities. The IvP Helm is the primary component of an additional set of capabilities implemented to form a full marine autonomy suite known as MOOS-IvP. This software and architecture are platform and mission agnostic and allow for a scalable nesting of unmanned vehicle nodes to form large-scale, long-endurance ocean sensing systems comprised of heterogeneous platform types with varying degrees of communications connectivity, bandwidth, and latency.

2.1 Introduction

The growing desire for autonomy in unmanned marine systems is driven by several trends, including increased complexity in mission objectives and duration, increased capability in on-board sensor processing and computing power, and an increase in the number of users and owners of unmanned vehicles. The MOOS-IvP project is an open-source project designed and developed in this context. It is an implementation of an autonomous helm and substantial support applications that aims to provide a capable autonomy system out of the box. It also has an architecture, software policy, documentation, and support network that allows this and the next generation

M.R. Benjamin (✉) • H. Schmidt • J.J. Leonard
Computer Science and Artificial Intelligence Laboratory, Massachusetts Institute of Technology, Cambridge, MA, USA
e-mail: benjamin@mit.edu; schmidt@mit.edu; leonard@mit.edu

P.M. Newman
Department of Engineering Science, University of Oxford, Oxford, UK
e-mail: pnewman@robots.ox.ac.uk

M.L. Seto (ed.), *Marine Robot Autonomy*, DOI 10.1007/978-1-4614-5659-9_2,

of scientists, with newer vehicles and mission ambitions, to be nimble to build innovative autonomy algorithms to augment an existing set of capabilities. This chapter describes the MOOS-IvP autonomy architecture and software structure.

MOOS-IvP is comprised of two open-source software projects. The mission oriented operating suite (MOOS) is a product of the Mobile Robotics Group at the University of Oxford and provides middleware capabilities in a publish-subscribe architecture as well as several applications ubiquitous in unmanned marine robotic and land robotic applications using MOOS. Additional MOOS applications, including the IvP Helm, are available in the MOOS-IvP project. IvP stands for interval programming and refers to the multi-objective optimization method used by the IvP Helm for arbitrating between competing behaviors in a behavior-based architecture.

The MOOS-IvP software is available on the web via anonymous read-only access [5]. It consists of more than 130,000 lines of C++, comprising about 30 distinct applications and over a dozen vehicle behaviors. It represents about 25 work years of effort or more from individual contributors. Autonomy configurations and missions in this environment have been tested in several thousands of hours of simulation and several hundred hours of in-water experiments on platforms including the Bluefin 21-inch UUV, the Bluefin 9-inch UUV, the Hydroid REMUS-100 and REMUS-600 UUVs, the Ocean Server Iver2 UUV, the Ocean Explorer 21-inch UUV, autonomous kayaks from Robotic Marine Systems and SARA Inc, the Kingfisher USV from Clearpath Robotics, and two USVs at the NATO Undersea Research Centre in La Spezia Italy, an unmanned RHIB from H-Scientific, and the USV-2600 from SeaRobotics.

2.1.1 Trends in Unmanned Marine Vehicles Relating to Autonomy

The algorithms and software described in this chapter have their genesis in unmanned underwater vehicles. Unlike unmanned sea-surface, ground and aerial vehicles, underwater vehicles cannot be remotely controlled; they *must* make decisions autonomously due to the low bandwidth in acoustic communications. Remote control, or teleoperation, in land, air, or surface vehicles may be viewed as a means to allow conservative, risk-averse operation with respect to the degree of autonomy afforded to the vehicle. In underwater vehicles, similar conservative tendencies are realized by scripting the vehicle missions to be as predictable as possible, comprised perhaps of a preplanned set of waypoints accompanied with depth and speed parameters. Sensors merely collected data to be analyzed after the vehicle was recovered from the water.

Advances in sensor technologies include greater capabilities, at lower cost, lower size, and lower power consumption. The same is true for the on-board computing components needed to process sensor data. Increasingly underwater

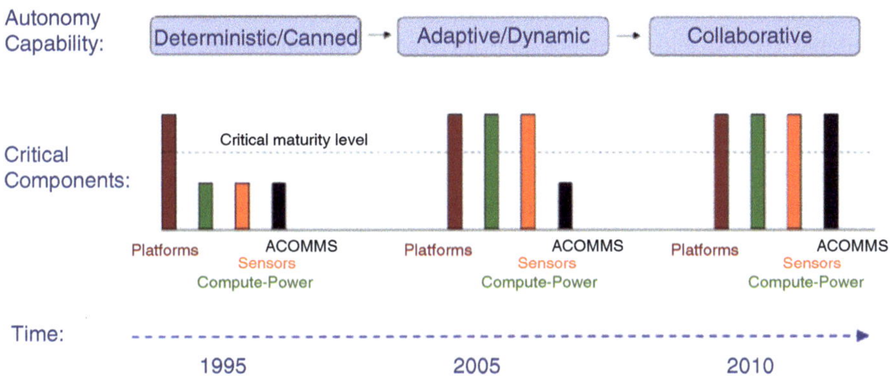

Fig. 2.1 Technologies and autonomy: A rough timeline and relationship between UUV autonomy and other critical UUV technologies. Critical components include (**a**) the platform itself in terms of reliability, cost, and endurance, (**b**) on-board computing power and sensor processing, (**c**) on-board sensors in terms of resolution, size, and cost, and (**d**) acoustic communications (ACOMMS). Each of these maturing technology trends affects what is possible and desired from the on-board autonomy system. The corresponding trend in autonomy is from deterministic vehicles acting independently, toward adaptive vehicles acting in collaboration

vehicles are able to see, hear, and localize objects and other vehicles in their environment and quickly analyze an array of qualities in water samples taken while underway. Likewise, the available mission duration at depth has grown longer due to improvements in inertial navigation systems, which have become cheaper, smaller, and more accurate and due to improvements in platform battery life. Each of these trends has contributed to making a UUV owner less satisfied with simply collecting the data and analyzing the results in a post-mission analysis phase. The notion of *adaptive autonomy* reflects the goal of using available in-stride information and analysis to the advantage of the mission objectives.

The chart in Fig. 2.1 conveys a rough timeline and relationship between the evolution of UUV autonomy capabilities and the evolution of other critical UUV technologies. The notion of *adaptive* in adaptive autonomy is a sliding scale and refers to the ability to allow increasing degrees of sensor information to affect in-stride autonomy decisions.

The notion of *collaboration* in collaborative autonomy may be viewed as a sliding scale as well. At one end of the spectrum are vehicles deployed alongside each other, executing a mission independently but each contributing to a joint mission. In this case, the collaboration occurs in the pre-deployment mission planning process. When at least periodic communication between deployed vehicles is feasible, real-time collaboration is possible, especially when each vehicle is able to adapt components of its mission to both its sensed environment *and* incoming communications from other vehicles. Advances in underwater acoustic communications (ACOMMS) in terms of reliability, range, flexibility in defining message sets, and bandwidth have enabled the development of adaptive, collaborative autonomy.

This trend also occurs in the context of declining cost and size of commercially available UUVs, making it possible for smaller organizations to operate several vehicles.

The MOOS-IvP autonomy architecture has been developed and refined in this context of migration to adaptive, collaborative autonomy. Mission structure is less defined in terms of a sequence of tasks but rather as a set of autonomy modes with conditions, events, and field commands defining the transitions between modes. The modes correlate to a set of one or more active behaviors, where each behavior may be its own substantial autonomy subcomponent. An autonomy system that includes the ability to adapt its mission to the environment, other collaborating vehicles, and periodic messages from within a field-control hierarchy will inevitably need to balance competing objectives in a way that reflects a singular mission focus. This chapter also discusses how multi-objective optimization is used at the behavior coordination level in the helm to achieve this design objective.

2.1.2 The Backseat Driver Design

The main idea in the backseat driver paradigm is the separation of vehicle *control* and *autonomy*. The control system runs on a platform's main vehicle computer and the autonomy system runs on a separate payload computer. A primary benefit is the decoupling of the platform autonomy from the vehicle hardware. The manufacturer provides a navigation and control system streaming vehicle position and trajectory information to the payload computer. In return, the main vehicle computer accepts a stream of autonomy decisions such as heading, speed, and depth. Exactly how the vehicle navigates and implements control is largely unspecified to the autonomy system running in the payload. The relationship is shown in Fig. 2.2.

The payload autonomy system consists of distinct processes communicating through a publish-subscribe database called the MOOSDB (mission oriented operating suite—database). One process is an interface to the main vehicle computer, and another key process is the IvP Helm implementing the behavior-based autonomy system. The MOOS community is referred to as the "autonomy system," since MOOS itself is middleware, and actual autonomous decision making, sensor processing, contact management, etc., are implemented as individual MOOS processes.

2.1.3 The Publish-Subscribe Middleware Design and MOOS

MOOS provides a middleware capability based on the publish-subscribe architecture and protocol. Each process communicates with each other through a single database process in a star topology (Fig. 2.3). The interface of a particular process is described by what messages it produces (publications) and what messages it consumes (subscriptions). Each message is a simple variable-value pair where

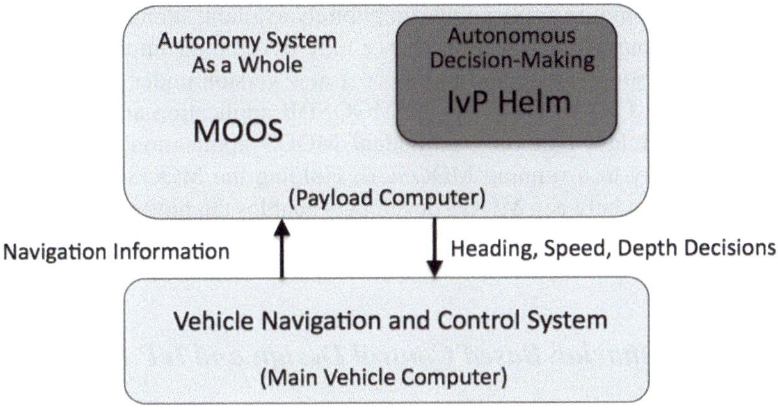

Fig. 2.2 The backseat driver paradigm: The key idea is the separation of vehicle autonomy from vehicle control. The autonomy system provides heading, speed, and depth commands to the vehicle control system. The vehicle control system executes the control and passes navigation information, e.g., position, heading, and speed, to the autonomy system. The backseat paradigm is agnostic regarding how the autonomy system implemented, but in this figure the MOOS-IvP autonomy architecture is depicted

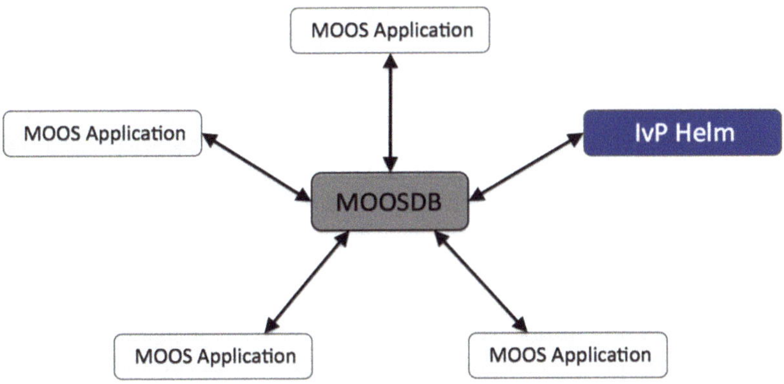

Fig. 2.3 A MOOS community: is a collection of MOOS applications typically running on a single machine each with a separate process ID. Each process communicates through a single MOOS database process (the MOOSDB) in a publish-subscribe manner. Each process may be executing its inner loop at a frequency independent from one another and set by the user. Processes may be all run on the same computer or distributed across a network

the values are either string or numerical values such as (STATE, "DEPLOY"), or (NAVSPEED, 2.2), or arbitrarily large binary packets. Allowing simple message types reduces the compile dependencies between modules and facilitates debugging since all messages are human readable.

The key idea with respect to facilitating code reuse is that applications are largely independent, defined only by their interface, and any application is easily replaceable with an improved version with a matching interface. Since MOOS

Core and many common applications are publicly available along with source code under an open-source GPL license, a user may develop an improved module by altering existing source code and introduce a new version under a different name. The term MOOS Core refers to (a) the MOOSDB application and (b) the MOOS application superclass that each individual MOOS application inherits from to allow connectivity to a running MOOSDB. Holding the MOOS Core part of the code-base constant between MOOS developers enables the plug-and-play nature of applications.

2.1.4 The Behavior-Based Control Design and IvP Helm

The IvP Helm runs as a single MOOS application using a behavior-based architecture for implementing autonomy. Behaviors are distinct modules describable as self-contained mini expert systems dedicated to a particular aspect of overall vehicle autonomy. The helm implementation and each behavior implementation expose an interface for configuration by the user for a particular set of missions. This configuration often contains particulars such as a certain set of waypoints, search area, and vehicle speed. It also contains a specification of mission modes that determine which behaviors are active under what situations and how states are transitioned. When multiple behaviors are active and competing for influence of the vehicle, the IvP solver is used to reconcile the behaviors (Fig. 2.4).

The solver performs this coordination by soliciting an objective function, i.e., utility function, from each behavior defined over the vehicle decision space, e.g., possible settings for heading, speed, and depth. In the IvP Helm, the objective functions are of a certain type, piecewise linearly defined, and are called IvP functions. The solver algorithms exploit this construct to rapidly find a solution to the optimization problem comprised of the weighted sum of contributing functions.

The concept of a behavior-based architecture is often attributed to [7]. Various solutions to the problem of action selection, i.e., the issue of coordinating competing behaviors, have been put forth and implemented in physical systems. The simplest approach is to prioritize behaviors such that the highest priority behavior locks out all others as in the subsumption architecture in [7]. Another approach, referred to as the potential fields, or vector summation approach (see [1, 10]) regards the average action between multiple behaviors to be a good compromise. These action-selection approaches have been used effectively on a variety of platforms, including indoor robots, e.g., [1, 2, 13, 14], land vehicles, e.g., [15], and marine vehicles, e.g., [6, 8, 11, 16, 17]. However, action selection via the identification of a single highest priority behavior and via vector summation have well-known shortcomings later described in [13, 14] and [15] in which the authors advocated for the use of multi-objective optimization as a more suitable, although more computationally expensive, method for action selection. The IvP model is a method for implementing multi-objective function-based action selection that is computationally viable in the IvP Helm implementation.

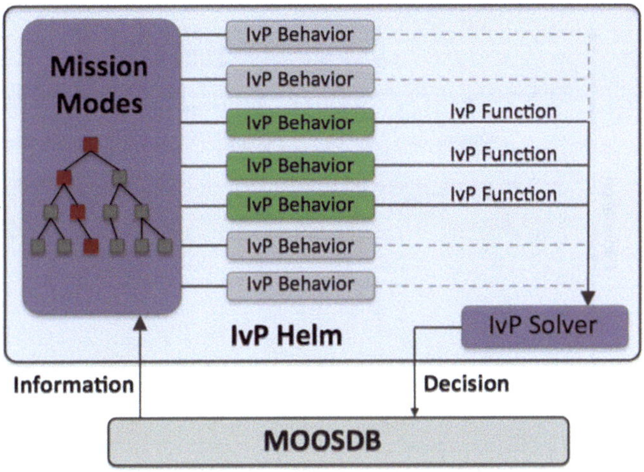

Fig. 2.4 The IvP Helm: The helm is a single MOOS application running as the process pHelmIvP. It is a behavior-based architecture where the primary output of a behavior on each iteration is an IvP objective function. The IvP solver performs multi-objective optimization on the set of functions to find the single best vehicle action, which is then published to the MOOSDB. The functions are built and the set is solved on *each* iteration of the helm—typically one to four times per second. Only a subset of behaviors are active at any given time depending on the vehicle situation and the state space configuration provided by the user

2.1.5 The Nested Autonomy Paradigm

For large-scale subsurface/surface ocean monitoring and observation systems (100+ square kilometers), no single unmanned platform has the ability in terms of sensing, endurance, and communications to achieve large-scale, long-endurance (several days to several weeks) system objectives. Even if multiple platforms are applied to the problem, effectiveness may be substantially diminished if limited to a single platform *type*. The *nested autonomy* paradigm, depicted in Fig. 2.5, is an approach to implementing a system of unmanned platforms for large-scale autonomous sensing applications. It is based in part on the objective of making seamless use of heterogeneous platform types using a uniform platform-independent autonomy architecture. It also assumes the platforms will have varying communications bandwidth, connectivity, and latency.

The *vertical* connectivity allows information to pass from sensors to the on-board sensor processing and autonomy modules, or from each node to other nodes in the cluster, or up to the field operator, and thus forms the basis for the autonomous *adaptive control* which is a key to the capability in compensating for the smaller sensor aperture of the distributed nodes. Similarly, the *horizontal* connectivity forms the basis for *collaboration* between sensors on a node (sensor fusion) or between nodes (collaborative processing and control).

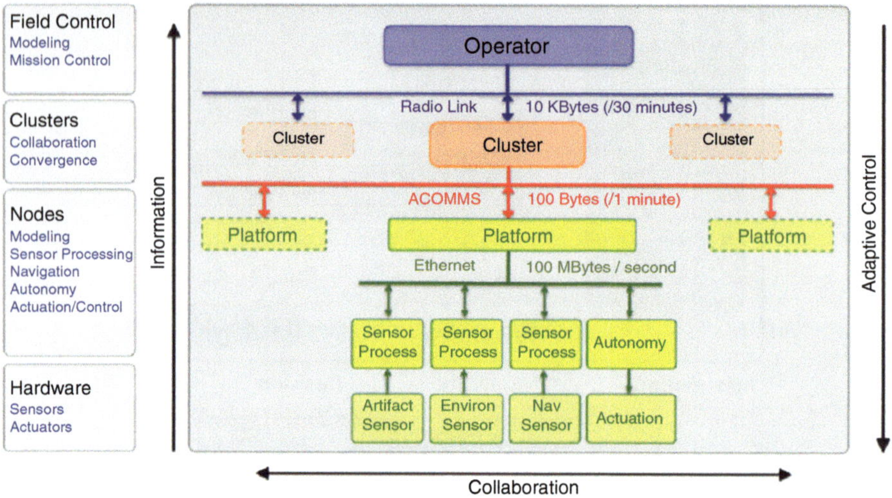

Fig. 2.5 The nested autonomy paradigm: Field-control operators receive intermittent information from field nodes as connectivity and bandwidth allow. Elements of clusters may serve a heterogeneous role as a gateway communications agent. Likewise, nodes receive intermittent commands and cues from field operators. Node autonomy compensates for and complements the sporadic connectivity to field control and other nodes in a cluster or network of clusters

The three layers of horizontal communication have vastly different bandwidths, ranging from 100 byte/min for the internode acoustic modem communications (ACOMMS) to 100 Mbyte/s for the on-board systems. Equally important, the layers of the vertical connectivity differ significantly in latency and intermittency, ranging from virtually instantaneous connectivity of the on-board sensors and control processes to latencies of 10–30 min for information flowing to and from the field-control operators. This, in turn, has critical implication to the timescales of the adaptivity and collaborative sensing and control. Thus, adaptive control of the network assets with the operator in the loop is at best possible on hourly to daily basis, allowing the field operator to make tactical deployment decisions based on, e.g., environmental forecasts and reports of interfering shipping distributions, etc. Shorter timescale adaptivity, such as autonomously reacting to episodic environmental events or a node tracking a marine mammal acoustically, must clearly be performed without operator intervention. On the other hand, the operator can still play a role in cuing forward assets in the path of the dynamic phenomenon, using the limited communication capacity, taking advantage of his own operational experience and intuition. Therefore, as much as a centralized control paradigm is infeasible for such systems, it is also unlikely that a concept of operations based entirely on nodal autonomy is optimal. Instead, some combination will likely be optimal, but in view of the severe latency of the *vertical* communication channels, the *nested autonomy* concept of operations described is heavily tilted toward autonomy.

The MOOS-IvP autonomy implementation is situated primary at the node level in the nested autonomy structure depicted in Fig. 2.5. However, aspects of the MOOS-IvP architecture are relevant to the larger picture as well. A key enabling factor to the nested autonomy paradigm is the platform independence of the node level autonomy system. The backseat driver design allows the decoupling of the vehicle platform from the autonomy system to achieve platform independence. The MOOS middleware architecture and the IvP Helm behavior-based architecture also contribute to platform independence by allowing an autonomy system to be comprised of modules that are swappable across platform types. Furthermore, collaborative and nested autonomy between nodes is facilitated by the simple modal interface to the on-board autonomy missions to control behavior activations.

2.2 A Very Brief Overview of MOOS

MOOS is often described as autonomy *middleware* which implies that it is a kind of glue that connects a collection of applications where the real work happens. MOOS does indeed connect a collection of applications, of which the IvP Helm is one. MOOS is cross platform standalone and dependency free. It needs no other third-party libraries. Each application inherits a generic MOOS interface whose implementation provides a powerful, easy-to-use means of communicating with other applications and controlling the relative frequency at which the application executes its primary set of functions. Due to its combination of ease-of-use, general extendibility, and reliability, it has been used in the classroom by students with no prior experience, as well as in many extended field exercises with substantial robotic resources at stake. To frame the later discussion of the IvP Helm, the basic issues regarding MOOS applications are introduced here. For further information on the original design of MOOS, see [12].

2.2.1 Inter-Process Communication with Publish/Subscribe

MOOS has a star-like topology as depicted in Fig. 2.3. Application within a MOOS community (a MOOSApp) has a connection to a single MOOS Database (called MOOSDB) lying at the heart of the software suite. All communication happens via this central server application. The network has the following properties:

- No peer-to-peer communication.
- Communication between the client and server is initiated by the client, i.e., the MOOSDB never makes an unsolicited attempt to contact a MOOSApp from out of the blue.
- Each client has a unique name.
- The client need not have knowledge.

Table 2.1 The contents of MOOS message

Variable	Meaning
Name	The name of the data
String value	Data in string format
Double value	Numeric double float data
Source	Name of client that sent this data to the MOOSDB
Auxiliary	Supplemental message information, e.g., IvP behavior source
Time	Time at which the data was written
Data type	Type of data (STRING or DOUBLE or BINARY)
Message type	Type of message (usually NOTIFICATION)
Source community	The community to which the source process belongs

- One client does not transmit data to another—it can only be sent to the MOOSDB and from there to other clients. Modern versions of the library sport a sub-one millisecond latency when transporting multi-MB payloads between processes.
- The star network can be distributed over any number of machines running any combination of supported operating systems.
- The communications layer supports clock synchronization across all connected clients and in the same vein can support "time acceleration" whereby all connected clients operate in an accelerated time stream—something that is very useful in simulations involving many processes distributed over many machines.
- Data can be sent in small bites as "string" or "double" packets or in arbitrarily large binary packets.

2.2.2 Message Content

The communications API in MOOS allow data to be transmitted between the MOOSDB and a client. The meaning of that data is dependent on the role of the client. However, the form of that data is not constrained by MOOS although for the sake of convenience MOOS does offer bespoke support for small "double" and string payloads.[1] Data is packed into messages which contain other salient information shown in Table 2.1.

Often it is convenient to send data in string format, for example the string `"Type=EST,Name=AUV,Pos=[3x1]3.4,6.3,0.2"` might describe the position estimate of a vehicle called "AUV" as a 3×1 column vector. This is human readable and does not require the sharing and synchronizing of headerfiles to ensure both sender and recipient understand how to interpret data (as is the case with binary data).

[1]Note that very early versions of MOOS only allowed data to be sent as strings or doubles—but this restriction is now long gone.

It is quite common for MOOS applications to communicate with string data in a concatenation of comma separated "name=value" pairs.

- Strings are human readable.
- All data becomes the same type.
- Logging files are human readable (they can be compressed for storage).
- Replaying a log file is simply a case of reading strings from a file and "throwing" them back at the MOOSDB in time order.
- The contents and internal order of strings transmitted by an application can be changed without the need to recompile consumers (subscribers to that data)—users simply would not understand new data fields but they would not crash.

The above are well-understood benefits of sending self-explanatory ASCII data. However, many applications use data types which do not lend themselves to verbose serialization to strings—think, for example about camera image data being generated at 40 Hz in full color. At this point the need to send binary data is clear and of course MOOS supports it transparently (and via the application pLogger supports logging and replaying it). At this point it is up to the user to ensure that the binary data can be interpreted by all clients and that any and all perturbations to the data structures are distributed and compiled into each and every client. It is here that modern serialization tools such as "Google Protocol Buffers" find application. They offer a seamless way to serialize complex data structures into binary streams. Crucially they offer *forwards* compatibility—it is possible to update and augment data structures with new fields in the comforting knowledge that all existing apps will still be able to interpret the data—they just won't parse the new additions.

2.2.3 Mail Handling—Publish/Subscribe—in MOOS

Each MOOS application is a client having a connection to the MOOSDB. This connection is made on the client side and the client manages a threaded machinery that coordinates the communication with the MOOSDB. This completely hides the intricacies and timings of the communications from the rest of the application and provides a small, well-defined set of methods to handle data transfer. The application can:

1. Publish data—issue a notification on named data
2. Register for notifications on named data
3. Collect notifications on named data—reading mail

Publishing Data: Data is published as a pair—a variable and value—that constitutes the heart of a MOOS message described in Table 2.1. The client invokes the `Notify(VarName, VarValue)` command where appropriate in the client code. The above command is implemented both for string, double and binary values, and the rest of the fields described in Table 2.1 are filled in automatically. Each notification results in another entry in the client's "outbox," which in older versions of MOOS is emptied the next time the MOOSDB accepts an incoming call from the client or in recent versions is pushed instantaneously to all interested clients.

Registering for Notifications: Assume that a list of names of data published has been provided by the author of a particular MOOS application. For example, an application that interfaces to a GPS sensor may publish data called GPS_X and GPS_Y. A different application may register its interest in this data by subscribing or registering for it. An application can register for notifications using a single method Register specifying both the name of the data and the maximum rate at which the client would like to be informed that the data has been changed. The latter parameter is specified in terms of the minimum possible time between notifications for a named variable. For example, setting it to zero would result in the client receiving each and every change notification issued on that variable. Recent versions of MOOS also support "wildcard" subscriptions. For example, a client can register for "*:*" to receive all messages from all other clients or "GPS_*:?NAV" to receive messages beginning with "GPS_" from any process with a four-letter name ending in "NAV."

Reading Mail: A client can enquire at any time whether it has received any new notifications from the MOOSDB by invoking the Fetch method. The function fills in a list of notification messages with the fields given in Table 2.1. Note that a single call to Fetch may result in being presented with several notifications corresponding to the same named data. This implies that several changes were made to the data since the last client–server conversation. However, the time difference between these similar messages will never be less than that specified in the Register function described above. In typical applications the Fetch command is called on the client's behalf just prior to the Iterate method, and the messages are handled in the user overloaded OnNewMail method.

2.2.4 Overloaded Functions in MOOS Applications

MOOS provides a base class called CMOOSApp which simplifies the writing of a new MOOS application as a derived subclass. Beneath the hood of the CMOOSApp class is a loop which repetitively calls a function called Iterate() which by default does nothing. One of the jobs as a writer of a new MOOS-enabled application is to flesh this function out with the code that makes the application do what we want. Behind the scenes this overall loop in CMOOSApp is also checking to see if new data has been delivered to the application. If it has, another virtual function, OnNewMail(), is called. This is the function within which code is written to process the newly delivered data.

The roles of the three virtual functions in Fig. 2.6 are discussed below. The pHelmIvP application does indeed inherit from CMOOSApp and overload these functions. The base class contains other virtual functions (OnConnectToServer() and OnDisconnectFromServer()) discussed in [12]. *The Iterate() Method:* By overriding the CMOOSApp::Iterate() function in a new derived class, the author

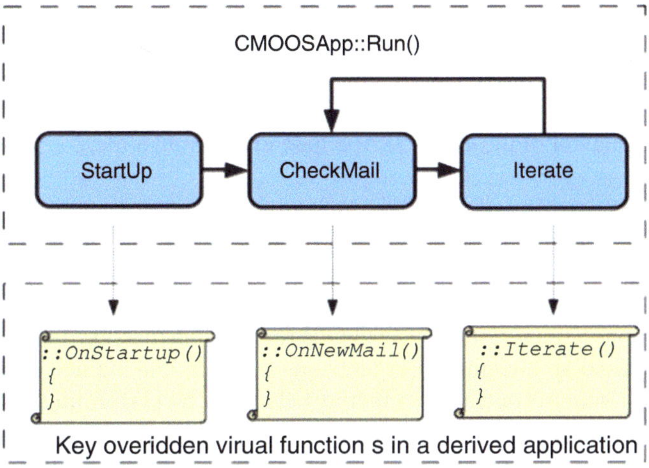

Fig. 2.6 Key virtual functions of the MOOS application base class: The flow of execution once `Run()` has been called on a class derived from `CMOOSApp`. The *scrolls* indicate where users of the functionality of CMOOSApp will be writing new code that implements whatever that is wanted from the new applications. Note that it is not the case (as the above may suggest) that mail is polled for—in modern incantations of MOOS it is pushed to a client a synchronously `OnNewMail` is called as soon as `Iterate` is not running

creates a function from which the work that the application is tasked with doing can be orchestrated. In the `pHelmIvP` application, this method will consider the next best vehicle decision, typically in the form of deciding values for the vehicle heading, speed, and depth. The rate at which `Iterate()` is called by the `SetAppFreq()` method or by specifying the `AppTick` parameter in a mission file. Note that the requested frequency specifies the maximum frequency at which `Iterate()` will be called—it does not guarantee that it will be called at the requested rate. For example, if you write code in `Iterate()` that takes 1 s to complete there is no way that this method can be called at more than 1 Hz. If you want to call `Iterate()` as fast as is possible simply request a frequency of zero—but you may want to reconsider why you need such a greedy application. *The OnNewMail() Method:* Just before `Iterate()` is called, the `CMOOSApp` base class determines whether new mail is present, i.e., whether some other process has posted data for which the client has previously registered, as described above. If new mail is waiting, the `CMOOSApp` base class calls the `OnNewMail()` virtual function, typically overloaded by the application. The mail arrives in the form of a list of `CMOOSMsg` objects (see Table 2.1). The programmer is free to iterate over this collection examining who sent the data, what it pertains to, how old it is, whether or not it is string or numerical data and to act on or process the data accordingly. In recent versions of MOOS it is possible to have `OnNewMail` called in a directly and rapidly in response to new mail being received by the backend communications threads. This architecture allows for very rapid response times (sub ms) between a client posting data and it being received and handled by all interested parties. *The OnStartup() Method:* This function is called just before the application enters into its own forever-loop depicted in Fig. 2.6. This is the application that implements the application's initialization code and in particular reads configuration parameters (including those that modify the default behavior of the CMOOSApp base class) from a file.

2.3 IvP Helm Autonomy

An autonomous helm is primarily an engine for decision making. The IvP Helm uses a behavior-based architecture to organize its decision making and is distinctive in the manner in which it resolves competing behaviors, by performing multi-objective optimization on their collective output using a mathematical programming model called interval programming. Here the IvP Helm architecture is described and the means for configuration given a set of behaviors and a set of mission objectives.

2.3.1 Influence of Brooks, Stallman, and Dantzig on the IvP Helm

The notion of a behavior-based architecture for implementing autonomy on a robot or unmanned vehicle is most often attributed to Rodney Brooks' subsumption architecture [7]. A key principle at the heart of Brooks' architecture, and arguably the primary reason its appeal has endured, is the notion that autonomy systems can be built *incrementally*. Notably, Brooks' original publication predated the arrival of open-source software and the Free Software Foundation founded by Richard Stallman. Open-source software is not a prerequisite for building autonomy systems incrementally, but it has the capability of greatly accelerating that objective. The development of complex autonomy systems stands to significantly benefit if the set of developers at the table is large and diverse. Even more so if they can be from different organizations with perhaps even the loosest of overlap in interest regarding how to use the collective end product.

The interval programming solution algorithm, as well as the term itself, was motivated by the mathematical programming model, linear programming, developed by George Dantzig [9]. The key idea in linear programming is the choice of the particular mathematical construct that comprises an instance of a linear programming problem—it has enough expressive flexibility to represent a huge class of practical problems, *and* the constructs can be effectively exploited by the simplex method to converge quickly even on very large problem instances. The constructs used in interval programming to represent behavior output (piecewise linear functions) were likewise chosen to have expressive flexibility for handling current and future behavior and due to the opportunity to develop solution algorithms that exploit the piecewise linear constructs.

2.3.2 Nontraditional Aspects of the IvP Behavior-Based Helm

While IvP Helm takes its motivation from early notions of the behavior-based architecture, it is also quite different in many regards. The notion of behavior independence to temper the growth of complexity in progressively larger systems is still a principle closely followed in the IvP Helm. Behaviors may certainly influence one another from one iteration to the next. However, within a single iteration, the output generated by a single behavior is not affected at all by what is generated by other behaviors in the same iteration. The only inter-behavior "communication" realized within an iteration comes when the IvP solver reconciles the output of multiple behaviors. The independence of behaviors not only helps a single developer manage the growth of complexity, but it also limits the dependency between developers.

Certain aspects of behaviors in the IvP Helm may also be a departure from some notions traditionally associated (fairly or not) with behavior-based architectures:

- Behaviors have state. IvP behaviors are instances of a class with a fairly simple interface to the helm. Inside they may be arbitrarily complex, keep histories of observed sensor data, and may contain algorithms that could be considered "reactive" or "plan-based."
- Behaviors influence each other between iterations. The primary output of behaviors is their objective function, ranking the utility of candidate actions. IvP behaviors may also generate variable–value posts to the `MOOSDB` observable by behaviors on the next helm iteration. In this way they can explicitly influence other behaviors by triggering or suppressing their activation or even affecting the parameter configuration of other behaviors.
- Behaviors may accept externally generated plans. Behavior input can be anything represented by a MOOS variable and perhaps generated by other MOOS processes outside the helm. It is allowable to have one or more planning engines running on the vehicle generating output consumed by one or more behaviors.
- Several instances of the same behavior are allowed. Behaviors generally accept a set of configuration parameters that allow them to be configured for quite different tasks or roles in the same helm and mission. Different waypoint behaviors, e.g., can be configured for different components of a transit mission. Or different collision avoidance behaviors can be instantiated for different contacts.
- Behaviors can be run in a configurable sequence. The `condition` and `endflag` parameters defined for all behaviors allow for a sequence of behaviors to be readily configured into a larger mission plan.
- Behaviors rate actions over a coupled decision space. IvP functions are defined over the Cartesian product of the set of vehicle decision variables. Objective functions for certain behaviors may only be adequately expressed over such a decision space. See, e.g., the function produced by the AvoidCollision behavior later in this chapter. This is distinct from the decoupled decision making

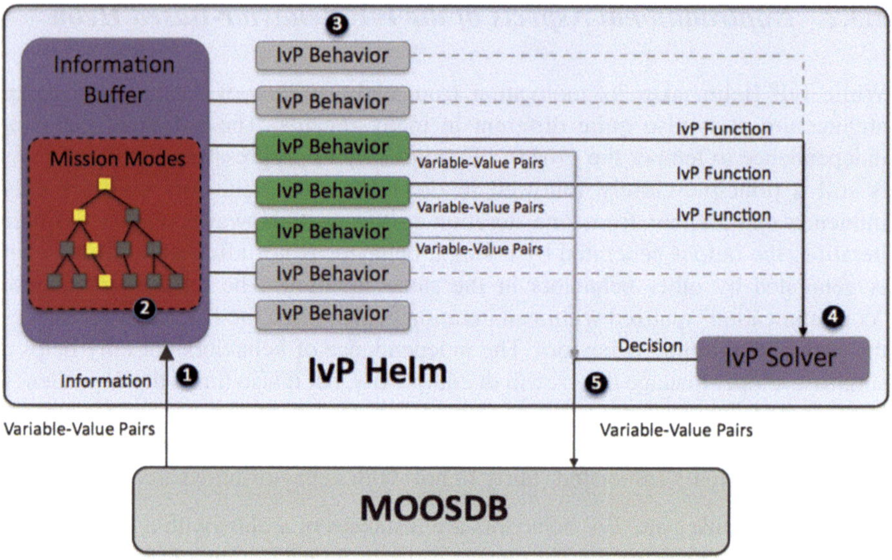

Fig. 2.7 The pHelmIvP iterate loop: (**a**) Mail is read from the MOOSDB. It is parsed and stored in a local buffer to be available to the behaviors. (**b**) If there were any mode declarations in the mission behavior file they are evaluated at this step. (**c**) Each behavior is queried for its contribution and may produce an IvP function and a list of variable–value pairs to be posted to the MOOSDB at the end of the iteration. (**d**) The objective functions are resolved to produce an action, expressible as a set of variable–value pairs. (**e**) All variable–value pairs are published to the MOOSDB for other MOOS processes to consume

style proposed in [13, 15]—early advocates of multi-objective optimization in behavior-based action selection.

The autonomy in play on a vehicle during a particular mission is the product of two distinct efforts, (1) the development of vehicle behaviors and their algorithms, and (2) mission planning via the configuration of behaviors and mode declarations. The former involves the writing of new source code, and the latter involves the construction of mission behavior files.

2.3.3 Inside the IvP Helm: A Look at the Helm Iterate Loop

As with other MOOS applications, the IvP Helm implements an `Iterate()` loop within which the basic function of the helm is executed. Components of the `Iterate()` loop, with respect to the behavior-based architecture, are described in this section. The basic flow, in five steps, is depicted in Fig. 2.7.

Step 1. Reading Mail and Populating the Information Buffer: The first step of a helm iteration occurs outside the `Iterate()` loop. As depicted in Fig. 2.6, a

MOOS application will read its mail by executing its `OnNewMail()` function just prior to executing its `Iterate()` loop if there is any mail in its in-box. The helm parses mail to maintain its own information buffer which is also a mapping of variables to values. This is done primarily for simplicity to ensure that each behavior is acting on the same world state as represented by the information buffer. Each behavior has a pointer to the buffer and is able to query the current value of any variable in the buffer or get a list of variable–value changes since the previous iteration.

Step 2. Evaluation of Mode Declarations: Next, the helm evaluates any mode declarations specified in the behavior file. Mode declarations are discussed in Sect. 2.3.4. In short, a mode is represented by a string variable that is reset on each iteration based on the evaluation of a set of logic expressions involving other variables in the buffer. The variable representing the mode declaration is then available to the behavior on the current iteration when it, e.g., evaluates its `condition` parameters. A condition for behavior participating in the current iteration could therefore read something like `condition = (MODE==SURVEYING)`. The exact value of the variable `MODE` is set during this step of the `Iterate()` loop.

Step 3. Behavior Participation: In the third step much of the work of the helm is realized by giving each behavior a chance to participate. Each behavior is queried sequentially—the helm contains no separate threads in this regard. The order in which behaviors are queried does not affect the output. This step contains two distinct parts for each behavior: (1) determination of whether the behavior will participate and (2) production of output if it is indeed participating on this iteration. Each behavior may produce two types of information as Fig. 2.7 indicates. The first is an objective function (or "utility" function) in the form of an IvP function. The second kind of behavior output is a list of variable–value pairs to be posted by the helm to the MOOSDB at the end of the `Iterate()` loop. A behavior may produce both kinds of information, neither, or one or the other, on any given iteration.

Step 4. Behavior Reconciliation: In the fourth step, the IvP functions are collected by the IvP solver to produce a single decision over the helm's decision space. Each function is an IvP function—an objective function that maps each element of the helm's decision space to a utility value. IvP functions are piecewise linearly defined, where each piece is an *interval* of the decision space with an associated linear function. Each function also has an associated weight and the solver performs multi-objective optimization over the weighted sum of functions (in effect a single objective optimization at that point). The output is a single optimal point in the decision space. For each decision variable the helm produces another variable–value pair, such as `DESIRED_SPEED = 2.4` for publication to the MOOSDB.

Step 5. Publishing the Results to the MOOSDB: In the last step, the helm publishes all variable–value pairs to the MOOSDB, some of which were produced directly by behavior, and some of which were generated by the IvP Solver.

2.3.4 *Hierarchical Mode Declarations*

Hierarchical mode declarations (HMDs) are an optional feature of the IvP Helm for organizing the behavior activations according to declared mission modes. Modes and sub-modes can be declared, in line with a mission planner's own concept of mission evolution, and behaviors can be associated with the declared modes. In more complex missions, it can facilitate mission planning (in terms of less time and better detection of human errors), and it can facilitate the understanding of exactly what is happening in the helm during the mission execution and in post-analysis.

2.3.4.1 Scripted Versus Adaptive Missions

A trend of unmanned vehicle usage can be characterized as being increasingly less of the shorter, scripted variety to be increasingly more of the longer, adaptive mission variety. A typical mission in our own lab five years ago would contain a certain set of tasks, typically waypoints and ultimately a rendezvous point for recovering the vehicle. Data acquired during deployment was off-loaded and analyzed later in the laboratory. What has changed? The simultaneous maturation of acoustic communications, on-board sensor processing, and longer vehicle battery life has dramatically changed the nature of mission configurations. The vehicle is expected to adapt to both the phenomena it senses and processes on board, as well as adapt its operation given field-control commands received via acoustic, radio, or satellite communications. Multi-vehicle collaborative missions are also increasingly viable due to lower vehicle costs and mature ACOMMS capabilities. In such cases a vehicle is not only adapting to sensed phenomena and field commands but also to information from collaborating vehicles.

Our missions have evolved from being a sequence task to be instead a set of modes—an initial mode when launched, an understanding of what brings us from one mode to another, and what behaviors are in play in each mode. Modes may be entered and exited any number of times, perhaps in sequences unknown at launch time, depending on what the vehicle senses and how they are commanded in the field.

2.3.4.2 Syntax of Hierarchical Mode Declarations

An example is provided showing of the use of HMDs with an example simple mission. This mission is referred to as the Bravo mission and can be found alongside the Alpha mission in the set of example missions distributed with the MOOS-IvP public domain software. The modes are explicitly declared in the Bravo behavior file to form the following hierarchy:

The hierarchy in Fig. 2.8 is formed by the mode declaration constructs on the left-hand side, taken as an excerpt from the `bravo.bhv` file. After the mode declarations

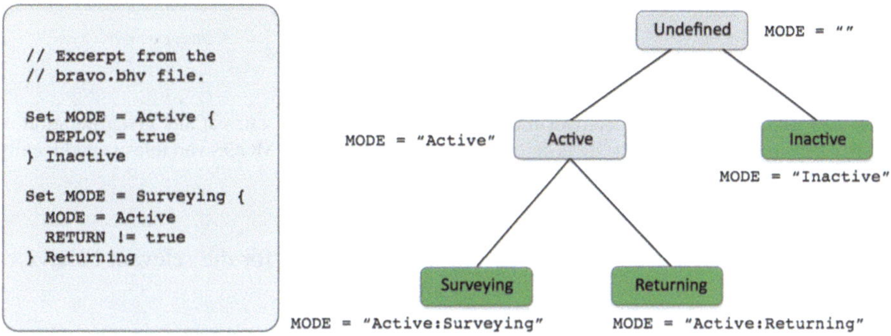

Fig. 2.8 Hierarchical modes for the Bravo mission: The vehicle will always be in one of the modes represented by a leaf node. A behavior may be associated with any node in the tree. If a behavior is associated with an internal node, it is also associated with all its children

are read when the helm is initially launched, the hierarchy remains static thereafter. The hierarchy is associated with a particular MOOS variable, in this case, the variable `MODE`. Although the hierarchy remains static, the mode is re-evaluated at the outset of each helm iteration based on the conditions associated with nodes in the hierarchy. The mode evaluation is represented as a string in the variable `MODE`. As shown in Fig. 2.8 the variable is the concatenation of the names of all the nodes. The mode evaluation begins sequentially through each of the blocks. At the outset the value of the variable `MODE` is reset to the empty string. After the first block in Fig. 2.8, `MODE` will be set to either `"Active"` or `"Inactive"`. When the second block is evaluated, the condition `"MODE=Active"` is evaluated based on how `MODE` was set in the first block. For this reason, mode declarations of children need to be listed after the declarations of parents in the behavior file.

Once the mode is evaluated, at the outset of the helm iteration, it is available for use in the run conditions of the behaviors (described below), via a string-matching relation that matches when one side matches exactly one of the components in the other side's colon-separated list of strings. Thus `"Active" == "Active:Returning"`, and `"Returning" == "Active:Returning"`. This is to allow a behavior to be easily associated with an internal node regardless of its children. For example, if a collision-avoidance behavior were to be added to the Bravo mission, it could be associated with the `"Active"` mode rather than explicitly naming all the sub-modes of the `"Active"` mode.

2.3.5 Behavior Participation in the IvP Helm

The primary work of the helm comes when the behaviors participate at each round of the helm `Iterate()` loop, by producing IvP functions, posting variable–value pairs to MOOS, or both. As depicted in Fig. 2.7, once the mode has been reevaluated

Fig. 2.9 Behavior states: A behavior may be in one of these four states at any given iteration of helm `Iterate()` loop. The state is determined by examination of MOOS variables stored locally in the helm's information buffer

taking into consideration newly received mail, it is time for the relevant behaviors to participate.

2.3.5.1 Behavior Conditions

On any single iteration a behavior may participate by generating an objective function to influence the helm's output over its decision space. Not all behaviors participate in this regard, and the primary criteria for participation is whether or not it has met each of its "run conditions." These are the conditions laid out in the behavior file of the form:

```
condition = <logic-expression>
```

Conditions are built from simple relational expressions, the comparison of MOOS variables to specified literal values or the comparison of MOOS variables to one another. Conditions may also involve Boolean logic combinations of relation expressions. A behavior may base its conditions on any MOOS variable such as:

```
condition = (DEPLOY=true) and (STATION_KEEP != true)
```

A run condition may also be expressed in terms of a helm mode such as:

```
condition = (MODE == LOITERING)
```

All MOOS variables involved in run condition expressions are automatically subscribed for by the helm to the MOOSDB.

2.3.5.2 Behavior Run States

On any given helm iteration a behavior may be in one of four states depicted in Fig. 2.9:

- Idle: A behavior is `idle` if it is not `complete` and it has not met its run conditions as described above in Sect. 2.3.5.1. The helm will invoke an idle behavior's `onIdleState()` function.
- Running: A behavior is `running` if it has met its run conditions and it is not `complete`. The helm will invoke a running behavior's `onRunState()` function thereby giving the behavior an opportunity to contribute an objective function.

- Active: A behavior is `active` if it is running and it did indeed produce an objective function when prompted. There are a number of reasons why a running behavior may not be active. For example, a collision avoidance behavior may opt to not produce an objective function when the contact is sufficiently far away.
- Complete: A behavior is `complete` when the behavior itself determines it is complete. It is up to the behavior author to implement this, and some behaviors may never complete. The function `setComplete()` is defined generally at the behavior superclass level, for calling by a behavior author. This provides some standard steps to be taken upon completion, such as posting of `endflags`, described below in Sect. 2.3.5.3. Once a behavior is in the `complete` state, it remains in that state permanently. All behaviors have a `DURATION` parameter defined to allow the behavior to be configured to time-out if desired. When a time-out occurs the behavior state will be set to `complete`.

2.3.5.3 Behavior Flags and Behavior Messages

Behaviors may produce a set of messages, i.e., variable–value pairs, on any given iteration (see Fig. 2.7). These messages can be critical for coordinating behaviors with each other and to other MOOS processes. This can also be invaluable for monitoring and debugging behaviors configured for particular missions. Behaviors do not post messages to the `MOOSDB`, they request the helm to post messages on their behalf. The helm collects these requests and publishes them to the `MOOSDB` at the end of the `Iterate()` loop. Variable–value pairs corresponding to state flags are set in the behavior file for each behavior. The following five flags are available:

- An `endflag` is posted once when or if the behavior enters the `complete` state. Multiple endflags may be configured for a behavior. By default, when a behavior is completed and has posted its endflag, it does not participate further. Its instance is destroyed and removed from the helm.
- An `idleflag` is posted on each iteration of the helm when the behavior is determined to be in the `idle` state. The variable–value pair representing the idleflag is given in the `idleflag` parameter in the behavior file.
- A `runflag` is posted on each iteration of the helm when the behavior is determined to be in the `running` state, regardless of whether it is further determined to be active or not. A `runflag` is posted exactly when an `idleflag` is not.
- An `activeflag` is posted on each iteration of the helm when the behavior is determined to be in the `active` state.
- An `inactiveflag` is posted on each iteration of the helm when the behavior is determined to be not in the `active` state.

A `runflag` is meant to "complement" an `idleflag`, by posting exactly when the other one does not. Similarly with the `inactiveflag` and `activeflag`. The situation is shown in Fig. 2.10. Behavior authors may implement their behaviors to post other messages as they see fit. For example, the waypoint

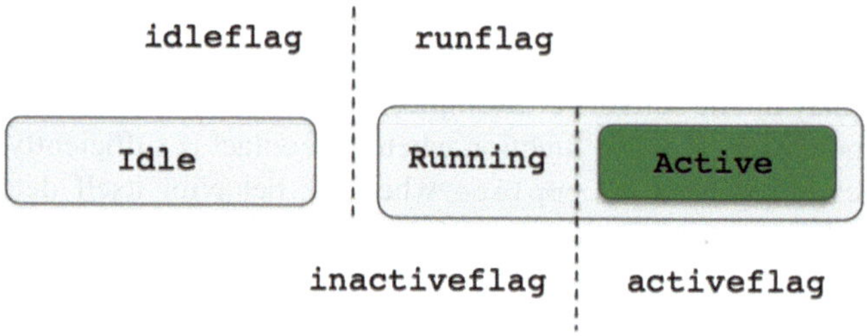

Fig. 2.10 Behavior flags: The four behavior flags `idleflag`, `runflag`, `activeflag`, and `inactiveflag` are posted depending on the behavior state and can be considered complementary in the manner indicated

behavior publishes a status variable, `WPT_STATUS`, with a status message similar to `"vname=alpha,index=0,dist=124,eta=62"` indicating the name of the vehicle, the index of the next point in the list of waypoints, the distance to that waypoint, and the estimated time of arrival.

2.3.6 Behavior Reconciliation Using Multi-objective Optimization

A unique aspect of the IvP Helm is the manner in which behaviors are reconciled when there are two or more behaviors competing for influence of the helm decision.

2.3.6.1 IvP Functions

IvP functions are produced by behaviors to influence the decision produced by the helm on the current iteration (see Fig. 2.7). The decision is typically comprised of the desired *heading*, *speed*, and *depth*, but the helm decision space could be comprised of any arbitrary configuration. Some points about IvP functions:

- IvP functions are piecewise linearly defined. Each piece is defined by an interval over some subset of the decision space, and there is a linear function associated with each piece (see Fig. 2.12).
- IvP functions are an *approximation* of an underlying function. The linear function for a single piece is the best linear approximation of the underlying function for the portion of the domain covered by that piece.
- IvP domains are discrete with an upper and lower bound for each variable, so an IvP function *may* achieve zero-error in approximating an underlying function by associating a piece with each point in the domain. Behaviors seldom need to do so in practice however.
- The Ivp function construct and IvP solver are generalizable to N dimensions.

- Pieces in IvP functions need not be of uniform size. More pieces can be dedicated to parts of the domain that are harder to approximate with linear functions.
- IvP functions need only be defined over a subset of the domain. Behaviors are not affected if the helm is configured for additional variables that a behavior may not care about. It is allowable, e.g., to have a behavior that produces IvP functions solely over the vehicle *depth*, even though the helm may be producing decisions over *heading*, *speed*, and *depth*.

IvP functions are produced by behaviors using the IvP Build Toolbox—a set of tools for creating IvP functions based on any underlying function defined over an IvP domain. Many, if not all, of the behaviors in this document make use of this toolbox, and authors of new behaviors have this at their disposal. A primary component of writing a new behavior is the development of the "underlying function," the function approximated by an IvP function with the help of the toolbox. The underlying function is a correlation between all candidate helm decisions, e.g., heading, speed, and depth choices, to a utility value from the perspective of what the behavior is trying to achieve.

2.3.6.2 The IvP Build Toolbox

The IvP Toolbox is a set of tools (a C++ library) for building IvP functions. It is typically utilized by behavior authors in a sequence of library calls within a behavior's (C++) implementation. There are two sets of tools—the *reflector* tools for building IvP functions in N dimensions and the *ZAIC* tools for building IvP functions in one dimension as a special case. The reflector tools work by making available a function to be approximated by an IvP function. The tools simply need this function for sampling. Consider the Gaussian function rendered in Fig. 2.11:

The "x″ and "y″ variables, each with a range of $[-250, 250]$, are discrete, taking on integer values. The domain therefore contains $501^2 = 251,001$ points or possible decisions. The IvP Build Toolbox can generate an IvP function approximating this function over this domain by using a uniform piece size, as rendered in Fig. 2.12a, b. The difference in these two figures is only the size of the piece. More pieces (Fig. 2.12a) result in a more accurate approximation of the underlying function but takes longer to generate and creates further work for the IvP solver when the functions are combined. IvP functions need not use uniformly sized pieces. By using the *directed refinement* option in the IvP Build Toolbox, an initially uniform IvP function can be further refined with more pieces over a sub-domain directed by the caller, with smaller uniform pieces of the caller's choosing. This is rendered in Fig. 2.12c. Using this tool requires the caller to have some idea where, in the sub-domain, further refinement is needed or desired. Often a behavior author indeed has this insight. For example, if one of the domain variables is vehicle heading, it may be good to have a fine refinement in the neighborhood of heading values close to the vehicle's current heading.

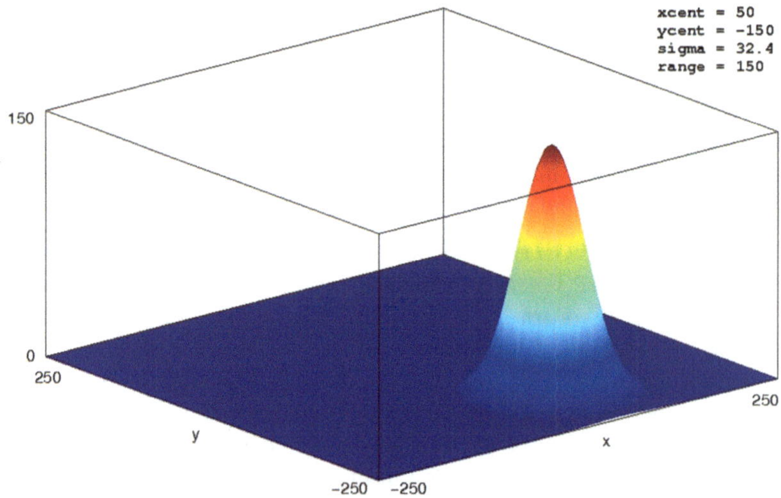

Fig. 2.11 A rendering of the function $f(x,y) = Ae^{-(\frac{(x-x_0)^2+(y-y_0)^2}{2\sigma^2})}$ where A = `range` = 150, σ = `sigma` = 32.4, x_0 = `xcent` = 50, and y_0 = `ycent` = -150. The domain here for x and y ranges from -250 to 250

In other situations, insight into where further refinement is needed may not be available to the caller. In these cases, using the *smart refinement* option of the IvP Build Toolbox, an initially uniform IvP function may be further refined by asking the toolbox to automatically "grade" the pieces as they are being created. The grading is in terms of how accurate the linear fit is between the piece's linear function and the underlying function over the sub-domain for that piece. A priority queue is maintained based on the grades, and pieces where poor fits are noted are automatically refined further, up to a maximum piece limit chosen by the caller. This is rendered in Fig. 2.12d.

The reflector tools work similarly in N dimensions and on multimodal functions. The only requirement for using the reflector tool is to provide it with access to the underlying function. Since the tool repetitively samples this function, a central challenge to the user of the toolbox is to develop a fast implementation of the function. In terms of the time consumed in generating IvP functions with the reflector tool, the sampling of the underlying function is typically the longest part of the process.

2.3.6.3 The IvP Solver and Behavior Priority Weights

The IvP solver collects a set of weighted IvP functions produced by each of the behaviors and finds a point in the decision space that optimizes the weighted combination. If each IvP objective function is represented by $f_i(\overrightarrow{x})$, and the weight

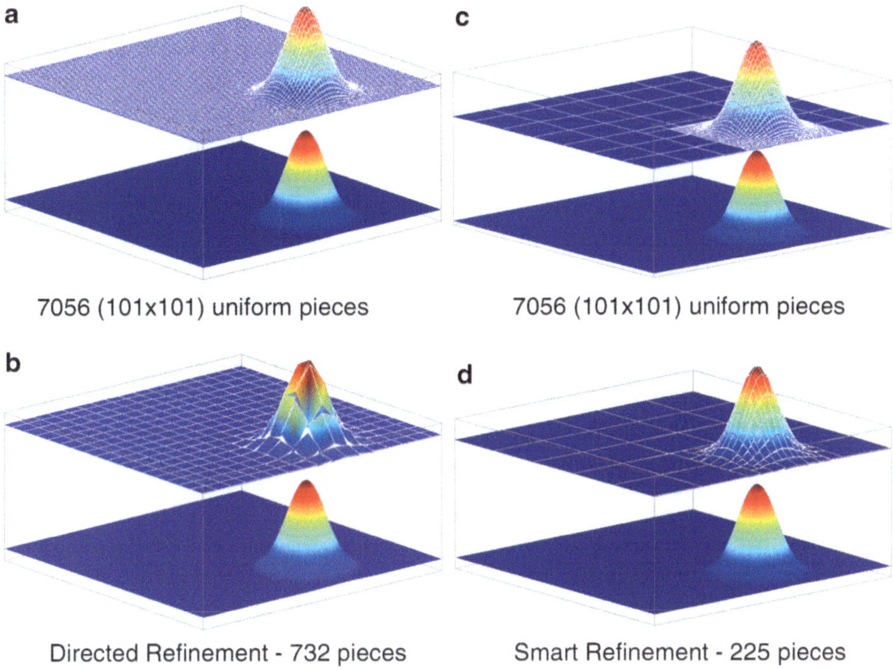

Fig. 2.12 A rendering of four different IvP functions approximating the same underlying function: The function in (**a**) uses a uniform distribution of 7,056 pieces. The function in (**b**) uses a uniform distribution of 1,024 pieces. The function in (**c**) was created by first building a uniform distribution of 49 pieces and then focusing the refinement on a sub-domain of the function. This is called directed refinement in the IvP Build toolbox. The function in (**d**) was created by first building a uniform function of 25 pieces and repeatedly refining the function based on which pieces were noted to have a poor fit to the underlying function. This is termed smart refinement in the IvP Build toolbox

of each function is given by w_i, the solution to a problem with k functions is given by

$$\mathbf{x}^* = \arg\max_{\mathbf{x}} \sum_{i=0}^{k-1} w_i f_i(\mathbf{x}).$$

The algorithm is described in detail in [3], but is summarized in the following few points.

- *The Search Tree*: The structure of the search algorithm is branch-and-bound. The search tree is comprised of an IvP function at each layer, and the nodes at each layer are comprised of the individual pieces from the function at that layer. A leaf node represents a single piece from each function. A node in the tree is realizable if the piece from that node and its ancestors intersect, i.e., share common points in the decision space.

- *Global Optimality*: Each point in the decision space is in exactly one piece in each IvP function and is thus in exactly one leaf node of the search tree. If the search tree is expanded fully, or pruned properly (only when the pruned out sub-tree does not contain the optimal solution), then the search is guaranteed to produce the globally optimal solution. The search algorithm employed by the IvP solver does indeed start with a fully expanded tree and utilizes proper pruning to guarantee global optimality. The algorithm does allow for a parameter for guaranteed limited back off from the global optimality—a quicker solution with a guarantee of being within a fixed percent of global optima. This option is not exposed in the configuration of the IvP Helm, which always finds the global optimum, since this stage of computation is very fast in practice.
- *Initial Solution*: A key factor of an effective branch-and-bound algorithm is seeding the search with a decent initial solution. In the IvP Helm, the initial solution used is the solution (typically `heading`, `speed`, `depth`) generated on the previous helm iteration. In practice this appears to provide a speedup by about a factor of two.

In cases where there is a "tie" between optimal decisions, the solution generated by the solver is nondeterministic. When the solver is used in the helm, the nondeterminism is mitigated somewhat by the fact that the solution is seeded with the output of the previous helm iteration as discussed above. In other words, all things being equal, the helm will default to producing a decision that matches its previous decision.

The setting of function priority weights occurs in one of three manners. First, many behaviors are designed with a *policy* for setting their own weights. For example, in a collision avoidance behavior, the weight varies between zero and a maximum weight depending on the relative position of the two vehicles. Second, weights are influenced by initial behavior configuration parameters regarding the priority weight policy. These are set during a pre-mission planning phase. Lastly, weights may be affected at runtime via the dynamic reconfiguration of behavior parameters to values different from those set at the time of mission launch. Such reconfiguration may be the result of a field-control message received from a remote operator or another platform, or the result of another on-board process outside the helm. In practice, users often make good use of simulation tools to confirm that parameter configurations and behavior weight-setting policies are in line with their expectations for the mission.

2.4 IvP Helm Behaviors

Helm behaviors derive part of their function from inheriting properties from a base class behavior implementation, and their unique capabilities are an extension of the base capability. The uniform base capability allows for mission configurations to be constructed in a simple predictable manner. Here we discuss (a) the base capabilities

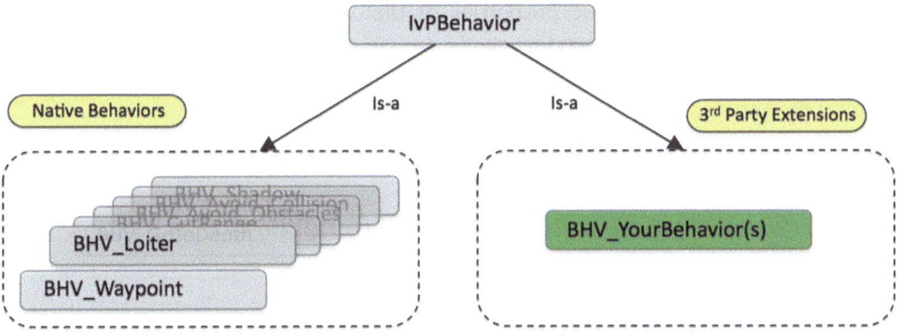

Fig. 2.13 Behavior inheritance: Behaviors are derived from the `IvPBehavior` superclass. The native behaviors are the behaviors distributed with the helm. New behaviors also need to be a subclass of the IvPBehavior class to work with the helm. Certain virtual functions invoked by the helm may be optionally but typically overloaded in all new behaviors. Other private functions may be invoked within a behavior function as a way of facilitating common tasks involved in implementing a behavior

of IvP behaviors, (b) how behaviors are handled generally by the helm in each iteration, (c) the hooks for creating a new third-party behavior, (d) an overview of standard behaviors that are distributed with the helm in the open domain, and (e) a more detailed look at a few representative behaviors.

2.4.1 Brief Overview

Behaviors are implemented as C++ classes with the helm having one or more instances at runtime, each with a unique descriptor. The properties and implemented functions of a particular behavior are partly derived from the `IvPBehavior` superclass, shown in Fig. 2.13. The is-a relationship of a derived class provides a form of code reuse as well as a common interface for constructing mission files with behaviors.

The `IvPBehavior` class provides three virtual functions which are typically overloaded in a particular behavior implementation:

- The `setParam()` Function: Parameter–value pairs are handled to configure a behavior's unique properties distinct from its superclass.
- The `onRunState()` Function: The primary function of a behavior implementation, performed when the behavior has met its conditions for running, with the output being an objective function and a possibly empty set of variable–value pairs for posting to the MOOSDB.
- The `onIdleState()` Function: What the behavior does when it has not met its run conditions. It may involve updating internal state history, generation of variable–value pairs for posting to the MOOSDB, or absolutely nothing at all.

This section discusses the properties of the `IvPBehavior` superclass that an author of a third-party behavior needs to be aware of in implementing new behaviors. It is also relevant material for users of the native behaviors as it details general properties.

2.4.2 *Parameters Common to All IvP Behaviors*

A behavior has a standard set of parameters defined at the `IvPBehavior` level as well as unique parameters defined at the subclass level. By configuring a behavior during mission planning, the setting of parameters is the primary venue for affecting the overall autonomy behavior in a vehicle. Parameters are set in the behavior file but can also be dynamically altered once the mission has commenced. A parameter is set with a single line of the form:

```
parameter = value.
```

In this section, the parameters defined at the superclass level and available to all behaviors are discussed. Each new behavior typically augments these parameters with new parameters unique to the behavior.

2.4.2.1 A Summary of General Behavior Parameters

The following parameters are defined for all behaviors at the superclass level.

NAME The name of the behavior—should be unique between all behaviors.

PRIORITY The priority weight of the produced objective function. The default value is 100. A behavior may also be implemented to determine its own priority.

DURATION The time in seconds that the behavior will remain running before declaring completion. If unspecified, the behavior will never time-out.

CONDITION This parameter specifies a condition that must be met for the behavior to be active. Conditions are checked for each behavior at the beginning of each control loop iteration. Conditions are based on current MOOS variables, such as `STATE = normal` or `(`$K \leq 4$`)`. More than one condition may be provided, as a convenience, treated collectively as a single conjunctive condition. The helm automatically subscribes for any condition variables.

RUNFLAG This parameter specifies a variable and a value to be posted when the behavior has met all its conditions for being in the *running* state. It is an equal-separated pair such as `TRANSITING`=true. More than one flag may be provided. These can be used to satisfy or block the conditions of other behaviors.

IDLEFLAG This parameter specifies a variable and a value to be posted when the behavior is in the *idle* state. It is an equal-separated pair such as `WAITING`=true. More than one flag may be provided. These can be used to satisfy or block the conditions of other behaviors.

ACTIVEFlAG This parameter specifies a variable and a value to be posted when the behavior is in the *active* state. It is an equal-separated pair such as `TRANSITING`=true. More than one flag may be provided. These can be used to satisfy or block the conditions of other behaviors.

INACTIVEFlAG This parameter specifies a variable and a value to be posted when the behavior is *not* in the *active* state. It is a equal-separated pair such as `OUT_OF_RANGE`=true. More than one flag may be provided. These can be used to satisfy or block the conditions of other behaviors.

ENDFLAG This parameter specifies a variable and a value to be posted when the behavior completes. The circumstances causing completion are unique to the individual behavior. However, if the behavior has a `DURATION` specified, the `completed` flag is set to true when the duration is exceeded. The value of this parameter is a equal-separated pair such as `ARRIVED_HOME`=true.

UPDATES This parameter specifies a variable from which updates to behavior configuration parameters are read from after the behavior has been initially instantiated and configured at the helm start-up time. Any parameter and value pair that would have been legal at start-up time is legal at runtime. This is one of the primary hooks to the helm for mission control—the other being the behavior conditions described above.

NOSTARVE The `NOSTARVE` parameter allows a behavior to assert a maximum staleness for one or more MOOS variables, i.e., the time since the variable was last updated.

PERPETUAL Setting the `perpetual` parameter to `true` allows the behavior to continue to run even after it has completed and posted its endflags.

2.4.2.2 Altering Behaviors Dynamically with the `UPDATES` Parameter

The parameters of a behavior can be made to allow dynamic modifications—after the helm has been launched and executing the initial mission in the behavior file. The modifications come in a single MOOS variable specified by the parameter `UPDATES`. For example, consider the simple waypoint behavior configuration below in Listing 1. The return point is the (0,0) point in local coordinates, and return speed is 2.0 m/s. When the conditions are met, this is what will be executed.

Listing 1—An example behavior configuration using the `UPDATES` parameter.

```
Behavior = BHV_Waypoint
{
  name      = WAYPT_RETURN
  priority  = 100
  speed     = 2.0
  radius    = 8.0
  points    = 0,0
  UPDATES   = RETURN_UPDATES
  condition = RETURN = true
  condition = DEPLOY = true
}
```

If, during the course of events, a different return point or speed is desired, this behavior can be altered dynamically by writing to the variable specified by the `UPDATES` parameter, in this case, the variable `RETURN_UPDATES` (line 7 in Listing 1). The syntax for this variable is of the form:

```
parameter=value # parameter=value # ... # parameter=value
```

White space is ignored. The "#" character is treated as special for parsing the line into separate parameter–value pairs. It cannot be part of a parameter component or value component. For example, the return point and speed for this behavior could be altered by any other MOOS process that writes to the MOOS variable:

```
RETURN_UPDATES = "points = (50,50) # speed = 1.5"
```

Each parameter–value pair is passed to the same parameter setting routines used by the behavior on initialization. The only difference is that an erroneous parameter–value pair will simply be ignored as opposed to halting the helm as done on start-up. If a faulty parameter–value pair is encountered, a warning will be written to the variable `BHV_WARNING`. For example:

```
BHV_WARNING = "Faulty update for behavior: WAYPT_RETURN."
```

Note that a check for parameter updates is made at the outset of helm iteration loop for a behavior with the call `checkUpdates()`. Any updates received by the helm on the current iteration will be applied prior to behavior execution and in effect for the current iteration.

2.4.3 IvP Helm Behaviors in the Public Domain

Below is a brief description of several commonly used behaviors in the IvP Helm for the marine vehicle domain. This is followed by a longer description of a few behaviors chosen for illustration.

1. The *AvoidCollision* behavior will maneuver the vehicle to avoid a collision with a given contact. The generated objective function is based on the calculated closest point of approach (CPA) for a given contact and candidate maneuvers. The safe-distance tolerance and policy for priority based on range is provided in mission configuration.
2. The *AvoidObstacles* behavior will maneuver the vehicle to avoid obstacles at known locations, each expressed as a convex polygons. The safe-distance tolerance and policy for priority based on range are configuration parameters.
3. The *CutRange* behavior will maneuver the vehicle to reduce the range between itself and a given contact. The generated objective function is based either on the calculated CPA for a given contact and candidate maneuver or based purely on a greedy approach of heading toward the contact's present position. The policies may be combined (weighted) by the user in mission configuration. This

behavior also has the ability to extrapolate the other vehicle's position from prior reports when contact reports are received intermittently.

4. The *GoToDepth* behavior will drive the vehicle to a sequence of specified depths and duration at each depth. The duration is specified in seconds and reflects the time at depth after the vehicle has first achieved that depth, where achieving depth is defined by the user-specified tolerance parameter. If the current depth is within the tolerance, that depth is considered to have been achieved. The behavior also stores the previous depth from the prior behavior iteration, and if the target depth is between the prior depth and current depth, the depth is considered to be achieved regardless of tolerance setting.
5. The *Loiter* behavior is used for transiting to and repeatedly traversing a set of waypoints. A similar effect can be achieved with the waypoint behavior, but this behavior assumes a set of waypoints forming a convex polygon. It also robustly handles dynamic exit and reentry modes when or if the vehicle diverges from the loiter region due to external events. It is dynamically reconfigurable to allow a mission control module to repeatedly reassign the vehicle to different loiter regions by using a single persistent instance of the behavior.
6. The *MemoryTurnLimit* behavior is used to avoid vehicle turns that may cross back on its own path and risk damage to towed equipment. It is configured with two parameters combined to set a vehicle turn radius pseudo-limit. This behavior is described more fully in Sect. 2.4.3.2.
7. The *OpRegion* behavior provides four different types of safety functionality: (a) a boundary box given by a convex polygon in the x-y or lat-lon plane, (b) an overall time-out, (c) a depth limit, and (d) an altitude limit. The behavior does not produce an objective function to influence the vehicle to avoid violating these safety constraints. This behavior merely monitors the constraints and may post an error resulting in the posting of an all-stop command.
8. The *PeriodicSpeed* behavior will periodically influence the speed of the vehicle while remaining neutral at other times. The timing is specified by a given period in which the influence is on and a period specifying when the influence if off. The motivation was to provide an ability to periodically reduce self-noise to allow for a window of acoustic communications.
9. The *PeriodicSurface* behavior will periodically influence the depth and speed of the vehicle while remaining neutral at other times. The purpose is to bring the vehicle to the surface periodically to achieve some event specified by the user, typically the receipt of a GPS fix. Once this event is achieved, the behavior resets its internal clock to a given period length and will remain idle until a clock time-out occurs.
10. The *Shadow* behavior will mimic the observed heading and speed of another given vehicle, regardless of its position relative to the vehicle.
11. The *StationKeep* behavior is designed to keep the vehicle at a given lat/lon or x,y station-keep position by varying the speed to the station point as a linear function of its distance to the point.
12. The *Trail* behavior will attempt to keep the vehicle at a given range and bearing to another specified vehicle. It may serve the purpose of formation-keeping.

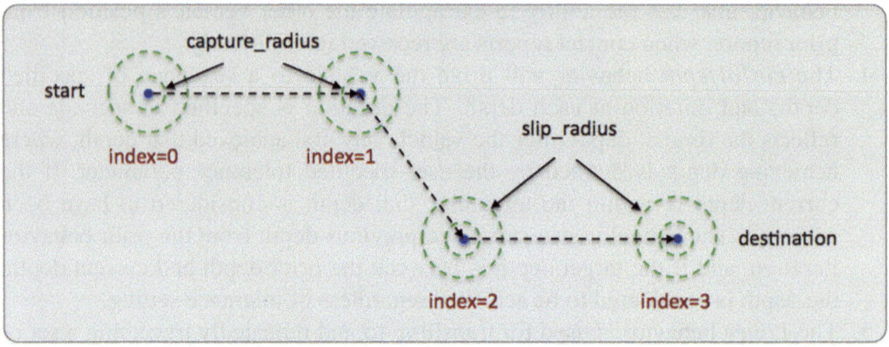

Fig. 2.14 The waypoint behavior: The waypoint behavior traverses a set of waypoints. A capture radius defines what it means to reach a waypoint, and a slip radius is specified to define what it means to be "close enough" should progress toward the waypoint be noted to degrade

It has the ability to extrapolate the other vehicle's position from prior reports when contact reports are received intermittently.

13. The *Waypoint* behavior is used for transiting to a set of specified waypoint in the $x - y$ plane. The primary parameter is the set of waypoints. Other key parameters are the inner and outer radius around each waypoint that determine what it means to have met the conditions for moving on to the next waypoint.

We now explore three of the above behaviors in detail.

2.4.3.1 The Waypoint Behavior

The waypoint behavior is used for transiting to a set of specified waypoint in the $x - y$ plane. The primary parameter is the set of waypoints. Other key parameters are the inner and outer radius around each waypoint that determine what it means to have met the conditions for moving on to the next waypoint. The basic idea is shown in Fig. 2.14.

The behavior may also be configured to perform a degree of track-line following, i.e., steering the vehicle not necessarily toward the next waypoint but to a point on the line between the previous and next waypoint. This is to ensure the vehicle stays closer to this line in the face of external forces such as wind or current. The behavior may also be set to "repeat" the set of waypoints indefinitely or a fixed number of times. The waypoints may be specified either directly at start-up or supplied dynamically during operation of the vehicle. There are also a number of accepted geometry patterns that may be given in lieu of specific waypoints, such as polygons and lawnmower pattern.

The Waypoint Behavior Configuration Parameters

The configuration parameters and variables published collectively define the interface for the behavior. The following are the parameters for this behavior, in addition to the configuration parameters defined for all behaviors, described in Sect. 2.4.2.

`POINTS:`	A colon-separated list of x,y pairs given as points in 2D space, in meters.
`SPEED:`	The desired speed (m/s) at which the vehicle travels through the points.
`CAPTURE_RADIUS:`	The radius tolerance, in meters, for satisfying the arrival at a waypoint.
`SLIP_RADIUS:`	An "outer" capture radius. Arrival declared when the vehicle is in this range and the distance to the next waypoint begins to increase.
`ORDER:`	The order in which waypoints are traversed—`"normal"` or `"reverse"`.
`LEAD:`	If this parameter is set, track-line following between waypoints is enabled.
`LEAD_DAMPER:`	Distance from track-line within which the `lead` distance is stretched out.
`REPEAT:`	Number of *extra* times waypoints are traversed. Or `"forever"`.
`CYCLEFLAG:`	MOOS variable–value pairs posted at end of each cycle through waypoints.
`VISUAL_HINTS:`	Hints for visual properties in variables posted intended for rendering.

Variables Published by the Waypoint Behavior

The MOOS variables below will be published by the behavior during normal operation, in addition to any configured flags. A variable published by any behavior may be suppressed or changed to a different variable name using the `post_mapping` configuration parameter, defined for all behaviors.

`WPT_STAT:`	A comma-separated string showing the status in hitting the list of points.
`WPT_INDEX:`	The index of the current waypoint. First point has index 0.
`CYCLE_INDEX:`	The number of times the full set of points has been traversed, if repeating.
`VIEW_POINT:`	Visual cue for indicating the waypoint currently heading toward.
`VIEW_POINT:`	Visual cue for indicating the steering point, if the `lead` parameter is used.
`VIEW_SEGLIST:`	Visual cue for rendering the full set of waypoints.

```
     WPT_STAT = vname=alpha,behavior-name=waypt_survey,
                index=1,hits=1/1,cycles=0,dist=30,eta=15
    WPT_INDEX = 3
  CYCLE_INDEX = 1
   VIEW_POINT = x=0,y=0,active=false,label=alpha's next waypoint,
                type=waypoint,source=alpha_waypt_return
 VIEW_SEGLIST = pts={60,-40:60,-160:150,-160:180,-100:150,-40},
                label=alpha_waypt_survey,vertex_color=yellow
```

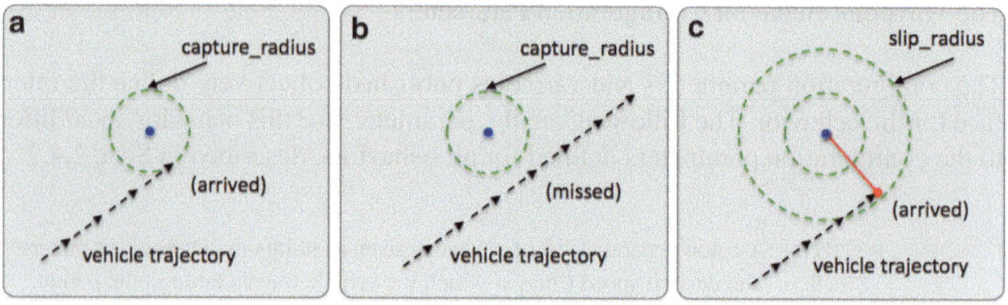

Fig. 2.15 The capture radius and slip radius: (**a**) A successful waypoint arrival by achieving proximity less than the capture radius. (**b**) A missed waypoint likely resulting in the vehicle looping back to try again. (**c**) A missed waypoint but arrival declared anyway when the distance to the waypoint begins to increase and the vehicle is within the slip radius

Specifying Waypoints: The `Points`, `Order`, and `Repeat` Parameters

The waypoints may be specified explicitly as a colon-separated list of comma-separate pairs or implicitly using a geometric description. The order of the parameters may also be reversed with the `order` parameter. An example specification:

```
points    = 60,-40:60,-160:150,-160:180,-100:150,-40
order     = reverse   // default is "normal"
repeat    = 3         // default is 0
```

A waypoint behavior with this specification will traverse the five points in reverse order (150, −40 first) four times (one initial cycle and then repeated three times) before completing. If there is a syntactic error in this specification at helm start-up, an output error will be generated and the helm will not continue to launch. If the syntactic error is passed as part of a dynamic update (see Sect. 2.4.2.2), the change in waypoints will be ignored and the a warning posted to the `BHV_WARNING` variable. The behavior can be set to repeat its waypoints indefinitely by setting `repeat="forever"`.

The `Capture_Radius` and `Slip_Radius` Parameters

The `capture_radius` parameter specifies the distance to a given waypoint the vehicle must be before it is considered to have arrived at or achieved that waypoint. It is the inner radius around the points in Fig. 2.14. The slip radius parameter specifies an alternative criteria for achieving a waypoint.

As the vehicle progresses toward a waypoint, the sequence of measured distances to the waypoint decreases monotonically. The sequence becomes non-monotonic when it hits its waypoint or when there is a near-miss of the waypoint capture radius. The `nm_radius` is a capture radius distance within which a detection of increasing distances to the waypoint is interpreted as a waypoint arrival. This distance would have to be larger than the capture radius to have any effect. As a rule of thumb, a distance of twice the capture radius is practical. The idea is shown in Fig. 2.15. The

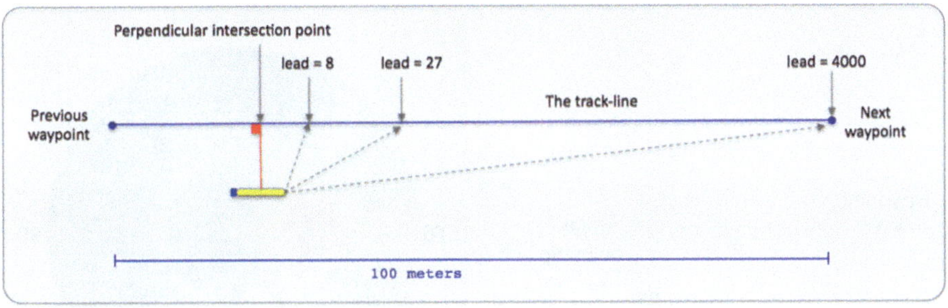

Fig. 2.16 The track-line mode: When in track-line mode, the vehicle steers toward a point on the track-line rather than simply toward the next waypoint. The steering point is determined by the `lead` parameter. This is the distance from the perpendicular intersection point toward the next waypoint

behavior keeps a running tally of hits achieved with the capture radius and those achieved with the slip radius. These tallies are reported in a status message.

Track-Line Following Using the `lead` Parameter

By default the waypoint behavior will output a preference for the heading that is directly toward the next waypoint. By setting the `lead` parameter, the behavior will instead output a preference for the heading that keeps the vehicle closer to the track-line or the line between the previous waypoint and the waypoint currently being driven to (Fig. 2.16).

The distance specified by the `lead` parameter is based on the perpendicular intersection point on the track-line. This is the point that would make a perpendicular line to the track-line if the other point determining the perpendicular line were the current position of the vehicle. The distance specified by the `lead` parameter is the distance from the perpendicular intersection point toward the next waypoint, and defines an imaginary point on the track-line. The behavior outputs a heading preference based on this imaginary steering point. If the lead distance is greater than the distance to the next waypoint along the track-line, the imaginary steering point is simply the next waypoint.

If the `lead` parameter is enabled, it may be optionally used in conjunction with the `lead_damper` parameter. This parameter expresses a distance from the track-line in meters. When the vehicle is within this distance, the value of the `lead` parameter is stretched out toward the next waypoint to soften, or dampen, the approach to the track-line and reduce overshooting the track-line.

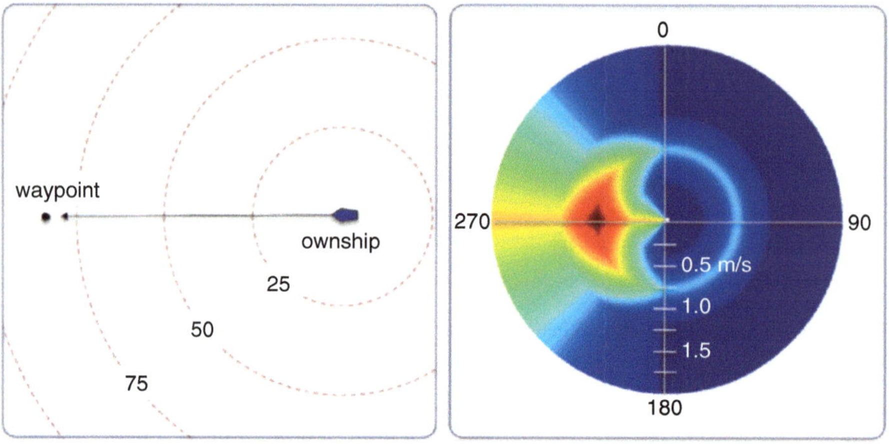

Fig. 2.17 A waypoint objective function: The objective function produced by the waypoint behavior is defined over possible heading and speed values. Depicted here is an objective function favoring maneuvers to a waypoint 270° from the current vehicle position and favoring speeds closer to the mid-range of capable vehicle speeds. Higher speeds are represented farther radially out from the center

The Objective Function Produced by the Waypoint Behavior

The waypoint behavior produces a new objective function, at each iteration, over the variables `speed` and `course/heading`. The behavior can be configured to generate this objective function in one of two forms, either by coupling two independent one-variable functions or by generating a single coupled function directly. The functions rendered in Fig. 2.17 are built in the first manner.

2.4.3.2 The MemoryTurnLimit Behavior

The objective of the Memory-Turn-Limit behavior is to avoid vehicle turns that may cross back on its own path and risk damage to the towed equipment. Its configuration is determined by the two parameters described below which combine to set a vehicle turn radius limit. However, it is not strictly described by a limited turn radius; it stores a time-stamped history of recent recorded headings, maintains a *heading average*, and forms its objective function on a range deviation from that average. This behavior merely expresses a preference for a particular heading. If other behaviors also have a heading preference, coordination/compromise will take place through the multi-objective optimization process. The following parameters are defined for this behavior:

`MEMORY_TIME`: The duration of time for which the heading history is maintained and heading average calculated.

TURN_RANGE: The range of heading values deviating from the current heading average outside of which the behavior reflects sharp penalty in its objective function.

The heading history is maintained locally in the behavior by storing the currently observed heading and keeping a queue of n recent headings within the MEMORY_TIME threshold. The heading average calculation below handles the issue of angle wrap in a set of n headings $h_0 \dots h_{n-1}$ where each heading is in the range $[0, 359]$.

$$\text{heading\_avg} = \text{atan2}(s, c) \cdot 180/\pi, \tag{2.1}$$

where s and c are given by

$$s = \sum_{k=0}^{n-1} \sin(h_k \pi/180)), \qquad c = \sum_{k=0}^{n-1} \cos(h_k \pi/180)).$$

The vehicle turn radius r is not explicitly a parameter of the behavior but is given by

$$r = v/((u/180)\pi),$$

where v is the vehicle speed and u is the turn rate given by u = TURN_RANGE/ MEMORY_TIME. The same turn radius is possible with different pairs of values for TURN_RANGE and MEMORY_TIME. However, larger values of TURN_RANGE allow sharper initial turns but temper the turn rate after the initial sharper turn has been achieved.

The objective function produced by this behavior looks effectively like a constraint on the value of the heading. Some typical functions are rendered from simulation in Fig. 2.18. This figure depicts the MemoryTurnLimit IvP function alongside the Waypoint IvP function as the vehicle traverses a set of waypoints, where the final waypoint (waypoint #3 in the figure) represents nearly a 180° turn with respect to the prior waypoint. The first frame in the figure depicts the vehicle trajectory *without* the influence of the MemoryTurnLimit behavior and shows a very sharp turn between waypoints #2 and #3. The MemoryTurnLimit behavior, shown in the last five frames of Fig. 2.18, is used to avoid the sharp turn in the first frame. This behavior was configured with TURN_RANGE=45 and MEMORY_TIME=20. The turn between waypoints #1 and #2 is not affected by the MemoryTurnLimit behavior because the new desired heading (180°) is within the tolerance of the heading history recorded up to the turning point. The same cannot be said when the vehicle reaches waypoint #2 and begins to traverse to waypoint #3. The recent heading history as it arrives at waypoint #2 reflects the time spent traversing from waypoint #1 and #2. The heading average given by (2.1) at this point (time = 98) is 180°, and the IvP function produced by the MemoryTurnLimit behavior restricts the vehicle heading to be $180 \pm 45°$. As the vehicle continues its turn toward waypoint #3, the heading average of the MemoryTurnLimit behavior and its IvP function evolve to eventually allow a desired heading that is consistent with the optimal point of the Waypoint IvP function as shown in the last frame at time = 170. The resulting turn between

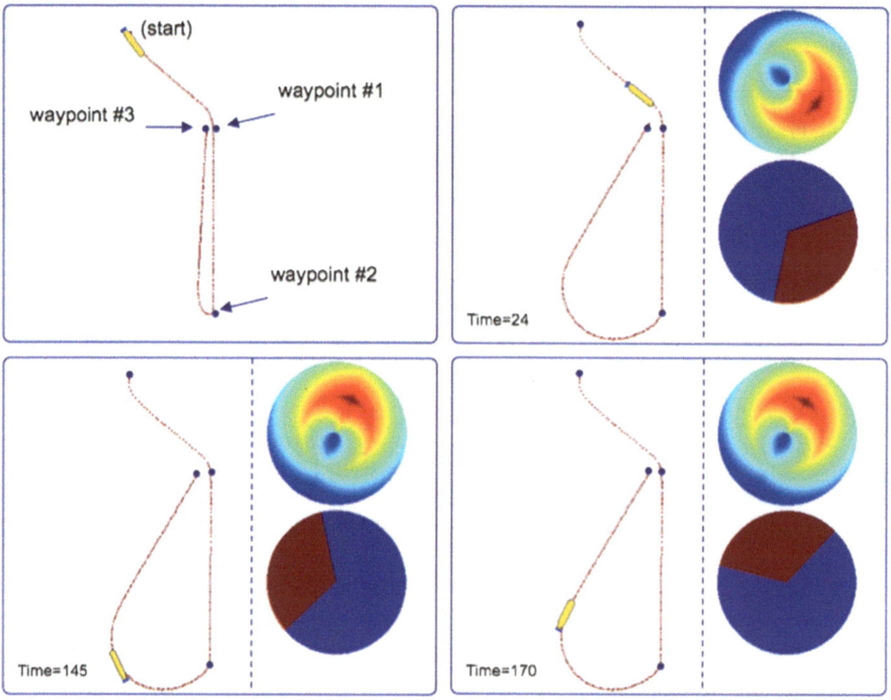

Fig. 2.18 The MemoryTurnLimit and waypoint behaviors: The MemoryTurnLimit behavior is used to avoid sharp vehicle turns between waypoints as shown, from simulation, in the first, *upper-left frame*. The MemoryTurnLimit behavior is configured with `TURN_RANGE=45` and `MEMORY_TIME=20`. In each frame, the IvP function of the waypoint behavior is shown in the *upper right* and that of the MemoryTurnLimit behavior is in the *lower right*. At time = 24, the vehicle is approaching waypoint #1 and its recent heading is 140°. After the vehicle reaches waypoint #2, the desired heading from the waypoint behavior is not in the range of allowed headings from the MemoryTurnLimit behavior. As the vehicle turns, e.g., time = 145, the heading history of the MemoryTurnLimit behavior evolves to eventually allow the peak desired heading of the waypoint behavior, time = 170

waypoints #2 and #3 is much wider than that shown in the first frame. Discussion of this behavior used in field experiments with a UUV towing a sensor array can be found in [4].

2.4.3.3 The AvoidCollision Behavior

The AvoidCollision behavior will produce IvP objective functions designed to avoid collisions (and near collisions) with another specified vehicle. The IvP functions produced by this behavior are defined over the domain of possible heading and speed choices. The utility assigned to a point in this domain (a heading–speed pair) depends in part on the calculated CPA between the candidate maneuver leg and the

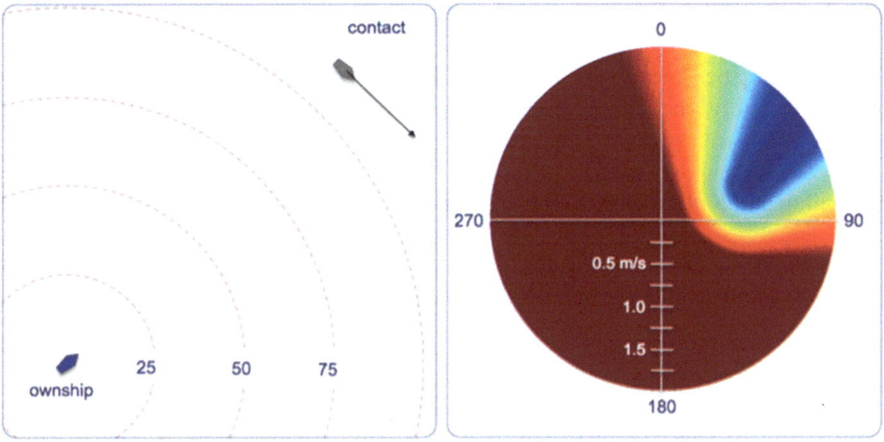

Fig. 2.19 The closest point of approach mapping: The function on the *right* indicates the relative change in calculated closest point of approach between ownship and contact position and trajectory shown on the *left*

contact leg formed from the contact's position and trajectory. Figure 2.19 shows the relationship $\text{cpa}(\theta, v)$ between CPA and candidate maneuvers (θ, v), where θ=heading and v=speed, for a given relative position between ownship and a given contact vehicle and trajectory. The IvP function generated by the AvoidCollision behavior applies a further user-defined utility function to the CPA calculation for a candidate maneuver, $f(\text{cpa}(\theta, v))$. The form of $f()$ is determined by behavior configuration parameters described below.

`COMPLETED_DIST`:	Range to contact outside of which the behavior completes and dies.
`MAX_UTIL_CPA_DIST`:	Range to contact outside which a considered maneuver has max utility.
`MIN_UTIL_CPA_DIST`:	Range to contact within which a considered maneuver will have min utility.
`PWT_INNER_DIST`:	Range to contact within which the behavior has maximum priority weight.
`PWT_OUTER_DIST`:	Range to contact outside which the behavior has zero priority weight.
`CONTACT`:	Name or unique identifier of a contact to be avoided.
`DECAY`:	Time interval during which extrapolated position slows to a halt.
`EXTRAPOLATE`:	If true, contact position is extrapolated from last position and trajectory.
`TIME_ON_LEG`:	The time on leg, in seconds, used for calculating closest point of approach.

The AvoidCollision Configuration Parameters

The following parameters are defined for this behavior, in addition to the parameters defined for all IvP behaviors discussed earlier. A more detailed description follows:

Spawning and Killing New Behavior Instances Dynamically

The AvoidCollision behavior may be configured as a *template* that spawns a new behavior for each new contact, using the below two configuration lines:

```
templating = spawn
updates    = COLL_AVOID_INFO
```

Configured in this manner, the following posting to the MOOSDB would suffice to spawn a new instance of the behavior:

```
COLL_AVOID_INFO = name=avd_zulu_002 # contact = zulu_002
```

The MOOS variable matches the MOOS variable specified in the `UPDATES` configuration parameter. The `"name=avd_zulu_002"` component specifies a unique behavior name and indicates to the helm that a new behavior is to be spawned upon receipt. When the behavior is spawned, all initial behavior parameters supplied in the behavior mission file are applied, and the `"contact=zulu_002"` component is applied as a further configuration parameter. The new AvoidCollision instance will remain with the helm until the contact goes out of the range specified by the `COMPLETED_DIST` parameter. The posting of the MOOS variable that triggers the spawning is done by a contact manager. In this case the contact manager is a separate MOOS application called pBasicContactMgr. Details of this are beyond the scope of this chapter.

Configuring the AvoidCollision Behavior Utility Function

The IvP function generated by this behavior is defined over the range of possible heading and speed decisions. Its form is derived in part from the calculation of the CPA calculated for a candidate maneuver leg, of a duration given by the `TIME_ON_LEG` parameter, which is set to 60 seconds by default. The *utility* of a given CPA value is determined further by a pair of configuration parameters. Distances less than or equal to `MIN_UTIL_CPA_DIST` are given the lowest utility, equivalent to an actual collision. Distances greater than or equal to `MAX_UTIL_CPA_DIST` are given the highest utility. These two parameters, and the raw CPA calculations, determine the form of the function. The magnitude, or weight, of the function is determined by the range between the two vehicles and two further configuration parameters. At ranges greater than `PWT_OUTER_DIST`, the weight is set to zero. At ranges less `PWT_INNER_DIST`, the weight is set to 100 % of the user-configured priority weight. The weight varies linearly when the range between vehicles falls somewhere between.

Closest Point of Approach Calculations and Caching

The production of IvP functions for this behavior is potentially CPU intensive compared to other behaviors, primarily due to the repeated calculations of CPA values. Recall from Sect. 2.3.6.2 that IvP functions built with the reflector tool are built by repeated sampling of the underlying function. This repeated sampling, and the existence of common partial calculations between samples, allows for caching of intermediate results to greatly speed up sampling for this behavior. This is implemented in a C++ class called a CPAEngine in separate utility library for use in other behaviors that reason about CPA, such as the CutRange and Trail behaviors.

Current ownship position is known and given by $(x,\ y)$, and the other vehicle's current position and trajectory is given by $(x_b,\ y_b,\ \theta_b,\ v_b)$. To compute the CPA distance for a given $\langle \theta, v, t \rangle$, first the time $t_{\min}$ when the minimum distance between two vehicles occurs is computed. The distance between the two vehicles at the current time can be determined by the Pythagorean theorem. For any given time t (where the current time is $t = 0$), and assuming the other vehicle stays on a constant trajectory, the distance between the two vehicles for any chosen $\langle \theta, v, t \rangle$ is given by

$$\text{dist}^2(\theta, v, t) = k_2 t^2 + k_1 t + k_0, \tag{2.2}$$

where

$$\begin{aligned}
k_2 &= \cos^2(\theta)v^2 - 2\cos(\theta)v\cos(\theta_b)v_b + \cos^2(\theta_b)v_b^2 + \sin^2(\theta)v^2 \\
&\quad -2\sin(\theta)v\sin(\theta_b)v_b + \sin^2(\theta_b)v_b^2, \\
k_1 &= 2\cos(\theta)vy - 2\cos(\theta)vy_b - 2y\cos(\theta_b)v_b + 2\cos(\theta_b)v_b y_b \\
&\quad +2\sin(\theta)vx - 2\sin(\theta)vx_b - 2x\sin(\theta_b)v_b + 2\sin(\theta_b)v_b x_b, \\
k_0 &= y^2 - 2yy_b + y_b^2 + x^2 - 2xx_b + x_b^2.
\end{aligned}$$

The stationary point is obtained by taking the first derivative with respect to t:

$$\text{dist}^2(\theta, v, t)' = 2k_2 t + k_1.$$

Since there is no "maximum" distance, this stationary point always represents the time of the CPA, and therefore $t_{\min} = -k_1/2k_2$. The value of $t_{\min}$ may be in the past, i.e., less than zero, if the two vehicles are currently opening range. On the other hand, $t_{\min}$ may occur after t, the time length of the candidate maneuver $\langle \theta, v, t \rangle$. Therefore the value of $t_{\min}$ is clipped by $[0, t]$. Furthermore, $t_{\min} = 0$ in the special case when the two vehicles have the same heading and speed (the only case where k_2 is zero). The actual CPA value is obtained by substituting $t_{\min}$ back into (2.2):

$$\text{cpa}(\theta, v) = \sqrt{k_2 {t_{\min}}^2 + k_1 t_{\min} + k_0}. \tag{2.3}$$

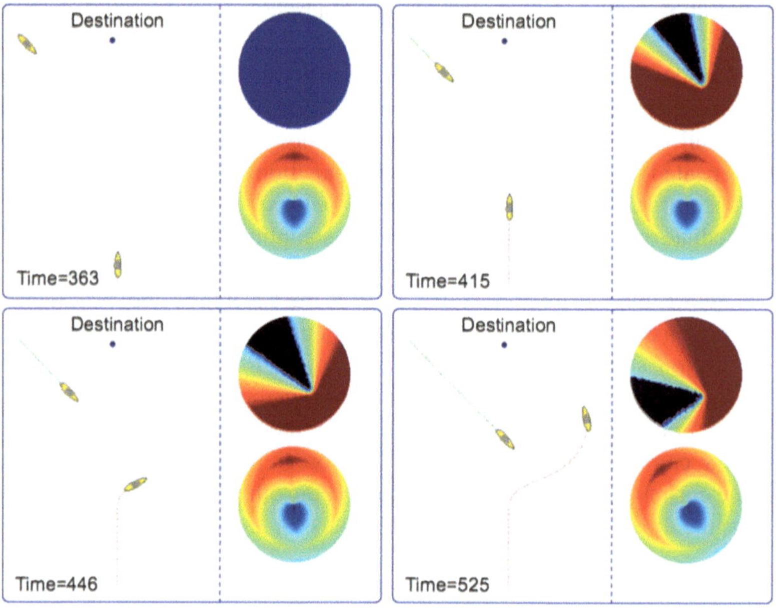

Fig. 2.20 The AvoidCollision and waypoint behaviors: The bravo vehicle maneuvers to a destination point with the waypoint behavior. The IvP objective function produced by the waypoint behavior is the lower function shown on the *right* in each frame. Beginning in the second frame, time=415, the AvoidCollision behavior becomes active and begins to produce IvP objective functions shown in the *upper right* of each frame. At each point in time, the helm chooses the heading and speed that represent the optimal decision given the pair of IvP functions

In the generation of a single IvP function with $\text{cpa}(\theta, v)$ as a component of each sample of the decision space, intermediate values (2.2) may be cached that have the same values of current vehicle position (x, y) and current position and trajectory of the other vehicle (x_b, y_b, θ_b, v_b). A further cache, normally of size 360, is typically used for terms involving θ, ownship heading.

The AvoidCollision behavior is shown in Fig. 2.20 in a simple scenario working with the waypoint behavior in simulation. In Fig. 2.20, the effect of the AvoidCollision behavior in two fielded UUVs is shown.

2.5 Conclusions

In this chapter two architectures were described in detail. The MOOS publish-subscribe middleware is both an architecture and mechanism for inter-process communication and process scheduling but also as an open-source software project, a collection of substantial applications for sensing, communications, autonomy, debugging, and post-mission analysis. The IvP Helm is a behavior-based architecture, unique in its use of the IvP model for multi-objective optimization for resolving competing autonomy behaviors. It is also an open-source project that includes many well-tested vehicle behaviors and autonomy tools for creating, debugging, and analyzing autonomy capabilities.

Acknowledgements The prototype of MOOS was developed by Paul Newman at MIT under the GOATS'2000 NURC Joint Research Program, with ONR support from Grant N-00014-97-1-0202 (Program Managers Tom Curtin, Code 322OM, Jeff Simmen, Code 321OA, Tom Swean, Code 321OE, and Randy Jacobson, Code 321TS). The development of the Nested Autonomy concept for environmental acoustic sensing and the MIT component of the GLINT'08 experiment was funded by the Office of Naval Research under the GOATS program, Grant N-00014-08-1-0013 (Program Manager Ellen Livingston, ONR Code 321OA). The development of the unified communication, command, and control infrastructure and the execution of the SWAMSI09 experiment was supported by ONR, Grant N-00014-08-1-0011 (Program Manager Bob Headrick, Code 321OA).

The IvP Helm autonomy software and the basic research involved in the interval programming model for multi-objective optimization has been developed under support from ONR Code 311 (Program Managers Don Wagner and Behzad Kamgar-Parsi). Prior prototype development of IvP concepts benefited from the support of the In-house Laboratory Independent Research (ILIR) program at the Naval Undersea Warfare Center in Newport, RI.

The NATO Undersea Research Centre (NURC) has supported the development of the MOOS-IvP Nested Autonomy concept by conducting 7 major field experiments, in which MIT LAMSS has been a partner, including GOATS'2000 and GOATS'2002, FAF'2003, FAF'2005, CCLNet'08, GLINT'08, and GLINT'09. Without the world-class seagoing experiment capabilities of NURC, with its state-of-the-art RVs, NRV Alliance and CRV Leonardo and their outstanding crew, and NURC's excellent engineering and logistics support, the Nested Autonomy concept and the underlying MOOS-IvP software base would not have reached the level of sophistication and robustness that it has achieved.

References

1. Arkin RC (1987) Motor schema based navigation for a mobile robot: an approach to programming by behavior. In: Proceedings of the IEEE conference on robotics and automation, Raleigh, NC, pp 264–271
2. Arkin RC, Carter WM, Mackenzie DC (1993) Active avoidance: escape and dodging behaviors for reactive control. Int J Pattern Recognit Artif Intell 5(1):175–192
3. Benjamin MR (2004) The interval programming model for multi-objective decision making. In: Technical report AIM-2004-021, Computer Science and Artificial Intelligence Laboratory, MIT, Cambridge, MA
4. Benjamin M, Battle D, Eickstedt D, Schmidt H, Balasuriya A (2007) Autonomous control of an unmanned underwater vehicle towing a vector sensor array. In: International conference on robotics and automation (ICRA), Rome, Italy
5. Benjamin M, Schmidt H, Leonard JJ. http://www.moos-ivp.org
6. Bennet AA, Leonard JJ (2000) A behavior-based approach to adaptive feature detection and following with autonomous underwater vehicles. IEEE J Oceanic Eng 25(2):213–226
7. Brooks RA (1986) A robust layered control system for a mobile robot. IEEE J Robotics Automation RA-2(1):14–23
8. Carreras M, Batlle J, Ridao P (2000) Reactive control of an AUV using motor schemas. In: International conference on quality control, automation and robotics, Cluj Napoca, Rumania
9. Dantzig GB (1948) Programming in a linear structure. Comptroller, US Air Force, Washington, DC

10. Khatib O (1985) Real-time obstacle avoidance for manipulators and mobile robots. In: Proceedings of the IEEE international conference on robotics and automation, St. Louis, MO, pp 500–505
11. Kumar R, Stover JA (2001) A behavior-based intelligent control architecture with application to coordination of multiple underwater vehicles. IEEE Trans Syst, Man, and Cybernetics - Part A: Cybernetics 30(6):767–784
12. Newman PM (2003) MOOS - a mission oriented operating suite. In: Technical report OE2003-07, MIT Department of Ocean Engineering
13. Pirjanian P (1998) Multiple objective action selection and behavior fusion. Ph.D. thesis, Aalborg University
14. Riekki J (1999) Reactive task execution of a mobile robot. Ph.D. thesis, Oulu University
15. Rosenblatt JK (1997) DAMN: a distributed architecture for mobile navigation. Ph.D. thesis, Carnegie Mellon University, Pittsburgh, PA
16. Rosenblatt JK, Williams SB, Durrant-Whyte H (2002) Behavior-based control for autonomous underwater exploration. Int J Inform Sci 145(1–2):69–87
17. Williams SB, Newman P, Dissanayake G, Rosenblatt JK, Durrant-Whyte H (2000) A decoupled, distributed AUV control architecture. In: Proceedings of 31st international symposium on robotics, Montreal, Canada, pp 246–251

Chapter 3
Towards Deliberative Control in Marine Robotics

Kanna Rajan, Frédéric Py, and Javier Barreiro

Abstract We describe a general purpose artificial-intelligence-based control architecture that incorporates in situ decision making for autonomous underwater vehicles (AUVs). The Teleo-reactive executive (`T-REX`) framework deliberates about future states, plans for actions, and executes generated activities while monitoring plans for anomalous conditions. Plans are no longer scripted a priori but synthesized onboard with high-level directives instead of low-level commands. Further, the architecture uses multiple control loops for a "divide-and-conquer" problem-solving strategy allowing for incremental computational model building, robust and focused failure recovery, ease of software development, and ability to use legacy or nonnative computational paradigms. Vehicle adaptation and sampling occurs in situ with additional modules which can be selectively used depending on the application in focus. Abstraction in problem solving allows different applications to be programmed relatively easily, with little to no changes to the core search engine, thereby making software engineering sustainable. The representational ability to deal with time and resources coupled with Machine Learning techniques for event detection allows balancing shorter term benefits with longer term needs, an important need as AUV hardware becomes more robust allowing persistent ocean sampling and observation. `T-REX` is in regular operational use at MBARI, providing scientists a new tool to sample and observe the dynamic coastal ocean.

K. Rajan (✉) • F. Py
Monterey Bay Aquarium Research Institute, CA, USA
e-mail: Kanna.Rajan@mbari.org; fpy@mbari.org

J. Barreiro
NASA Ames Research Center, Moffett Field, CA, USA
e-mail: javier.barreiro@nasa.gov

M.L. Seto (ed.), *Marine Robot Autonomy*, DOI 10.1007/978-1-4614-5659-9_3,

3.1 Introduction

Autonomous platforms in the marine environment have had a substantial and immediate impact for both civil and military applications. Autonomous underwater vehicles (AUVs) have not only altered the ability of oceanographers to reach beyond the surface and make high-resolution observations but also impacted the engineering methodologies behind sampling, control, and robotics itself [1–6]. Yet challenges remain, principally in making such robotic devices more attuned to the environment, to sample large swaths at the mesoscale ($>50\,km^2$) and doing so systematically and adaptively. Further, over the last two decades, substantial improvements in hardware including propulsion and battery technologies have outpaced software especially in algorithmic methods in sampling and control and in systematic software engineering development. Large-scale high-resolution observations have become increasingly tractable using autonomous platforms both on and below the surface. This, in turn, have lead to periodic calls for sustained exploration and sampling [7], illustrating the high scientific value placed on such data. The complexity of oceanic processes and the chronic under-sampling of the coastal oceans, especially of dynamic features studied over a variety of spatial and temporal scales, has made it challenging for robotic platforms to autonomously track and sample. A prominent example of such a process is coastal algal blooms which are patchy and could cover large coastal zones. Persistent observation of such dynamic events dictates that our robotic assets track and sample such patches which can evolve rapidly due to inherent biogeochemical activity, as well as advection and diffusion of the water mass it resides in.

At the Monterey Bay Aquarium Research Institute (MBARI), a unique interdisciplinary collaboration between biologists, ecologists, geneticists, and roboticists on the CANON (Controlled, Agile, and Novel Observation Network) program [8] has driven autonomous system exploration and control. Fundamental questions related to sampling strategies have driven the use cases for Advanced concepts in autonomy, specifically to Artificial Intelligence (AI) Planning and Execution. Current work is also looking at how other fields in machine intelligence, especially Machine Learning, can be leveraged to impact the science of engineering systems in marine robotics and as a consequence oceanography itself.

In this chapter, we propose an alternative view to decision making using in situ automated planning [9]. More importantly, we demonstrate that earlier control paradigms (primarily driven by *reactive* control approaches) can be augmented to provide targeted observation and sampling as well as a more nuanced and balanced consideration of mission objectives, environmental conditions, and available resources. Planning[1] (or deliberation or projection in action space) is important to balance current needs of a robot with future desires or goals taking into account time and available resources in situ on the robot. Planning without time or action planning in and of itself is not adequate; dealing with time ensures that the balance between

[1] We use the terms "planning" and "deliberation" interchangeably in this chapter.

now and the future is handled systematically. Yet, plans have to be tied to robotic action and in dynamic real-world environments. Further, plans not only project future states which are likely to change, but they also need to factor in substantial environmental uncertainty.[2] In the oceanographic context, use of environmental cues is often hampered by poor predictive skill [10] of ocean models and little to no availability of a priori synoptic views of the survey area. Planning therefore provides a "buffer" against the stochastic environment.

To date AUV control architectures have primarily been driven by reactive Subsumption-based approaches [11]. They have been adequate for routine straight-line survey trajectories which have played a valuable role in the acceptance of AUVs as a new tool for ocean science and military applications. However, scaling of problem domains in addition to dealing with event response and discovery in the water column challenges *how* controllers are written, maintained, and upgraded over the life cycle of the platform. Dealing with off-nominal endogenous or exogenous conditions including sensor failures stretches how AUVs are currently being utilized; one consequence, for instance, is any off-nominal condition results in a *fail-safe* surfacing mode to radio for help. Software design using these reactive approaches not only makes it hard for sustaining engineering development, but these approaches are also not adequately adaptive for emerging ocean science problems. Despite vigorous defense by Brooks early on [12, 13], autonomous systems do require a model if not of the world then of the *physics* of the vehicle when it interacts with the world.

Criticism of AI planning, on the other hand, also well articulated by Brooks, was targeting the performance of in situ deliberation. Terrestrial robotic platforms were well known to frequently stop and "plan," assuming in the meantime that the world would remain static [14]. In the marine robotics world, such a situation is not permissible given the need to make continuous observations without stopping a non-holonomic robotic platform such as an AUV. However, algorithmic and implementational advances by the AI community over the years have produced a range of planners with demonstrated embedded capabilities [15–20]. Our work has been influenced by and in turn influenced a wide range of activities in the field of AI Planning and Plan Execution with demonstrated real-world capabilities in the space domain [19, 21–23] for NASA. Lessons learned have been applied to the world of marine robotics which we bring together in this chapter. Important lessons that were derived from this rich legacy included the following:

- Projecting plans (or using *deliberation*) can and should balance near-term objectives with end-term or even evolving goals.
- Dealing with time and resources is essential in real-world planning.
- A rich representation that can deal with co-temporal events for modeling complex systems is important.

---

[2]It is important to remember that these principles are broad much as Dwight Eisenhower is reputed to have said *In preparing for battle I have always found that plans are useless, but planning is indispensable* and *Failing to plan is planning to fail*.

- Fast solvers for incremental causal reasoning can be built as a basis for dealing with dynamic controllable events where replanning is tightly integrated.
- Planning should be tightly interleaved with execution to enable responsiveness to events which impact plan failure. This is also important for platform adaptivity.
- A model-based approach should separate control formulation of the platform which describes operational characteristics from the domain. The controller can then be rigorously tested and certified even as the domain model evolves to fit evolving requirements and/or diverse applications.

We have developed, tested, and deployed the Teleo-reactive executive (T-REX) an on-board adaptive control system that integrates AI-based planning and state estimation in a hybrid executive [24–26]. State estimation allows the system to keep track of world evolution as time advances based on a formal model distributed through the architecture. Onboard planning and execution enables adaptation of navigation and instrument control based on estimated state. It further enables goal-directed commanding within the context of projected mission state and allows for replanning for off-nominal situations and opportunistic science events. The framework in addition to being used on an AUV, is general enough to be used for controlling a personal robot [27–29] and is deployed on a European planetary rover testbed [30]. While it is possible to integrate probabilistic estimation on top of this framework [31], such representation is not explicitly handled by the framework and is out of scope of this chapter. Deliberation and reaction are integrated systematically over different temporal and functional scopes within a single agent and a single model that covers the needs of high-level mission management, low-level navigation, instrument control, and detection of unstructured and poorly understood phenomena. T-REX is deployed on MBARI's *Dorado* AUV shown in Fig. 3.1, which to the best of our knowledge is the only operational marine platform anywhere being used for routine scientific surveys with onboard plan synthesis.

This chapter is organized as follows. We start by placing this effort in the context of the overall research in AI Planning in Sect. 3.2 followed by some foundational concepts in our planning framework in Sect. 3.3 which includes a brief overview of AI Planning in Sect. 3.3.1. In Sect. 3.4 we go deeper into the EUROPA framework itself, the basis of all deliberation within T-REX. A key portion is Sect. 3.4.2 which details the internal plan representation within a situated agent like T-REX. Modeling (Sect. 3.4.4), inference (Sect. 3.4.5), and search (Sect. 3.4.6) articulate important details associated with EUROPA critical to T-REX. With these foundational elements of deliberation out of the way, we transition to T-REX itself in Sect. 3.5, bring out architectural details in Sect. 3.5.2, deal with the execution cycle in Sect. 3.5.3, highlight how T-REX deliberates in Sect. 3.5.4, and finally delving deeper into the key concept of synchronization in Sect. 3.5.4.1. Further details on deliberation within T-REX (Sect. 3.5.4.2) and how planning and execution are interleaved are detailed in Sect. 3.5.4.3. We offer results from our experiments at sea in Sect. 3.6, some insights into the near-term future goals of our research in Sect. 3.7, and conclude with Sect. 3.8.

Fig. 3.1 The MBARI *Dorado* AUV on its support vessel the R/V *Zephyr*

3.2 Background and Related Work

A distinct problem in marine robotics has been the use of the overloaded term "autonomy". Not only does the notion transcend different disciplines within engineering in this domain (e.g., the control engineering community sees it distinctively from those in AI) but users of marine platforms in oceanography as well as in maritime defense conflate the methodology of use (tethered vs. untethered) with their control. In this chapter, our notion of "autonomy" refers to not only the notion of dealing with the *sense, plan, and act* paradigm in robotics but also the need to perform *computational search* between different action outcomes, an idea central to the field of AI. Consequently, our literature survey here is targeted towards a more focused and (to this chapter) relevant part of the field of computation.

With few exceptions, most AUV control systems have been variations of reactive approaches. Typically, waypoint-based predefined mission designs are uploaded to the AUV; specialized code fragments called *behaviors* are designed for the specific mission and a choice of behaviors for the mission is used on the computational stack [32]. Swapping in and out of these behaviors using conditionals or cleverer and quantitative approaches such as [33] forms the basis for adaptation and safety in the vehicle. Where such swapping does not aid adaptation, individual behaviors are tweaked to generate some measure of responsiveness to the surrounding environment [34] to generate interesting observations for oceanographers, yet technically naive in terms of scalability and systematicity. Benjamin et al. [35] is clear about the extent of adaptation in stating:

> *The primary difficulty often associated with behavior-based control concerns action selection—namely how to ensure the chosen action really is in the best overall interest of the robot or vehicle.*

Such techniques have proved their mettle in the first round of use of AUVs; they have provided operators a simple way to handle control complexity of the vehicle while returned data at far higher resolution than would have been possible using traditional ship-based methods and at substantially reduced amortized costs of platform operation and ownership over multiple years. AUV operators and clients for their data products have reasons to be well satisfied with results thus far. However, as hardware and more cost-effective sensors become more robust to the environment and as mission durations increase (as demonstrated by glider experiments such as [36]), sustained presence in the ocean requires the ability to deal with off-nominal conditions, which balances the needs of high-level mission management, low-level navigation, instrument control, and detection of unstructured and poorly understood phenomena. Earlier reactive methods are unable to perform such a balancing act without skewering the foundations of the software development methodology leading to code bloat or worse. Carreras et al. [37] provides a good overview of AUV control architectures essentially based on reactive control methods.

Deliberative techniques for robotic control on the other hand contrast in providing a distinct set of advantages:

1. Action selection is driven axiomatically based on a systematic assessment of a range of causal conditions. An action is selected for instantiation only when there is causal support in the form of a constraint in a deterministic model of the vehicle's operation.
2. The system is goal directed which forms an objective towards which the system is expecting to converge providing a foci in action selection.
3. Scaling to different (computational) problem size is dependent on incremental model building rather than being concerned about nuanced interactions between behaviors.
4. Systematic software engineering methodologies such as spiral development [38] can further be used for scaling up the task as demonstrated by [39].
5. When interactions between actions and/or behavior fragments do (and must) occur there are explicit constraints encoded in the model that must be computationally satisfied contrasting with carefully crafted strategies to instantiate behaviors on a stack. Such explicit recording of constraints in the model in turn aids long-term software maintenance and sustaining engineering.

Traditionally generalized planning has been considered computationally intensive (the general planning problem can be worse than NP-complete [9]). A major reason for this belief had to do with the role of the planner (which was assumed to be generative) and its place in the architecture (infrequently called to replan the *entire* mission in off-nominal conditions) within a *sense-plan-act* paradigm. This can limit the reactivity of a robot especially when the environment could change at a rate faster than the planner can plan.

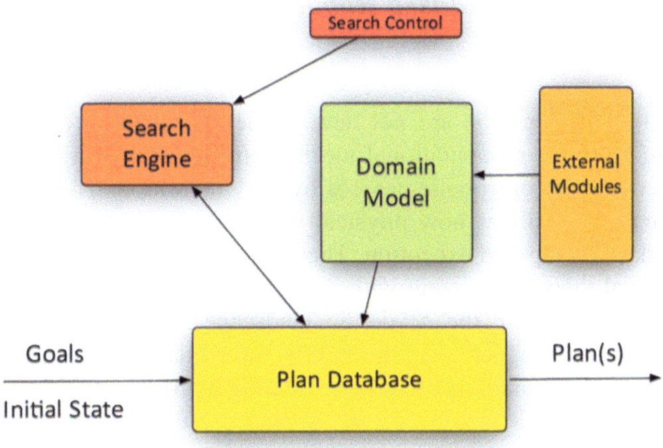

Fig. 3.2 A general architectural block diagram for an AI-based planner

The architecture of a typical AI planner is along the lines of Fig. 3.2. Given a *domain model* which encodes the platform constraints in a higher-level language, an *initial state*, a high-level objective or end *goal*, the *plan database* (or `plandb`) is where all assertions are tracked and maintained. Theorem proving in the form of satisfaction of axioms embodied in the domain model occurs within the `plandb`. It is the job of the *search engine* to provide the inferential mechanism for state-space exploration potentially aided by heuristics encoded for *search control*. In some complex domains [19] *external modules* can provide additional domain expertise which can augment the model. The end result is plan formulation using a multitude of approaches either using forward or backward search or using more generalized *partial order* methods [9].

There are costs associated with such inference-based control which are not inconsequential. Foremost among them is the steep learning curve that comes with the design of the domain model. With limited support tools necessary for knowledge capture and design of the constraints often in stylized higher-level languages (see [40] for an example), it requires the modeler to think in alternative ways to do problem solving. More importantly it is often necessary for the modeler to be exposed to the internal representation of the planner and how it performs search. The level of expertise is often well above what would be expected of a typical well-rounded computer science graduate student. Yet in our experience, the cost of model design in such systems is well balanced against the cost of re-engineering new mission scripts for new deployment in more "classical" script-based controllers.

However, this is mild criticism, given the end benefits and the scientific and engineering goals associated with adaptive and persistent observation for marine robots whether for civil or military applications. Given the general problem solving nature of such deliberative systems, there is also a substantial overall benefit for the field of robotics.

3.2.1 Evolution of AI-Based Robotic Planning: A Necessary Digression

Planning and plan execution are not new to the overall field of robotics. Early motivation of building computational mechanisms for decision making were intended to be deployed on robotic devices by very definition. The seminal volume [41] very clearly articulates how physical manifestation of robots could be controlled by task planning and execution. The very first planner [42] was an exercise in computational theorem proving with early state-space planners popularized by `STRIPS` [43] dominating the academic landscape. An early demonstration of planning as a situated agent[3] was the highly influential `Shakey` the robot [14]. Subsequent work driven largely by defense funding in the United States sequestered the discipline in academic labs principally applied for algorithmic development and if embedded then in testing within the confines of structured indoor settings. The leap to applying them for real-world problem solving is relatively new and driven largely by NASA applications pioneered by [19, 21, 23, 44–46].

The dominant approach for building agent control systems utilize a three-layered architecture (see [47] for a well thought out rationale and survey), notable examples of which include `IPEM` [48], `ROGUE` [16], the LAAS Architecture [17], the Remote Agent Experiment `RAX` [19], and the Autonomous Spacecraft Experiment `ASE` [49] (see [50] for a survey).

Representationally, the first paper on managing time systematically within AI Planning frameworks was [51]; subsequent efforts by Boddy and others [52, 53] refined the notion of using a temporal database and the conceptualization of temporal intervals. Using these concepts Muscettola and others [54–57] developed the notion of plan/schedule envelopes using the notion of state variable instantiated timelines. Simultaneously work by Dechter et al. [58] resulted in efficient temporal constraint propagation which systematically defined the notion of simple temporal networks (STNs) and path and arc consistency algorithms. This work neatly put together earlier efforts by Mackworth and Freuder [59, 60] and the Constraints community on propagation algorithms. In parallel work in the UK resulted in the design of `O-Plan` [61], which leveraged the notion of constraints and metric time, but within a hierarchical task network (HTN) representation. These efforts were soon after Veres paper on planning and time [62] which had a profound effect on the AI Planning community. In particular the work of Muscettola [44] was embraced by NASA for telescope scheduling. Around the same time period, researchers at LAAS came up with an architecture [17] which encapsulated the `IxTeT` planner [55] which dealt with time and resources. It featured advanced concepts for a temporal planner with the notion of chronicles plan recognition for partial plan fragments and early use of systematic resource constraints.

[3] By *situated* we emphasize that an agent is embedded within a physical robot.

3.2.2 The Remote Agent Experiment and Beyond

While HSTS [44] was being implemented as a ground-based planning tool for decision support for US-based space observations (EUVE and Cassini) [63], the opportunity to fly onboard NASAs New Millennium Deep Space 1 came about. The design of the Remote Agent Experiment RAX [19, 21, 39, 45, 64–66] was a direct offshoot of this effort where HSTS (written in LISP) was flown on board and successfully commanded the spacecraft for a week in May 1999.[4] There were a number of significant software engineering lessons learned with the HSTS technology as implemented for the RAX experiment:

1. Constraint-based representations were not only sufficient for plan synthesis but also valuable during debugging and development as a means of building a viable domain model.
2. Domain models when separated from the search engine as articulated by the model-based approach [67] ensured that sufficient effort would be targeted on human validation of the model building phase while ensuring that search engine stability across applications and domains resulted in lower cost to deploy.
3. Flexible temporal representation generating a range of plans as against a single plan was robust for robotic plan execution.
4. If planning and execution were disconnected (as in RAX), dispatchabilty [68] and controllability [69] issues within temporal plans need be addressed. In RAX, a post-processing step was added to counter these effects. It was clear that execution needed access to the planners database and be able to propagate execution time constraints prior to command dispatching.

It was with the demonstration of RAX 65 million miles in outer space, that temporal reasoning using planning methods came to the forefront of AI Planning for situated agents. Lessons learned from RAX led to the development of IDEA [70–73] with the central theme of using planners collectively for problem solving. Another critical theme was iteratively and incremental plan repair of an anytime plan [74] as proposed by CASPER [18].

This last lesson in particular had a lasting impact with the observation that planning and execution are strongly interrelated. This core concept was behind the next generation of the Remote Agent, called IDEA (intelligent deployable execution agents), where planning and execution were *intertwined* within a single domain model that spanned the most abstract (for high-level planning) to the least (for diagnosis). IDEA [70, 71] agents were expected to interact to achieve plan formulation; however, there was no systematic framework for formally governing these interactions. IDEA was also computationally heavy and required substantial effort to customize for an application domain. It did move to a retooled version of HSTS, called EUROPA [75, 76] while being ported to the C++ language.

[4]This was the first (and to our knowledge only) software ever to be written in LISP to be flown in space.

While IDEA was being deployed as a coordinating executive for earth analog rover deployments [77] a separate development was undertaken to deploy the EUROPA planner for NASAs Mars exploration rovers (MER) mission. The MAPGEN (Mixed-initiative Activity Plan GENerator) planner [22, 23, 78, 79] was designed to be used as a decision support in the mixed-initiative mode. MAPGEN continues to be used to this day to command the twin rovers on the Red Planet.

The autonomous sciencecraft experiment (ASE) [49, 80] was another autonomy demonstration in space, this time for a more observable vehicle, the EO-1 in Earth orbit. It encapsulated a classical three-layered architecture like RAX. Goals could not only be sent by a ground segment, but also by an onboard science driver which could encapsulate a range of machine-learned feature detectors which could trigger the planner. However, ASE was not an integrated system; the CASPER planner [18] would generate plans which were executed by a separate commercially available rule-based system [81]. CASPER demonstrates what is called as a *continuous* paradigm for incremental planning and iterative plan repair; this methodology takes the overall mission plan and selectively plans for the near term at a more detailed level. Robust software engineering is also not an important aspect targeted by ASE given the disparate models between the planner and the rule-based executive. With such a system, real-time updates from the environment could not be accommodated since the planner is a monolithic computational environment. The rate of updates from the science drivers is roughly on the order of an earth orbit on the EO-1 vehicle. As a result, the ASE is only an incremental update on RAX and is not suitable for domains where complex automated reasoning and asynchronous environmental observations need to be factored in deliberation. TCA [15] is another framework for robot control and has similar motivations to RAX and ASE. Reactivity and deliberation are also key to TCA along with a systematic approach to development led to its design. However, it suffers from three key weaknesses. First it has a weaker representational framework of task trees which require representation of tasks within cleanly formulated hierarchies. Constraints between branches within a task hierarchy are allowed; timing between leaves of different trees is however not possible. Second, its temporal framework does not deal with flexibility; time points are fixed, and while partial orders are possible, it does deal with execution time uncertainty. Third, while TCA shares the notion of using domain expertise of different planners, it does so thru a centralized message passing mechanism. While this allows disparate code bases to talk to each other through IPC (an inter-process communication or message passing mechanism), failure in the centralized controller is tantamount to a system crash.

The ReSSAC project at ONERA [20] is motivated by controlling aerial UAV platforms for autonomous search and rescue. Like TCA, the project uses a supervisor to control planning and at the same time decouples the deliberation and reactive components from the lower-level functional layer. This deliberate separation is partly because of the use of optimizing MDPs (Markov decision processes [82]) for planning which uses probabilistic state transitions to determine the next course of action based on perceived state. To overcome the typical problem of state-space explosion with MDPs, a local heuristic is used in order to generate reachable goal states incrementally. This work however does not deal with metric time and comes with a modestly simple model.

CIRCA [83] is an effort to bring decision making for real-world problems with the augmented need to have *verifiable* controllers synthesized in situ to ensure that during state-space exploration, there are no erroneous transitions to dangerous states by asynchronously generating test-action pairs (TAPs). These are annotated production rules consisting of a set of tests (or preconditions) and a set of actions to take if all the tests return true. TAPs are synthesized and scheduled by CIRCA and provide a viable way to deal with a dynamically changing world. CIRCA however differentiates between reasoning about time and reasoning in real time with the implication that reasoning necessarily requires substantial computation for state-space exploration which negates the real-time (and thus real-world) impact.

A key and early concern that dominated planning was that of scalability. The planning cycle in many approaches was (and is) monolithic, often making fast reaction times impractical when necessary for embedded robotic agents. Many of these systems (which were three-layered) also utilized very different techniques for specifying each layer in the architecture resulting in duplication of effort and a diffusion of knowledge. The work we highlight in this chapter builds on the approach used by IDEA [70, 71] in utilizing a collection of controllers, each interleaving planning and execution within a common framework. IDEA, however, provided no support for conflict resolution between controllers, nor did it provide an efficient algorithm for integrating current state within a controller, relying instead on a possibly exponential planning algorithm. Efficient synchronization of state in a partitioned structure is fundamental to making the approach effective in practice.

Our system, T-REX, the focus of this chapter, is an agent control structure formally defined as a composition of coordinated control loops where sensing, planning, and acting result from concurrent deliberating tasks within a formal framework. T-REX was designed after carefully evaluating lessons learned from IDEA to which it owes substantially the notion of partitioned problem solving. Partitioning in T-REX is systematic and methodologically principled where we manage the information flow within the partitioned structure to ensure consistency in order to direct the flow of goals and observations in a timely manner. The resulting control structure improves scalability since many details of each controller can be encapsulated within a single control loop. Furthermore, partitioning increases robustness since controller failure can be localized to enable graceful system degradation, making this an effective *divide-and-conquer* approach to solving the overall control problem.

A recent variation of the key idea of such controller composition derived from T-REX is ROAR [84] where lower-level functional modules that control sensors are organized within a graph hierarchy. One reason they cite such a structure as being essential is in dealing robustly to failure; an off-nominal condition will allow rapid graph traversal to identify alternative ways in which sensing tasks can be executed and therefore aiding platform responsiveness.

3.3 Foundational Concepts

T-REX is an adaptive, Artificial Intelligence based controller and provides a general framework for building reasoning systems for real-world autonomous vehicles. At MBARI T-REX is used for AUV control; another instantiation of the system is being used for control of a terrestrial personal robot [27, 28]. The development of T-REX has been targeted at surveying a number of oceanographic features which are dynamic and spatiotemporally unpredictable. Typically this requires our AUVs to balance the goals of spatial coverage while opportunistically following features of scientific interest and to do so while being aware of its own resource limitations (e.g., the battery state of charge) and overall proximity to other observational assets for obtaining scientific ground truth. We are interested in representational frameworks that allow robots to pursue long-term objectives while choosing short-term gain and can gracefully deal with exogenous or endogenous change.

To enable this responsiveness to external observations, the agent has to be able to synthesize plans in situ and to replan without human intervention. T-REX uses a temporal constraint-based planner with a demonstrated legacy of having flown on NASA space missions. Our autonomy architecture brings three key innovations for AUV control: in situ plan synthesis, the use of flexible plan representations with a sound theoretical basis, and systematic compositional control with the use of partitioned networks and doing so with a demonstrated capability at sea for novel ocean observation methods.

The next few sections will cover EUROPA in a bottom-up fashion, going from its general theoretical framework to the specific representation, modeling, and algorithms implemented by it. A brief introduction to constraint programming and constraint-based planning is provided followed by the specific representation that EUROPA uses to undertake constraint-based planning. Subsequently, we provide an overview of EUROPA's architecture and how it can be embedded in a planning application such as T-REX. Finally, EUROPA's modeling, inference, and search mechanisms are covered in detail so that the reader can appreciate how T-REX can perform deliberation while being responsive to events in the real world.

3.3.1 Automated Planning

> [Planning] *is an abstract, explicit deliberation process that chooses and organizes actions by anticipating their expected outcomes. This deliberation aims at achieving as best as possible some prestated objectives. Automated planning is an area of Artificial Intelligence (AI) that studies this deliberation process computationally.* – from **Automated Planning Theory and Practice** by Ghallab, Nau and Traverso [9]

Planning as an activity is ingrained in human decision making [85] and its generalization into a computational framework demonstrating machine intelligence has for long been targeted by pioneers in computer science [41]. Early work in this

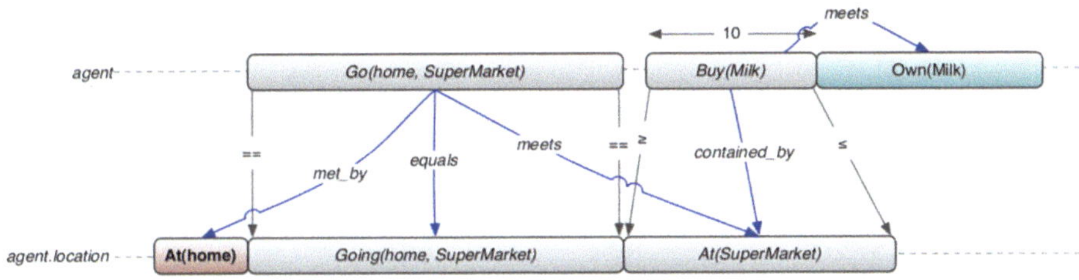

Fig. 3.3 A potential plan for a robot shopping for groceries

context focused on logical theorem proving by computer [42] which has formed the basis of much of the discipline. The emphasis in these early efforts was to logically derive the existence proof of a trajectory of actions for a hypothetical robot (moving blocks, chess pieces, furniture, etc.) to achieve a specific goal given specific initial conditions. Two important properties of *soundness*[5] and *completeness*[6] along with derived results for algorithmic complexity [86, 87] were a defining cornerstone for understanding how different computational paradigms could actually be tractable first in simulation and subsequently in actual robotic platforms. In this chapter our emphasis is on AI Planning and plan execution on an embedded marine robotic platform. The techniques we highlight are general purpose as noted in Sect. 3.1.

To articulate the fundamentals of automated planning briefly and use that to motivate the mechanisms we use in our specific form of planning we start with an example:

> In the near future, a personal robot sets out to buy a gallon of milk. This involves a number of tasks: obtain keys, obtain wallet, start car, drive to store, find and obtain milk, purchase milk, etc. The embedded planner has to have a "model" of the world in which it "lives" and has to use the task primitives in this model to structure the actions so it achieves its goal. Constraints control when certain tasks can or cannot occur. For example, the robot must obtain the keys and wallet *before* driving to the store and pick up the milk *before* purchasing it.

For such a robot the milk-buying plan at the store might look like that shown in Fig. 3.3.

At the core of the T-REX framework is the deliberation engine, EUROPA, which has a rich legacy from NASA missions [19, 21–23, 45]. EUROPA is a versatile constraint-based temporal planner which continues to be deployed on a diverse set of applications including recently, planning for the International Space Station [76]. We first motivate this section with some problem domains this planner has handled.

[5]A planning algorithm is *sound* if invoked on a problem *P* returns a plan which is a solution for *P*.

[6]A planning algorithm is *complete* if invoked on a solvable problem *P* is guaranteed to return a solution.

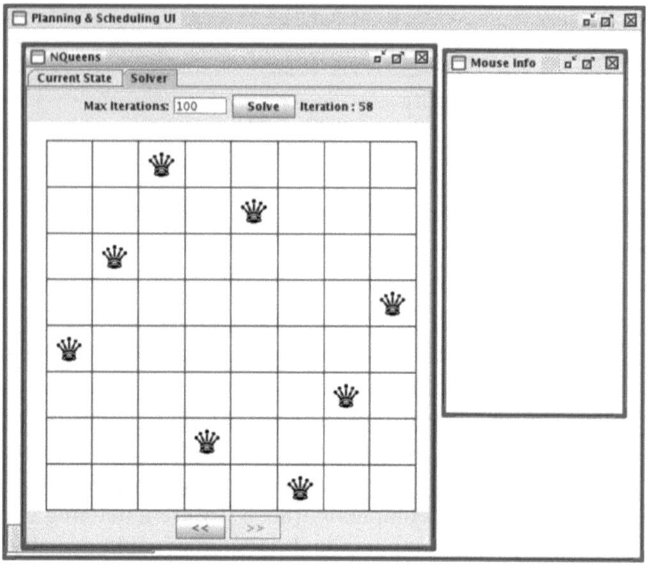

Fig. 3.4 An N-Queens solution generated by EUROPA

Constraint Satisfaction: A canonical problem in dealing with constraints is the N-Queens problem in which chess queens must be placed on an $N \times N$ chessboard so no queen can attack the other.

If we define $N \times N$ variables Q_{rc}, $r \in [1,N]$, $c \in [1,N]$, $Q_{rc} = 1$ if cell r,c in the chessboard is occupied by a queen, 0 otherwise. Then the following constraints need to be satisfied:

$$\text{Sum}(Q_{rc}) = \begin{cases} 1 \ \forall r & \text{(only one queen per row)} \\ 1 \ \forall c & \text{(only one queen per column)} \end{cases}$$

$$\text{Sum}(Q_{r+i,c+i}) = 1$$

$$\text{Sum}(Q_{r-i,c-i}) = 1 \qquad \text{(only one queen on each diagonal)}$$

The problem can be solved by finding assignments for all variables Q_{rc} that satisfy the above constraints. Figure 3.4 is a solution found by EUROPA using a compact version of that formulation and a specialized search procedure.

Scheduling: The resource-constrained project scheduling problem (RCPSP) [88] consists of a set of activities that must be scheduled in a way that satisfies minimum and/or maximum temporal separation constraints. The activity schedule must also respect fixed limits on the availability of resources required to perform each activity. In addition to satisfying temporal and resource constraints, it is common for the user to want to minimize makespan [9] so that the entire project is finished as early as possible. Figure 3.5a shows an example of a solution provided by EUROPA for an RCPSP instance with ten activities, five resources, and 30 temporal constraints.

Fig. 3.5 EUROPA-based solutions to scheduling and planning problems. (**a**) A EUROPA solution to an RCPSP [88] problem. (**b**) A EUROPA solution to a Shopping Agent problem domain where the agent needs to buy bananas, milk, and a drill

Planning: The Shopping Agent problem [89] is a variation of the problem with which we started this chapter. An agent needs to purchase a set of products (milk, drill, etc.) that are available at specific locations (supermarket, hardware store, etc.). The agent is subject to temporal (must complete tasks by specific deadlines) and resource (fuel, carrying capacity, etc.) constraints and it needs to synthesize actions that need to be performed to find and acquire the required items, as well as when to perform each of those actions. Figure 3.5b shows a solution produced by EUROPA for such a problem instance.

As the examples above show, a planning problem (actions to achieve a goal) may embed a scheduling problem (what resources are necessary to achieve stated goals) and both planning and scheduling may embed a constraint satisfaction problem. The relationships between planning, scheduling, and constraint satisfaction have been examined [90] and have lead to use of constraint reasoning in EUROPA as a core building block.

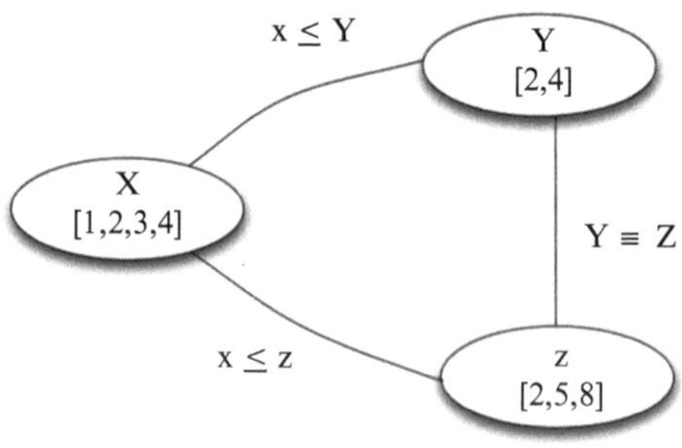

Fig. 3.6 Variables and constraints represented as a constraint network before arc consistency is enforced

3.3.2 *Constraint Programming*

Constraint satisfaction programming (also known as constraint programming or CP) is an AI discipline that provides a generic framework for representing, solving, and making logical inference on constraints. A complete treatment of this discipline is given in [91, 92] with concise introductions provided in [93, 94].

A constraint programming problem consists of a set of variables $V = x_1, \ldots, x_N$, where each variable takes values from a domain $d_1, \ldots, d_N$; in this chapter we will deal only with discrete and finite domains. Given a defined conjunctive set of constraints [95] on the variables: $C = \{c_1(x_1, \ldots, x_N), \ldots, c_K(x_1, \ldots, x_N)\}$, the objective is to find one or more value assignments to V where all constraints are satisfied. To solve a problem, CP techniques use logical inference to perform constraint propagation, which consists of two operations:

1. **Bounds propagation** inferring upper and lower variable bounds. For example, from the constraints $x_1 + x_2 \leq 2$ & $x_i \geq 0$, we can infer $[0, 2]$ bounds for x_1 and x_2.
2. **Domain reduction** inferring a valid set of values for a variable. For example, for constraints `allDifferent`(x_1,x_2,x_3), $x_i \geq 0$ & $x_i \leq 4$, if $x_1 = 1$ and $x_2 = 3$ we can infer that the valid and reduced domain for x_3 is $\{2, 4\}$.

A set of constraints and variables that describe a problem domain are typically represented as a *constraint network*, where the variables are nodes and constraints are arcs between the variables. A pair of variables x and y is said to be *arc consistent* if for every value in x's domain, there is a value in y's domain that is consistent with all the constraints that connect x and y. A naive approach to ensuring arc consistency consists of cycling through all the variable pairs and performing constraint propagation until there are no variable domain changes. The `AC-3` algorithm [59], a popular implementation, improves over the naive approach by ignoring constraints whose variables were not affected since the last iteration. Figure 3.6 shows a set of constraints on three variables (x, y, and z) represented as a constraint network and their domains before arc consistency are enforced. After the `AC-3` algorithm is executed, the domains for the variables would be pruned to $x = \{2\}$, $y = \{2\}$, and $z = \{2\}$.

Arc consistency and other more stringent consistency definitions can be used to efficiently prove that a CP problem is *unsatisfiable*. However, they generally cannot prove that a CP problem is satisfiable and provide specific variable assignments that constitute solutions. To do so, search algorithms like global backtracking [96] or local search are used. These use consistency as an efficient way to prune the search space [97]. In theory, solving a CP problem is NP-hard [9], but often in practice heuristic methods provide viable ways to generate efficient solutions.

CP is usually implemented as part of a programming language and constraints are usually represented as objects [98]. Any constraint that the user is able to reason about can be introduced into the system and as long as the constraint propagation protocols specified by the host CP system are enforced, it will be indistinguishable from any other "primitive" CP constraint, such as $\leq$ or $\geq$.

3.3.3 Constraint-Based Planning

The most common planning formulations use propositional logic, where the state of the world is represented by a set of propositions (statements that can be true or false), and operators change the truth values of these propositions [95]. Although these formulations are powerful for descriptive semantics in automated planning, there are classes of problem domains that are difficult to represent. In particular, it is hard to represent time, resources, mutual exclusion, and concurrency. Constraint programming is a framework that lends itself well to represent and reason about these elements [55] with the consequence that there has been substantial effort in building automated planners that leverage CP techniques. The most common approach in constraint-based planning has been to encode the entire planning problem as a CP problem, where action choices and relationships are represented as variables and/or constraints [99–102]. The advantage of using a pure CP formulation is that standard solvers can be then used to solve the planning problem. However, this approach also has significant drawbacks:

1. Given that action choices and relationships (rules) are expressed through variables, the domain descriptions that result from this approach are not intuitive and therefore difficult to understand and debug.
2. If the structure of a planning model (actions, conditions, effects, dependencies at the action level) is not explicitly maintained by the CP planner, the search algorithms are deprived of critical information to make better decisions. If that structure is maintained (for instance, by internally marking variables that represent action choice and relationships between actions), it is still hard to write search algorithms as any planning-specific insight has to be translated into the CP representation of variables and constraints.

These drawbacks have been addressed by a number of frameworks like `Descartes` [103, 104], `IxTeT` [105, 106] and `CAIP` (constraint-based interval and attribute planning) [44, 75]. They take advantage of a CP representation to deal with time and resources, while maintaining an explicit representation of the elements relevant to planning (action choices, causal relationships, etc.). This combination results in a planning approach that can deal with the kind of constraints found in real-world problems while allowing for intuitive problem descriptions and a natural way to express search algorithms and heuristics. EUROPA is one such implementation of the CAIP framework; others include [18, 107]. In EUROPA planning problems are specified in terms of:

Intervals: These are entities that have a temporal scope (that is, a beginning and an end in time) and a set of attributes. Intervals are used to represent actions and states in the planning domain.

Variables and Constraints: Variables represent interval attributes and constraints enable a direct representation of temporal, resource, and any other restrictions that a valid plan must comply with.

Domain Configuration Rules: This is an explicit representation of the conceptual relationships between actions and states in the problem domain, for instance, for a Shopping Agent to be able to purchase a product, it must be at a location where the product is available.

To generate plans, the CAIP framework uses a plan-space search approach [9], where a *partial plan*[7] is evolved until all open goals are achieved and all inconsistencies are removed. The specific search algorithm used by EUROPA will be covered in further detail below.

3.4 The EUROPA Planner

A Brief Note on EUROPA History and Notation

The origin of what we will call EUROPA in this chapter is derived from the HSTS planner [44] used on a number of ground-based demonstrations for space telescope scheduling as noted in Sect. 3.2.2. HSTS was also the LISP-based planner that was flown as part of the Remote Agent RAX experiment [19]. Concurrent to the RAX deployment, a C++ version of the planner was implemented at NASA Ames. This went on to be used for the MAPGEN system and continues to be used to date in the mission-critical uplink process for the command/control of the MER mission [23]. A vastly re-factored, higher performance open-source version was named EUROPA_2 [108] and is the subject of this chapter and used by T-REX. Note, however, in using the name EUROPA, we refer to the *fundamental ideas as well as the planning paradigm.*

[7] A partial plan is a subsequence of a plan which can be refined into a plan structure.

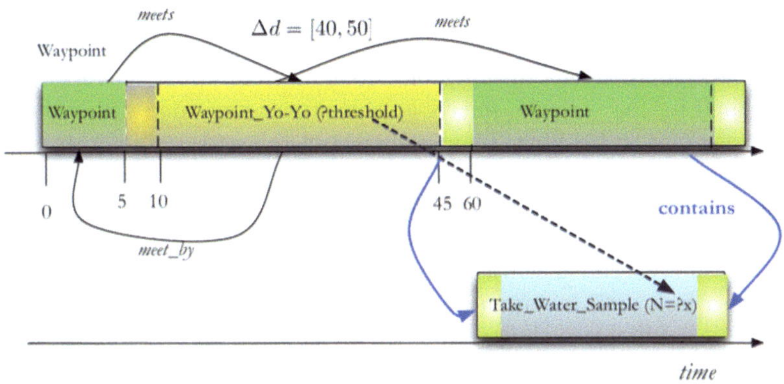

Fig. 3.7 Tokens with flexible temporal intervals and parametric constraints between tokens. This example shows the triggering of a water sampler based on a feature threshold while the vehicle Yo–Yo's. The `Waypoint_Yo--Yo` token has a flexible duration start and end times

3.4.1 Introduction

EUROPA uses a *domain model* (see Fig. 3.2) written in a declarative language, together with initial conditions and goals (also in that specified language), to construct a set of temporal relations that must be true at start time. These models include assertions about the physics of the vehicle, i.e., how it responds to external stimulus and internally driven goals. By propagating these relations forward using STNs [58] and applying goal constraints, EUROPA can select a set of conditions that should be true in the future, where some of these conditions will correspond to actions the agent must take. The planner can backtrack and try another path during the search process if a goal cannot be reached while being capable of discarding unachievable goals.

Traditionally robot execution has relied on dispatching commands at precise times. Such linear sequences of precisely timed commands give no ability to adjust execution on the basis of sensory information. Although some commands can issue tests on sensor readings, these tests have the objective of verifying whether expected execution conditions are occurring. If not, the state of the system is declared off-nominal and execution of the sequence of commands is interrupted. More recently, executives have been proposed and implemented that significantly broaden the way robots can be commanded [17, 19]. For example, the Remote Agent executive interpreted a *temporally flexible plan* which represents each start time as a flexible timepoint variable backed by an explicit network of bounded delay constraints between such variables. These timepoints [52] represent temporal intervals signifying a change in state and are of the form $[lb, ub]$ (where $lb, ub \in N$) are temporal lower and upper bounds, respectively. Instead of specifying a fixed integer, time is represented as an interval. Figure 3.7 shows an example of a flexible partial plan.

Unlike a traditional fixed time-tagged command sequences, such flexible plans leave room for adaptation at execution time. When the executive considers when to start a task, it propagates information through the constraint network, computes a time bound for the variable, selects an actual execution time within the bound, and starts the task at that time. Temporally flexible plans therefore express a *range of possible outcomes* of the robot's interaction with the environment within which the executive can elect at run time, the most appropriate for actual execution. The fact that constraints are explicitly represented ensures that through constraint propagation the executive will respect global limits expressed in the plan (e.g., don't start a task until a certain condition has been satisfied but still satisfying some overall deadline). Such flexibility is critical when dealing with dynamic and uncertain conditions for robotic action in real-world environments, where precise timing of an action might be indeterminate. Further, the advantage of flexibility can be contrasted with the consequences of the intrinsic inflexibility of traditional command sequences. Because they are inflexible, sequences are brittle and therefore must necessarily be designed considering worst case scenarios.

EUROPA uses a *state variable* representation [26,44,45] to describe the evolution of state over time. The instantiated history of such an evolution over a temporal horizon we call as a *timeline* and one which represents a single thread in the execution of a concurrent system. At any given time each thread can execute a single procedure, with each timeline consisting of a sequence of procedures which encapsulate and describe state evolution. We call these instantiated atomic entities *tokens*. A token therefore describes a procedure invocation, the state variables on which it can occur, the parameter values of the procedure, and the timepoints defining the interval (see Fig. 3.7). We allow encapsulation of uncertainty within these tokens with a range of start and end times and parameters, all of which are encoded as variables. A constraint solver in turn manipulates these variables defined in an EUROPA domain model. For example, consider a scientific need to take water samples 100 m from an environmental hotspot while an AUV is performing a Yo–Yo sawtooth pattern[8] in the water-column. Two samples are needed if the feature's signal is above a threshold, one otherwise. The token that is capturing the sensory threshold has a parametric constraints to the token which triggers the requisite water sampler. In addition, the start time of the water sampling procedure is highly dependant on the variability of subsea currents and actual speed of the vehicle. Therefore a number of values are possible for the start times and duration of the sampling, all of which are valid combinations for desired outcomes.

[8]Variability in the water column is along the vertical dimension. Since the Dorado platform can only move forward, the Yo–Yo pattern is the most efficient mechanism for studying water column properties.

3.4.2 EUROPA Plan Representation

EUROPA follows the representation outlined by CAIP augmented with elements that are necessary for an efficient implementation of constraint-based planning. We briefly describe the main elements, followed by a detailed description for each of these elements, and how together they constitute a rich representational paradigm for inference in Sects. 3.4.4–3.4.6.

1. **Tokens** Temporally scoped entities that correspond to intervals in CAIP that are used to represent state and actions in time. A **token state model** is defined to support an efficient implementation of plan-space search algorithms.
2. **Domains, variables, and constraint** are used to represent restrictions in the problem domain in terms of a CP formulation.
3. **Domain rules** are used to describe internal and external relationships on and between tokens, as prescribed by CAIP.
4. **Objects** are used to represent problem domains in a more modular and scalable way. **Timelines and resources** are built-in implementations of object types that appear frequently in real-world domains.

At the atomic level, the entities in the EUROPA plan representation are:

Variables Values that need to be represented to describe the problem domain over which constraints are specified. In the Shopping Agent example, the times at which the agent needs to leave or be back, or executes a purchase, are all instances where *variables* representing time is used. As noted in Sect. 3.3.3, a variable can take values from a domain. In EUROPA, domains are defined over a specific data type; the primitive data types supported are *String*, *Boolean*, *Numeric*, and *Object* (see below). Further, two types of variable domains can be defined over a data type *enumerated* and *interval* (used only for numeric data types, they are specified by a $[lowerbound, upperbound]$ pair, for instance: $[1, 10]$, $[5.0, 100.0]$). It is this latter ability to represent and reason over numeric intervals that forms the basis of representing flexible plans.

Constraints Valid plans in most problem domains have to satisfy a number of restrictions, for instance, a Shopping Agent may only be able to perform one task at a time, may have limited time or an energy or fuel budget to perform a task. These restrictions are represented by *constraints*, which can be defined over any combination of variables in a domain. EUROPA provides a *constraint library* that implements a number of useful constraints (temporal, resource, relational) while allowing the possibility to augment with new constraints if required.

Objects The items we wish to describe and refer to in a domain are considered objects. In the Shopping Agent example, we might consider the agent, products, and locations all to be *objects*. Objects have state and behavior. For example, a Shopping Agent can have its location, a bag that contains the products it has already purchased (the bag can in turn be another object), and behavior, as in going to a location to look for a product, perform a purchase, return home, etc., all as objects. Those that have similar state and behavior can be described generically in terms of object types or classes. EUROPA allows the definition of classes similar to object-oriented programming languages, where a class can have attributes (represented by variables) and classes can be arranged in a single-parent hierarchy through inheritance. What is different, however, is that to describe state and behavior for planning, one needs to build on the formalism of first-order logic.

Tokens In first-order logic, a predicate defines a relation between objects and properties [95]. In EUROPA, we define such relations between variables whose domains are sets of objects and sets of properties to describe state and behavior. For example, one might use a predicate At(a, l) to indicate that agent a is at location l or a predicate Buy(a, p) to indicate that agent a is acting to buy product p. Note that a is a variable which may have a number of possible values in a problem with multiple agents. Similarly, l and p are variables whose values are the set of possible positions and products, respectively. EUROPA can be used to create partial plans, where the domains of variables a and l can have more than one possible value. Or they can be grounded representations, where single values will be specified for each variable as shown in the N-Queens problem.

In general, for an executable plan, it is insufficient to define predicates that describe required state or behavior without also specifying some temporal extent over which each of those predicates hold. Predicates that are always true can be thought to hold from the beginning to the end of the planning horizon. However, in practice, the temporal extent of interest must be defined with timepoints to represent its start and end. In such a representation, one writes At(a, s, e, l) to indicate that agent a is at location l from time s to time e. In fact, this pattern of usage to describe both state and behavior of objects is prevalent enough that EUROPA encapsulates into the token representation. Tokens have built-in variables to indicate the object to which it primarily applies as well as the timepoints over which it holds. A token, therefore, is an instance of a predicate that represents an object's state or behavior and is defined over a temporal extent. In EUROPA, all predicate instances are tokens. In the same way that objects are described generically by classes, in EUROPA tokens can be described generically by token types. Every token has five built-in variables:

1. *Start*: The beginning of the temporal extent over which the token predicate is defined.
2. *End*: The end of the temporal extent over which the predicate is defined.
3. *Duration*: The duration of the temporal extent. The constraint $start + duration = end$ is enforced automatically.
4. *Object*: The set of objects to which a token might apply. In a grounded plan representation each token applies to a specific object, reflecting the intuition that tokens are used to describe some aspect of an object (i.e., state or behavior) in time. However, in a partial plan, it is possible that the commitment to a specific object may not yet have been made.
5. *State*: Tokens can be ACTIVE, INACTIVE, MERGED, or REJECTED. The state variable captures the token's current state and its reachable states

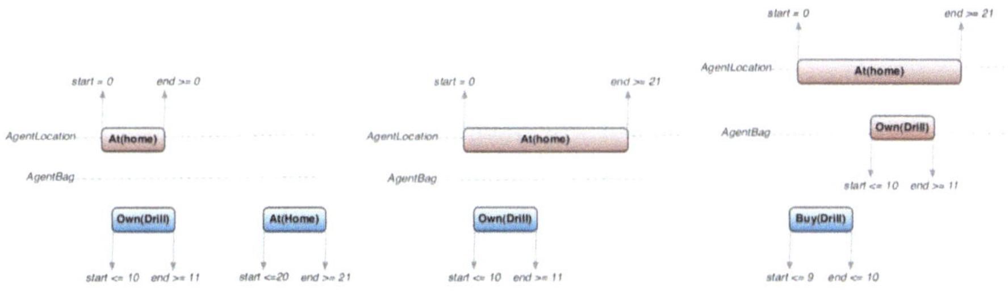

Fig. 3.8 Evolution of a timeline from the shopping agent example. (**a**) Timelines representing initial state and goals for a shopping agent. (**b**) State of timelines after satisfying a goal to be home by time 20 on merging with initial state. (**c**) State of timelines after satisfying a goal to own a drill by time 10 by activating corresponding token

through further restriction. This is required to support CAIP's approach to planning where goals and subgoals can be satisfied by either adding new intervals (tokens) to the plan or by merging with existing intervals.

Built-In Object Types In a domain model, there may be some classes that don't have any time-dependent state or behavior, and therefore those classes will not have any token types associated with them. For instance, in the Shopping Agent model, the set locations and products are static for the agent's purposes so one may only want to declare them but not associate any state or behavior with them. However, a common case in a nontrivial domain model is that objects will have many associated tokens in order to describe their state and behavior throughout a plan. There are some object types that are used so often in domain descriptions that EUROPA provides a built-in implementation for them:

1. *Timelines*: Often objects in a domain must be described by exactly one token for every given timepoint in the plan. Any instances of a class derived from EUROPA's built-in timeline class will induce ordering requirements among its tokens to ensure no temporal overlap may occur among them [44]. Figure 3.8a and others in this chapter illustrate how timelines are used to represent state in a plan.
2. *Resources*: Metric resources, for example, the energy of a battery or the capacity of a cargo hold, are objects with an explicit quantitative state in time and with a circumscribed range of changes that can occur to impact that state, i.e., produce, consume, use, change. Resources are a common enough requirement for EUROPA users that built-in object types (with their corresponding token types to denote production, consumption, etc.) are provided. Instances of classes derived from a resource will induce ordering requirements on their tokens in order to ensure that the level of the resource remains within specified limits.

Token State Model In EUROPA's representation, goals are temporally scoped states that need to be achieved, or actions that must be performed. As a result,

EUROPA internally represents goals as tokens. Goals can be posted explicitly, for example, when a user asks the Shopping Agent to own a specific product by a specific time, or implicitly when subgoals are created as a consequence of domain rules. For instance, the Shopping Agent must get to a location where a product is available before it is able to purchase it. In this case the original goal of purchasing the product spawns a *subgoal* of being at a specific location without which the purchase will not occur as defined in the domain model.

Looking at how a simple Shopping Agent goal would be addressed by EUROPA helps illustrate how token states support the planning process. Let's assume the planner starts with a partial plan that places the Shopping Agent at home at time 0 with no products in possession. Let's also assume that goals are posted for the Agent to own a drill by time 10 and to be home by time 20 as shown in Fig. 3.8a. Note that the tokens `Own(Drill)` and `At(Home)` that represent the goals are not immediately placed on the `AgentBag` or `AgentLocation` timelines, since EUROPA must first determine whether it can achieve those goals and how. When the goals are posted and before any decisions are made, the `Own(Drill)` and `At(Home)` tokens are created in an `INACTIVE` state. EUROPA then considers the partial plan and evaluates two possible alternatives:

1. If there are any tokens in the plan that are compatible with the new goal, EUROPA may attempt to satisfy the goal by *merging* it with any of them (if there is more than one it can try them one at a time or in some heuristic order), in this case, the state of the token would change to `MERGED`.
2. Instead of merging with existing tokens in the plan, EUROPA may decide to *insert* a new token into the plan, in which case the state would change to `ACTIVE`.

When a token is merged, all restrictions imposed on the merged token are transferred to the active token upon which it is merged. Merging a token requires finding a target `ACTIVE` token that is compatible. Two tokens are merge-compatible if they are instances of the same predicate and there are no empty intersections between corresponding token variables. The effect of merging is illustrated in Fig. 3.8b. The goal to be home by time 20 is satisfied by merging with the initial state that indicated that the agent is home at time 0. The merge is possible because the tokens are of the same `At` type and intersecting the domains for all of their variables does not yield an empty domain (representing an inconsistency). In this case the token variables are the location, which is `Home` in both cases and the temporal variables start, end, and duration, which can be intersected without causing an inconsistency. After the token representing the goal is merged onto the token representing the initial state, the resulting token represents that the agent must be not only `At(Home)` by time 0 but also `At(Home)` at time 20. This is reflected in the new constraints on the token's start and end variables as depicted in Fig. 3.8b.

When a token is activated, all domain rules associated with its type are evaluated, which may lead to new subgoals being generated. In our example, there could be a domain rule stating that if a Shopping Agent doesn't already own a

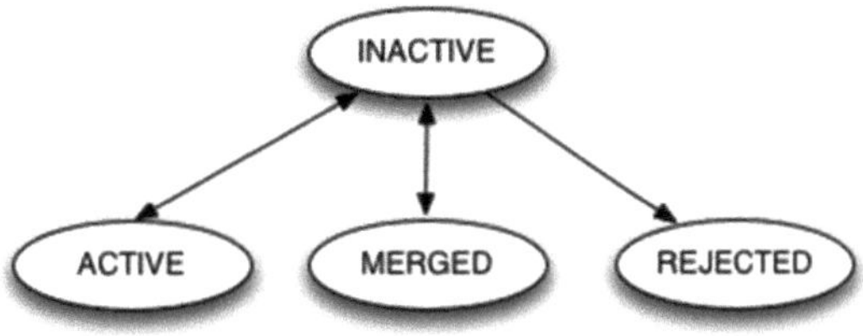

Fig. 3.9 Possible transitions for a token state

product, then it must Buy it before it can Own it. In such a case, activating the Own(Drill) token will cause a new Buy(Drill) to be created in an INACTIVE state, which in turn could spawn its own subgoals, like being at the location where the product can be purchased before being able to buy it, and so on, until all domain rules are satisfied. As Fig. 3.8c shows the decision to merge the goal to be At(Home) by time 20 with the initial state was not appropriate since the agent will have to get away from Home to be able to buy the drill which is inconsistent with the constraints that resulted after the merge. As a result that decision will have to be retracted and the alternative to activate or merge with a different token will need to be considered.

If after looking at all possibilities, neither activating nor merging work, a token may be REJECTED. This is possible only for goals posted by the end user and specified as optional. Subgoals generated as a result of token activation cannot be rejected. If a subgoal cannot be satisfied through activation or merging then it will cause the original activation decision that spawned it to be retracted.

As can be seen from the above example, during the search for a plan, a token's state can go from INACTIVE to ACTIVE, MERGED, or REJECTED as shown in Fig. 3.9.

Rules In order for a plan to be valid, it must comply with all rules pertinent to the relevant application domain. Rules govern the internal and external relationships of a token. For example, consider a parameterized action Go(origin,destination) which describes a Shopping Agent traveling from one location (*origin*) to another (*destination*). The parameters are instantiated on a token as variables whose domain of values is the set of all possible locations in a given problem. A rule governing an internal relationship among token variables might stipulate that a transition must involve a change in location. This can be expressed as a constraint on the definition of a predicate of the form: *origin != destination*. A further stipulation could be that the agent must be located at *origin* before travel can occur and the agent must be located at *destination* when travel is completed. This is an example of a rule governing an external relationship among tokens. It specifies a requirement that tokens of the predicate At that represent the location state for the Shopping Agent precede and succeed tokens of the predicate Go that represent traveling. Figure 3.10 illustrates the entities and relations involved in specifying such a rule on a Go action. The token on which the rule applies (Go) is referred to as the *master*. Each At and Going token required by the master is referred to as a *slave*. Application of a rule

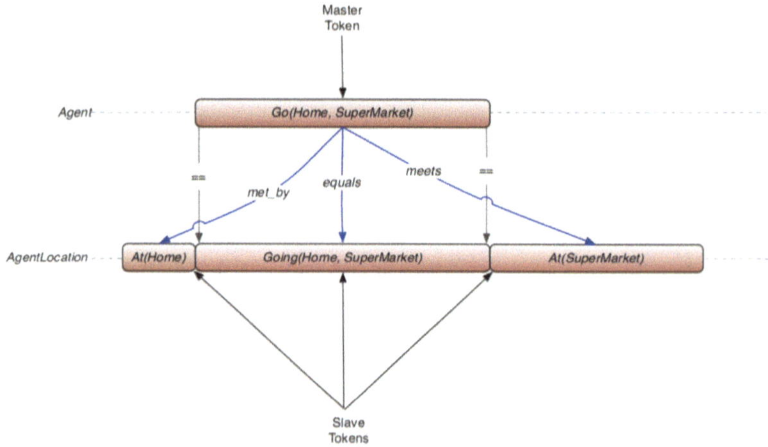

Fig. 3.10 Master and slave tokens generated as a result of a domain rule

on a token (i.e., a master) can thus cause slave tokens, variables, and constraints to be introduced. The capability of a domain rule to cause a slave token to be created is a key vehicle through which planning occurs. Semantically, a rule imposes a requirement for supporting slave tokens to be in the plan in order for the master token to be valid.

3.4.3 Architecture

EUROPA is implemented as a library that encapsulates all elements previously described and available via modeling and application programming interfaces (APIs). A user will typically use modeling facilities to inject a problem domain description and the API to extract and manipulate plans and to obtain information about plan flaws and constraint violations at any stage of the planning process.

Figure 3.11 shows the main architectural components in EUROPA and their relationships. The **constraint reasoning engine** (CRE) manages domain types, variables, and constraints that define relationships among them. It also provides an efficient arc consistency mechanism [59] designed so that specialized reasoning algorithms for new constraints can be easily and efficiently plugged in. The **plan database** (PDB) manages object and token types. It also maintains the state of partial plans in terms of object, variable and token instances. This constitutes the data foundation on which end user applications and automated problem solvers are built. The **rules engine** (RE) adds the semantics of domain rules that describe dependencies between tokens. The **solver module** provides the framework for flaw detection and resolution that is needed to implement search algorithms for

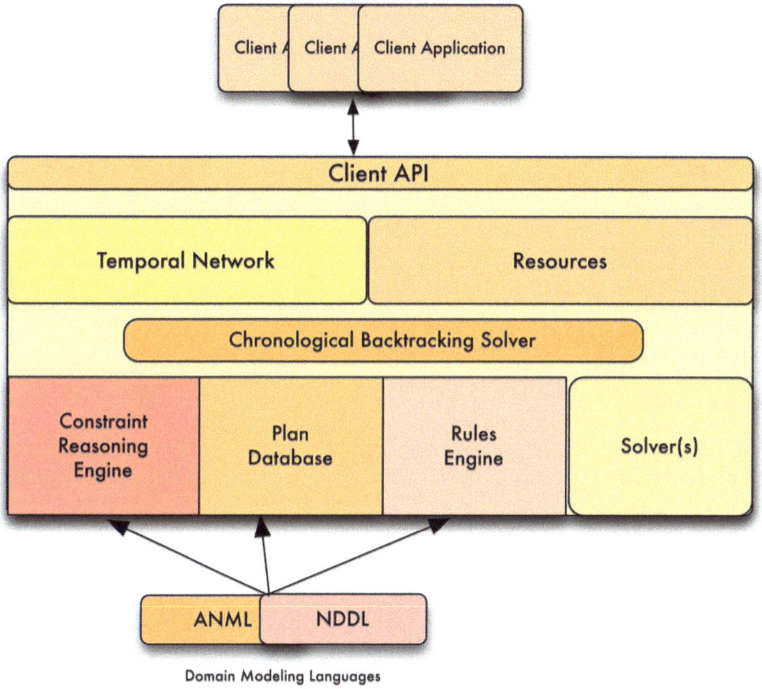

Fig. 3.11 The EUROPA software architecture

planning. The **API layer** provides access to services from all the different modules in a programmatic way so that EUROPA can be easily embedded into specialized applications.

Describing problem domains and problem instances in terms of objects, variables, and domain rules is cumbersome to the extent of being impractical; this motivated the creation of modeling languages that support those definitions in a much more natural and compact way. NDDL (new domain description language [pronounced "noodle"]) [40] is the primary modeling language for EUROPA described in Sect. 3.4.4. Another, ANML [109], is a language under development that provides a more intuitive syntax and out of scope for this chapter.

In an effort to minimize the amount of work that a user needs to perform to solve specific problems, EUROPA also bundles extension modules that have been found to be useful in many domains:

1. *Temporal Network*: Extensions to the CRE module to reason about temporal constraints.
2. *Resources*: Extensions to the CRE, PDB, and **solver module** to provide reasoning and search for metric resources.
3. *Chronological Backtracking Solver*: A general purpose built-in solver to provide the ability to plan out of the box. Problems that require challenging search may require specialized solvers to be built as noted in Sect. 3.4.6.

```
class LightBulb extends Timeline
{
    predicate On  {}
    predicate Off {}
}
class LightSwitch extends Timeline
{
    LightBulb myBulb_;

    LightSwitch(LightBulb b)
    {
        myBulb_ = b;
    }
    action turnOn  { duration=1; }
    action turnOff { duration=1; }
}
LightSwitch::turnOn
{
    met_by(condition object.myBulb_.Off);  // Bulb must be Off to be turned On
    meets(effect object.myBulb_.On);   // Must be turned on through the switch
}
LightSwitch::turnOff
{
    met_by(condition object.myBulb_.On);   // Bulb must be On to be turned Off
    meets(effect object.myBulb_.Off);      // Must be turned off through the switch
}
// Problem instance : turning the light off
LightBulb bulb1 = new LightBulb();
LightSwitch switch1 = new LightSwitch(bulb1);

// At time 0, the bulb is on
fact(bulb1.On initialCondition);
eq(initialCondition.start,0);

// We want the bulb to be off by time 10
goal(bulb1.Off goal1);
lt(0,goal1.start);
lt(goal1.start,10);
```

Fig. 3.12 NDDL modeling example for a light bulb and switch control

Before going into the details of each module, a simple example illustrates how they interact at an abstract level.

Figure 3.12 is an NDDL model (covered in more detail in Sect. 3.4.4) of a light bulb and a switch that controls it along with a problem instance that requires the light bulb to be off by time = 10, given an initial state with the bulb on at time = 0. The NDDL module allows EUROPA to ingest this model and the problem description. Once the NDDL statements are parsed, different object types, object instances, and constraints are instantiated inside the PDB and CRE modules where they are available for the user and the solver modules to query and manipulate programmatically. This instantiation of the initial state constitutes the initial plan that EUROPA holds for this particular problem instance.

The user can then invoke functionality available in the solver module to modify or extend this initial plan in a way that eliminates any flaws (violated constraints, unsatisfied goals, oversubscribed resources). When actions are added to a plan, the RE ensures that domain rules expressed in the model are enforced when EUROPA creates new action or predicate instances. For instance, if a LightSwitch::turnOn action is added to the plan, the RE ensures that requirements for the existence of the conditions and effects of that action (i.e., the corresponding LightBulb must be off before and on after the action) are posted through the PDB and CRE modules.

3.4.4 *Modeling in* EUROPA

EUROPA can ingest descriptions of models, plans, and goals and thereby can be applied to different domains and problems, simply by providing different models and goals. Consequently the expressiveness of the language for models, goals, and plans is of great importance. For EUROPA, NDDL allows the user to specify the different components of a problem domain in a precise and concise manner. Among the main features that make NDDL a powerful tool for describing problem domains are that it is object oriented. It supports classes and inheritance in similar fashion to popular object-oriented languages like Java and C++. It also supports polymorphism for some planning components. Using object classes and instances is a time-tested approach to naturally describe problem domains [110]. It offers declarative constructs to define constraints and causality dependencies between actions and effects and it offers procedural constructs to populate EUROPA's PDB with specific problem instances expressed in terms of objects, variables, constraints, facts, and goals.

To articulate the most important elements of NDDL, we show how the Shopping Agent problem could be modeled in Fig. 3.13. The main elements of a problem domain can be mapped to EUROPA's representation in the following manner:

Object Types: NDDL supports an object-oriented approach, where common data and behavior can be abstracted as a class. In this case, we have the *Product* class represent the products that the Shopping Agent needs, *Location* represent the places where products can be found, and *ProductLocation* represents the association arising when a product can be purchased at a location. The Shopping Agent itself is represented by an *Agent* class, which relies on the *AgentLocation* class to keep track of where the agent is at any point in time. NDDL supports a single-rooted class hierarchy so that state and behavior can be extended through inheritance. We take advantage of that feature by making *AgentLocation* inherit from EUROPA's built-in *Timeline* class which does not allow predicates to overlap in time for the same object instance. This feature is used, for instance, to ensure that an agent can only be at one place at a time in our example.

```
// Locations (Home, SuperMarket, etc.)
class Location {
  string name;
  Location(string _name){
    name = _name;
  }
}
// Products (Milk, Banana, etc.)
class Product {
  string name;
  Product(string _name) {
   name = _name;
  }
}
// ProductLocations (Banana can be found at SuperMarket, for example)
class ProductLocation {
   Location location;
   Product product;

   ProductLocation(Location _location, Product _product){
     location = _location;
     product = _product;
   }
}
// Use build-in timeline functionality to enforce that an agent cannot
// be at more than one place at one time
// Go to more than one place at one time
// Go somewhere and be At somewhere at the same time
class AgentLocation extends Timeline{
  predicate At {
    Location loc;
  }
  predicate Going {
    Location from;
    Location to;
   }
}
// In addition to having a location timeline, the agent can buy and
// own things. Note that the actions and the location predicates can
// be concurrent so they cannot be on the same timeline.
class Agent {

  AgentLocation location;
  Agent() {
   location = new AgentLocation();
  }
  action Buy {
   Product product;
  }
  action Go {
   Location from;
   Location to;
  }
  predicate Own {
   Product product;
  }
}
```

Fig. 3.13 NDDL model of the shopping agent example

```
Agent::Go {
  met_by(condition object.location.At origin);
  eq(from, origin.loc);

  equals(effect object.location.Going going);
  eq(going.from, from);
  eq(going.to, to);

  meets(effect object.location.At destination);
  eq(to, destination.loc);
}

Agent::Buy {
  // A Buy takes 10 time units
  eq(10, duration);

  // initialized to all locations
  ProductLocation possibleStores;

  // limit possibleStores variables to ones that provide what we need to buy
  eq(product, possibleStores.product);

  // We must be At a location during our Buy, and that location must have the
  // product we want available:
  contained_by(condition object.location.At currLocation);
  eq(currLocation.loc, possibleStores.location);

  meets(effect Own purchase);
  eq(purchase.product,product);
}
```

Fig. 3.14 NDDL rules for the shopping agent

Variable Types and Instances: All object attributes are represented as variables in this model. For instance, it uses integer variables to represent the built-in temporal parameters for each token (duration, start, and end times), string variables to represent *Location* and *Product* names, and object reference variables to represent the structural relationships between different objects (for instance, each agent maintains a variable to keep track of its *AgentLocation*).

Predicate Types: Some classes like *Product*, *Location*, and *ProductLocation* are static and so don't have any predicate or action types associated with them. The *AgentLocation* class is used to keep track of state and so has `At` and `Going` predicate types associated with it. The *Agent* class keeps track of the products it purchases through the `Own` predicate; having a separate *AgentBag* class with its own predicates would have been an equally valid modeling choice.

Action Types: In our example, *Agent* is solely the class that has behavior associated with it. Consequently, it defines action types to represent when the agent is taking action to `Go` to a *Location* or to `Buy` a *Product*.

Domain rules can be defined in NDDL in the context of a predicate or an action. In Fig. 3.14, the main elements of a rule definition are as follows:

Variables: A token type (predicate or action) is part of a class; therefore, class attributes (see `object.location` above), token type parameters (see `[from, to]` in `Agent::Go`, or `[duration, product]` in `Agent::Buy`), and locally declared variables (see `possibleStores` in `Agent::Buy`) can all be used in the definition of a rule.

Constraints: Restrictions on values that variables can take within a rule are expressed through constraints. In the example above, the `eq` constraint is used; in general, any constraint available in `EUROPA`'s library can be used.

Subgoals: In `Agent::Go`, subgoals are used to specify that the Shopping Agent's location must transition from origin to destination in a temporally consistent way. The `[meets, equals, metby]` operators (see Sect. 3.4.5) are used to specify subgoals, along with the temporal relationship each token representing a subgoal has with the token representing the `Agent::Go` action. Also, `[condition, effect]` annotations can be used when defining a subgoal inside an action. `EUROPA` does not intrinsically associate semantics with these annotations, but they can be extremely useful for search modules when creating plans.

With the specification for the Shopping Agent domain above, a particular problem instance could be stated as in Fig. 3.15.

This problem instance consists of **Variables** and **Objects**, **Facts**, **Goals**, and **Constraints**. In our example we define temporal variables that are used to specify the planning horizon (`horizonStart` and `horizonEnd`) and objects are used to specify locations, products, product-location associations, and the Shopping Agent that we want to plan for. Facts, represented as tokens in the `PDB`, are used to specify the current state of the world and do not need any support. In this example we use facts to specify that the Shopping Agent is initially at home. Goals are also represented as tokens in the `PDB`. Different from facts, goals trigger inference in `EUROPA` which determine whether flaws (unresolved subgoals, overloaded resources, etc.) need to be resolved; if so, search modules can be invoked to resolve those flaws as part of the deliberation process. Finally, constraints are restrictions on and between facts and goals. In the example above, we use temporal constraints to specify when facts hold and when goals need to be accomplished. `before` and `after` are `Allen Algebra` [111] operators, which `EUROPA` supports to specify temporal relations between tokens (specifically between their start and end timepoints). The algebra is particularly effective in dealing with interval arithmetic; its elements are summarized in Fig. 3.16.

3.4.5 Automated Inference

In `EUROPA`, all variables and constraints from a problem domain description are stored inside the `CRE`. It provides a framework where any number of constraint types can collaborate to perform inference and prune variable domains using bounds prop-

```
// Allocate instances
Location Home = new Location("Home");
Location SuperMarket = new Location("SuperMarket");
Location HardwareStore = new Location("HardwareStore");

Product Banana = new Product("Banana");
Product Milk = new Product("Milk");
Product Drill = new Product("Drill");

ProductLocation bananaLocation = new ProductLocation(SuperMarket, Banana);
ProductLocation milkLocation = new ProductLocation(SuperMarket, Milk);
ProductLocation drillLocation = new ProductLocation(HardwareStore, Drill);

Agent agent = new Agent();

// Indicate that the database is closed - no new objects can be created
// this allows EUROPA to perform more efficient inference
close();

// We start the day at Home:
fact(agent.location.At atHomeForBreakfast);
atHomeForBreakfast.loc.specify(Home);

// Goals for all of the agent's needs, buy if needed
goal(agent.Own gotMilk);
gotMilk.product.specify(Milk);
goal(agent.Own gotBanana);
gotBanana.product.specify(Banana);
goal(agent.Own gotDrill);
gotDrill.product.specify(Drill);

// Make sure agent is home for dinner
goal(agent.location.At atHomeForDinner);
atHomeForDinner.loc.specify(Home);

// Agent has all day to satisfy goals:
gotMilk after atHomeForBreakfast;
gotMilk before atHomeForDinner;

gotBanana after atHomeForBreakfast;
gotBanana before atHomeForDinner;

gotDrill after atHomeForBreakfast;
gotDrill before atHomeForDinner;

// Force things to happen within our planning horizon:
int horizonStart = 0;
int horizonEnd = 100;
leq(horizonStart, atHomeForBreakfast.start);
leq(atHomeForDinner.end, horizonEnd);
```

Fig. 3.15 NDDL instance and goal specification for the shopping agent

Temporal Relation	Inverse Temporal Relation	Topology
T1 before (d D) T2	T2 after (d D) T1	
T1 starts_before (d D) T2	T2 starts_after (d D) T1	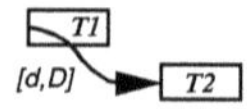
T1 ends_before (d D) T2	T2 ends_after (d D) T1	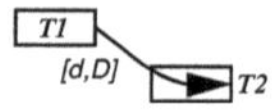
T1 starts_before_end (d D) T2	T2 ends_after_starts (d D) T1	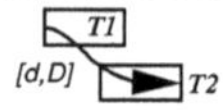
T1 contains((a A) (b B)) T2	T2 contained_by ((a A) (b B)) T1	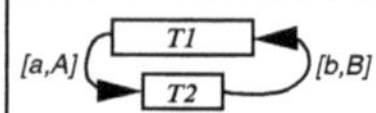
T1 parallels((a A) (b B)) T2	T2 paralleled_by ((a A) (b B)) T1	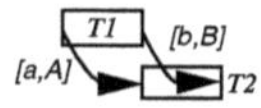

Fig. 3.16 Temporal relations defined within the planner are based on Allen Algebra [111] relations shown above

agation and domain reduction techniques [91]. EUROPA uses a propagator pattern which keeps track of a set of constraint instances of the same type and is responsible for ensuring variables are updated through inference whenever a relevant change occurs (variable domain modifications or constraint addition/deletion).

In a planning problem, variable or constraint changes can come directly from an end user (by introducing new activities or goals and relating them to the rest of the plan through constraints, by changing the start or end time of an activity, etc.) or from an automated planner (by expanding subgoals, introducing precedence constraints to resolve mutual exclusion conflicts, etc.). When these changes occur, each propagator is only able to ensure consistency for the constraints that are associated with it. To maintain global consistency, the CRE implements a variant of the AC-3 algorithm [59] where propagators are invoked repeatedly until the constraint network returns to a quiescent state.

3.4.5.1 Constraint Formulation

An EUROPA user can introduce new constraint types by implementing specialized constraint and propagator classes. However, EUROPA provides a built-in constraint library with a range of arithmetic, temporal, set, and resource constraint types so that a large number of problem domains can be modeled. Tables 3.1–3.4 show the main constraint types included in EUROPA's constraint library. Temporal and resource constraints are especially important to model real-world planning domains.

EUROPA uses variables to explicitly represent timepoints for plan actions and states. Constraints among timepoints provide a natural way to express domain

Table 3.1 Calculation constraints: these constraints enforce that one variable (usually a) is constrained by a calculation done on the remaining variables

Constraint	Syntax	Description	Variable types
DistanceSquares	DistanceSquares(a,b,c)	$c = \text{sqrt}(a+b)$ if a and b are singleton	Numeric intervals
CalcDistance	CalcDistance(a,b,c,d,e)	a is Euclidean distance between points (b,c) and (d,e)	Numeric intervals
Sin	Sin(a,b)	$a = \sin(b)$	Numeric intervals
AllDiff	AllDiff$(a,b,\ldots)$	Restrict all domains so the intersection of any pair of domains is empty	Comparable
EqualMaximum	EqualMax$(a,b,\ldots)$	$a = \max(b,\ldots)$	Numeric
EqualMinimum	EqualMin$(a,b,\ldots)$	$a = \min(b,\ldots)$	Numeric
CountZeros	CountZeros$(a,b,\ldots)$	a is the count of the rest that can be zero	Numeric
CountNonZeros	CountNonZeros$(a,b,\ldots)$	a is the count of the rest that can be non-zero	Numeric
DiffSquare	DiffSquare$(a,b,\ldots)$	$c = (a-b)^2$ if a and b are singleton	Numeric intervals
Card	Card$(a,b,\ldots)$	a must be greater than or equal the count of the other variables that are true	Numeric

Table 3.2 Set constraints: these constraints are likely to be used for set comparisons, etc.

Constraint	Syntax	Description	Variable types
SubsetOf	SubsetOf(a,b)	a is a subset of b	Comparable
MemberImply	MemberImply (a,b,c,d)	If a is a subset of b, then require that c is a subset of d	A and b comparable, c and d comparable
Lock	Lock(a,b)	Restrict a's domain to be contained in b's domain	Comparable

Note, however, that many of the other constraints could be used for sets (i.e., comparable variables) just as these constraints could be applied to numeric domains as well

Table 3.3 Object hierarchy constraints: these constraints are imposed on objects and tokens

Constraint	Syntax	Description	Variable types
CommonAncestor	CommonAncestor (a,b,c)	a and b must be contained by the same object in c (or by c itself)	a,b and c are all objects
HasAncestor	HasAncestor(a,b)	a must be contained by some object in b	a and b are objects

These constraints are used to assert which object one or more tokens is contained by. Most often, the commonAncestor constraint is used to subgoal across timelines which must share a contained object in common

Table 3.4 Miscellaneous constraints

Constraint	Syntax	Description	Variable types
Unary	NA	Restrict a variable's domain; given a variable and a domain, intersect them. Note that this constraints is only available internally and not exposed in NDDL	Anything
or	or$(a, b \ldots)$	At least one of the variables must be true	Numeric
absVal	absVal(a, b)	$a.lb \geq 0$, $a.ub = \max(\text{abs}(b.lb))$, $b.lb \geq -a.lb$, $b.ub \leq a.ub$	Numeric

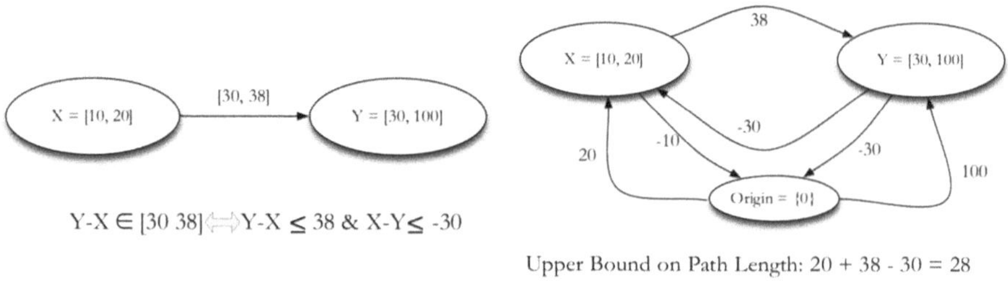

Fig. 3.17 An illustration of a STN and its corresponding distance graph. (**a**) A simple temporal network (STN). Nodes correspond to variables and their domains and arc's the *constraint* or distance between them. (**b**) The resulting distance graph from (a)

axioms. For example, in order to state that activity *A* must occur before activity *B*, we can say that the end timepoint of *A* is *leq* the start timepoint of *B*. Dechter et al. [58] proposed that constraints among timepoints can be grouped together to form a STN. Such a network can be transformed into a distance graph (DG) where the outward arc from a node represents the maximum distance from the source node to the target node. They also show that shortest path algorithms (see [87] for a review) could be used to propagate values in the network and discover a negative cycle: namely a path from a node to itself that has a path length less than 0. If such a cycle exists, the network is provably inconsistent. It was further shown that a single-source shortest path algorithm was sufficient to detect a negative cycle and provide sufficient propagation to yield a backtrack-free search. EUROPA's propagator for temporal constraints is an implementation of such a single-source algorithm. Figure 3.17 illustrates an STN with two variables and a single constraint along with the resulting DG.

Most real-world planning problems need to deal with plans made up of actions that typically depend upon limited resources (e.g., fuel, storage capacity, battery state of charge, data buffers). In EUROPA's representation, resource object types come in three forms with related constraints. **Reservoir** resources can have both `consume(resource,time,quantity)` and `produce(resource,time,quantity)` constraints; production and consumption is instantaneous. **Reusable** resources, on the other hand, have constraints that use a quantity of the resource for a specific duration

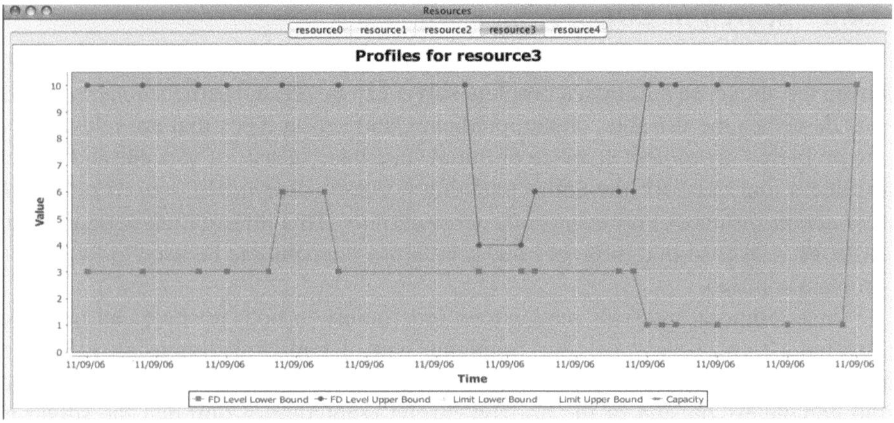

Fig. 3.18 An example of resource envelopes [112, 113] computed by EUROPA

(`uses(resource,start,duration,quantity)`), which are then subsequently released. **Unary** resources are reusables with capacity = 1, which accounts for runtime savings in comparison to the generic reusable class. Configuration options are offered such that in addition to the usual constraint violations when a resource is completely depleted, plan flaws can be generated whenever a resource level is outside a specific bound or when instantaneous production or consumption exceeds a specific threshold.

Determining whether a resource is the source of a plan flaw or a constraint violation in general is substantially more difficult in EUROPA's flexible representation than just computing a profile of the resource and checking against some limits. With temporal flexibility, a number of resource profiles could be applicable depending on precisely when each production/consumption event occurs out of a range of possible values that are in the domain of the variables that represent the temporal extent for the event. Muscettola [112, 113] propose an algorithm where the variables that represent the time for production/consumption events are placed in a temporal network, similar to the network used by the temporal propagator; constraints on the resource can then be seen as quantities *flowing* through that network. Computing the maximum flow then, yields the maximum resource usage and thus a lower bound on the resource level; computing a minimum flow will similarly yield an upper bound. They show that such upper and lower bound envelopes on the resource level can be computed in polynomial time [87]. Resource envelopes can be useful when trying to evaluate plan brittleness in the face of changing activity durations or start times. They can also be useful when searching for plans; however, they are relatively expensive to compute. Therefore their ability to prune the search space must be weighted against the ability to search faster [114, 115]. Figure 3.18 shows an example of such a resource computation.

3.4.6 *Search in* EUROPA

Given the above, an automated problem solver can be created using a *domain model* that describes the variable, object, predicate, and action types that are relevant for the problem—a *problem instance* or initial state that consists of variable and object instances that exist for the entire planning horizon—*temporally scoped predicate and action* instances and *temporally scoped goals*. This information is retained in EUROPA's PDB so that inference and search mechanisms can be used to look for a problem solution.

Since temporal intervals may extend indefinitely in both positive and negative directions (e.g., $[-\infty,+\infty]$), for problems with a temporal dimension it is also common to specify a *planning horizon* when invoking a solver. Any decisions that fall completely outside of the horizon can be ignored, ensuring that the solver is making as few commitments as necessary and reducing plan generation complexity. The *planning horizon* is also useful when dealing with recurrent activities. Consider, for instance, a model where a medical checkup has to occur every 3 months. The user can model a recurring activity and then use the horizon to ensure that a finite number of medical checkup activities are generated as part of a plan.

3.4.6.1 Flaw Resolution

EUROPA provides a built-in solver that performs *Plan Space Planning* [9]; the initial state is considered a partial plan that needs to be refined towards a solution that achieves stated goals. The operations to refine the partial plan PP at any time are:

1. Find flaws of PP, i.e., conditions that prevent it from being a solution plan.
2. Select one such flaw.
3. Select a resolver for the flaw.
4. Refine PP by applying the resolver.
5. If an inconsistency is found, try another resolver.
6. If all possible resolvers for a particular flaw fail, return failure; otherwise continue until resolving all flaws.

The initial partial plan is the state of EUROPA's PDB after the initial state has been instantiated. This results in a set of variable, object, and token instances. Inference then takes place to detect flaws in the partial plan. The three kinds of flaws that can be detected are:

Unbound Variable: It is a variable in the partial plan whose domain is not a singleton. Unbound variables are resolved by value selection from their domain.

Open Condition: An open condition is an INACTIVE token which can be generated by explicitly posted goals or during token activation when rules may cause the creation of INACTIVE slave tokens as described in Sect. 3.4.2. Open conditions can be resolved in three ways:

Merging: An `INACTIVE` token is merged with a matching `ACTIVE` token already in the plan. A token a is said to match another token $a?$ iff a and $a?$ unify [89] and the temporal constraints involving a are satisfied by $a?$ Thus, a and $a?$ can be considered the same token without the need to introduce a in the plan. Consequently, the domain rules associated with a are not introduced into the planning process, since they are already triggered when $a?$ was instantiated in the plan.

Activation: A token a is introduced in the current plan associating it with the appropriate timeline, without choosing a specific time slot; if its timing causes a problem, a threat flaw will be triggered (see below). The domain rules associated with a are applied and the subgoal activities resulting from those domain rules are introduced as `INACTIVE` tokens. This results in a number of open condition flaws corresponding to the new subgoal activities.

Rejection: Some goals may be optional; if so its corresponding flaw may be resolved by discarding the `INACTIVE` token.

Threat: Once a token has been placed in the partial plan it may impact other tokens indirectly through possible overlapping requirements on objects. Recall, for example, that a token may belong to objects (e.g., timelines) which require a total order over their tokens. If any two tokens could possibly overlap (though not necessarily), then they pose a threat to each other in terms of achieving an extension of the current partial plan which is complete and consistent. Similarly, threats may arise where tokens share a common resource and their current state might yield extensions of the current partial plan which are inconsistent. Threats are resolved by imposing ordering constraints among tokens.

Open conditions and threats allow flaw detection and resolution at a higher level of abstraction (i.e., in terms of objects and tokens) than that of binding variables as is common in constraint satisfaction problems. This is advantageous when a solver applies heuristics for ordering choices since it provides a richer context in which to make decisions. Furthermore, it aids in reducing the amount of work done by a solver so that only the necessary refinements are made, ensuring a partial plan with sufficient flexibility. For example, one can bypass committing unbound timepoint variables since threats will force a solver to impose ordering restrictions on these variables. Thus, the planning process may yield partially ordered plans for which all possible extensions are provably valid.

The Plan Space Planning algorithm described can be implemented in multiple ways depending on the approach chosen for flaw and resolver selection and for backtracking. `EUROPA`'s built-in solver implements a chronological backtracking algorithm that is summarized in Algorithm 1.

The algorithm takes as input a partial plan p and returns true if a complete and consistent refinement of p could be found (or if p is initially complete and consistent) and false otherwise. Briefly the algorithm is described as:

Line 2: Propagate the constraints to test for inconsistency. If inconsistent, then return false since no refinements to p can yield a consistent plan.

Line 3: Choose a flaw from the set of available flaws.

Algorithm 1: bool *solve*(*plan*): To deliberate and generate partial plans

```
   Input: PartialPlan plan
   Output: true if a plan solution was found, false otherwise
1  begin
2      if isInconsistent(plan) then return false;
3      Flaw flaw ← chooseFlaw(plan) ;
4      if ∅ = flaw then return true ;
5      DecisionPoint decision ← makeDecisionPoint(flaw, plan) ;      // potential
       branching point(s)
6      ; while decision.hasNext() do
7          PartialPlan pp ← decision.executeNext() ;
8          if solve(pp) then return true ;
9          else decision.undo() ;// possible backtracking point
10     end
11     return false ;
12 end
```

Line 4: If there are no flaws, then p is complete and terminate with success.
Line 6: Formulate a decision point which is a branch in the search space. Each choice is a particular refinement operation and the `DecisionPoint` collects all possible refinement operations for the given flaw.
Line 6: Keep trying until the chosen flaw is resolved, or until there are no further resolvers to try.
Line 7: A new partial plan is obtained by application of a refinement operator. Note that the ordering over refinement operators to select is a nondeterministic step.
Line 8: Recursive call to solve the new planning problem. If successful, then success.
Line 9: Otherwise, retract the last refinement operation and move on to try the next.
Line 11: Arriving here implies all options to resolve the flaw have been exhausted, including the case where no options were available initially. Thus the problem is insoluble.

Algorithm 1 provides for a sound and complete search, assuming that no flaws or available refinement operators are pruned unnecessarily. Dead ends in the search are discovered through constraint propagation, a vehicle for evaluating the consistency of a partial plan and also for filtering infeasible values from consideration prior to commitment. Consistency testing is initiated by the `isInconsistent` procedure. The algorithm permits a heuristically controlled search by applying orderings for `chooseFlaw` and `makeDecisionPoint` resulting in a chronologically backtracked search. Heuristics can be implemented by applying ordering/filtering policies for `chooseFlaw` and `makeDecisionPoint`, for instance, by dealing with earliest flaws first, or by making planning decisions (open condition flaws) before scheduling decision (threats).

It should be emphasized that while this algorithm is commonly employed by applications that embed `EUROPA`, it is only one of many that could be implemented.

Algorithm 2: *solveNQueens(ce, maxIter)*: An N-Queens solver

```
Input: ConstraintEngine ce
Input: int maxIter
begin
    int curIter ← 0 ;
    while (ce.getViolations() > 0) & (curIter < maxIter) do
        Variable queenToMove ← getQueenWithMaxViolation(ce) ;
        SortedSet<int> moves ← getMoves(ce, queenToMove, curPos) ;
        bool moved ← false ;
        foreach int newPos of moves do
            moved ← tryMove(queenToMove, curPos, newPos);
            if moved then break ;
        end
        if !moved then forceMove(queenToMove, curPos, moves.first()) ;
        checkSolution() ;          // Evaluate if a new best solution
        curIter++ ;
    end
end
```

As an example, consider the solver in Algorithm 2 for the N-Queens problem, where an integer variable V_i represents the queen in column i; $V_1 = 3$ implies that the variable on column 1 is placed on row 3. Non-attack constraints are posted as arithmetic constraints among the V_i variables to ensure that all the rows and diagonals contain only one queen. The `CRE` can be used to compute the non-attack constraint violations, with `ce.getViolations()` a call on the `EUROPA` API. The `CRE` can also be used to pick the queen that has the most violations in `getQueenWithMaxViolation()` and to look ahead and see what moves are possible for a queen that would remove its violations in `getMoves()`.

In similar fashion, the constraint violation and flaw reporting mechanisms that `EUROPA` provides can be used as building blocks that specialized or general purpose search algorithms can use to construct plans.

3.5 Planning and Execution with `T-REX`

3.5.1 Introduction

A prime motivation behind the `T-REX` architecture was to design a controller which could bring automated planning closer to low-level control of the vehicle while ensuring reactivity. Traditionally planning has been deemed computationally expensive with planner performance restricting robot reactivity [9, 72]. In control architectures where task planning is embedded, planning consequently remains at an abstract level with the assumption that it would otherwise impede system reactiveness and that the environment can change at a faster rate than the planner can plan for. In such a situation, the agent may thrash [9] if the internal state of the plan gets out of sync with the actual state of the world.

An important consequence of this has been that the planning problem managed in situ has traditionally remained fairly detached from low-level control. Most adaptations at the lower-level reactive layers in contrast are managed by specialized components (e.g., `LAAS Architecture` [17,116] and `CLARATy` [117]) that rely on different representations for modeling local control behavior or worse do not have any explicit formal agent behavior model. The consequent different techniques for specifying each layer in the architecture results in duplication of effort and a diffusion of knowledge leading to significant design and integration issues [39]. `IDEA` mitigated most of these issues by tightly integrating planning and execution in a single representational and computational framework with a unified declarative model. Our approach builds on this design methodology albeit in a substantially systematic manner.

The `T-REX` architecture contributes to agent architectures over and beyond `IDEA` in the following significant ways:

- *Partitioning* the global planning problem into multiple decision loops called *reactors*. Each reactor has its own scope both functionally (one, e.g., focusing on the conversion of high-level waypoints into low-level commands) and temporally as each reactor declares its planning look-ahead (π) and expected latency (λ) before being able to produce a plan. These parameters depend on where a reactor is located in a dependency graph such as the one shown in Fig. 3.20a. A reactor closer to the hardware in the hierarchy is expected to deliberate less and be highly responsive to observations from the world. This implies lower values of π and λ (see Sect. 3.5.2). In the current design, π and λ are application dependent and fixed at design time.
- *Coupling* a tighter integration between the planning process and world evolution with the use of the *same* plan database. By doing so, the planner is better informed in its deliberation despite a larger latency than the changing world.

Systematic partitioned inference is a major contribution of `T-REX` which is covered in more detail in [26, 118]. Coupled interaction between planning and the world, on the other hand, has a substantial impact on how planning can be integrated in such frameworks. In this chapter we show how using the `EUROPA` planner can be coupled to the `T-REX` framework by first linking concepts in temporal planning with the framework. Note that in the following treatise, we call the formal semantic framework of `T-REX` as the "agent." Reactors constitute an entity within this agent and are stand-alone blocks of inference. The scope, detail, and number of reactors are application specific and outside the scope of this chapter.

---

[9] In such an event even as the planner works towards goal satisfaction, the initial conditions leading to achieve that goal are no longer valid. The planner then tries to achieve a different goal which too has to be discarded similarly. And so on.

3.5.2 Architectural Concepts in T-REX

The T-REX architecture is structured around the notion of EUROPA *state variables* as the fundamental basis of interaction. Additionally it comes with a well-defined ownership model associated with each such state variable. In this framework each reactor is seen as a "black box" providing its own *internal* state variable which it is responsible for in order to maintain its value at every instant of time. Each such instance is represented by a *tick* with a fixed duration; in our implementation of T-REX on the Dorado AUV, we use a tick of 1 s. In order to identify its state a reactor can subscribe to *external* state variables owned by other reactors, for which it will receive updates when the owner reactor makes changes to its *internal* state due to deliberation or an exogenous state update.

As the execution frontier, τ advances, the T-REX agent propagates the current state information across all reactor state variables through the *synchronization* process. The sequencing of the reactors synchronization is identified by traversing a reactor dependency graph from the least dependent—ones with no *external* state variables—to the most, i.e., which typically do not have *internal* state variables. Conversely each reactor can request a future change on one of its *external* state variables which will be transformed as an *internal* goal to the owner reactor(s). Such *goal dispatching* is expected to produce a deliberation phase for the receiving reactor which can cyclically, in turn, produce a plan impacting its *external* state variables' future that will propagate down the hierarchy following the same mechanism. Figure 3.20a shows an example of such a dependency graph implemented as a directed acyclic graph.

In such a process, one challenge is to avoid being overwhelmed by goals that may threaten the ability of a reactor to process and integrate in its evolving partial plan. For that reason, each reactor specifies the following two parameters that are applied towards the scope of the agent lifetime:

- λ is the specified latency of this reactor, which is an indicator of the *maximum expected time* for a reactor to produce a plan.[10]
- π is the planning *look-ahead* of the reactor, which is indicative of the number of ticks in the future this reactor is looking towards deliberation.

Using this information along with the dependency relation between reactors, T-REX defines the notion of execution latency of a state variable as follows:

Definition 3.1. The *execution latency* of a *state variable s* which is internal to the reactor *r* is the maximum expected delay before a goal on this state variable can be included in its plan and executed by the reactor *r*. It is defined recursively as

[10]Note that while it is highly recommended to select a sound value, a failure to produce a plan within a chosen value for this parameter is not considered as a critical failure of the reactor.

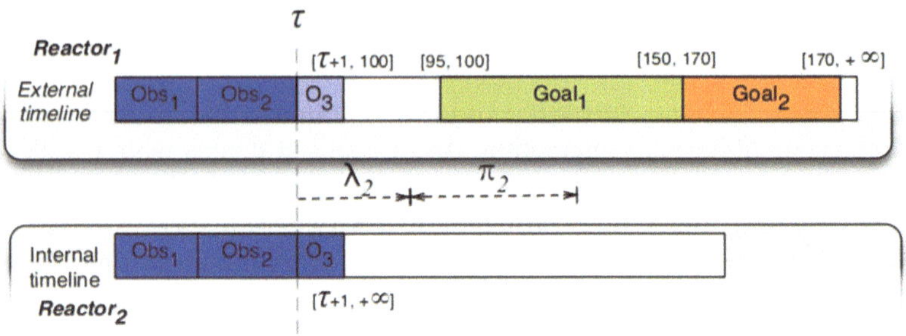

Fig. 3.19 Dispatching a token occurs only when the request start time overlaps a window. $\texttt{Goal}_1$ therefore gets dispatched from $\texttt{Reactor}_1$ to $\texttt{Reactor}_2$ but not $\texttt{Goal}_2$ to avoid early commitment. Tokens from $\texttt{Reactor}_1$ are "dispatched" for execution to $\texttt{Reactor}_2$ only when necessary while meeting the conditions in Definition 3.2. τ signifies the execution frontier as time moves from *left* to *right* in the illustration

$$\lambda_{\text{exec}}(s) = \lambda(r) + \max_{e \in \text{exts}(r)} \lambda_{\text{exec}}(e)$$

Where $\text{exts}(r)$ is the set of the state variables which are *external* to r.

For intertwined plan execution, the concept of execution latency is critical at the agent level since it can impact the planning/deliberation process on external state variables of a reactor. For simplicity we abstract it out as being equivalent to the reactor latency implying that we assume that all *external* timelines of this reactor have $\lambda_{\text{exec}} = 0$. At the T-REX agent level this execution latency is used to define the *dispatch window* of the same state variable (see [26] for formal proofs).

Definition 3.2. For every timeline s *internal* to a reactor r, the *dispatch window* of this timeline at the execution frontier τ is the time interval

$$\Pi(s) = [\tau + \lambda_{\text{exec}}(s), \tau + \lambda_{\text{exec}}(s) + \pi(r)].$$

This window is used by T-REX to identify when to dispatch a request to a *reactor* as a new goal. This happens only if the request start time—represented as the interval of possible start times—intersects this window. Requests further in the future are buffered by the agent until this time window advances for an overlap. This provides an elegant mechanism to request a change from the *external* state variable of a reactor while avoiding flooding the owner reactor of this state variable and avoiding early commitment. The choice of both the latency and look-ahead of a reactor can be delicate as one is dependent on the other. Latency is dependent on exogenous parameters such as the computer power of the host and the number of reactors needing to deliberate at once, among others. As the design stands, these variables should be evaluated carefully by the system designer balancing agent efficacy. Figure 3.19 illustrates an example between two reactors where one token is dispatched when overlapping the dispatch window while another is not.

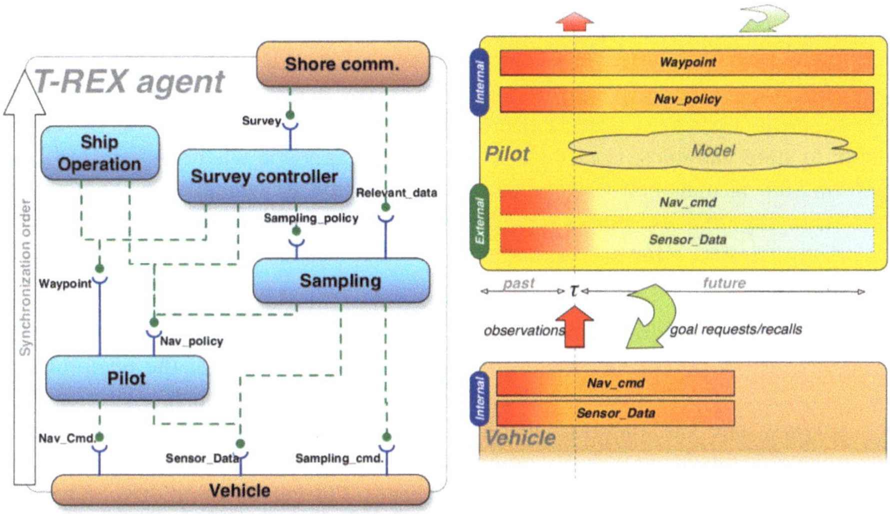

Fig. 3.20 T-REX reactors and shared state variables. (**a**) A T-REX agent is composed of multiple reactors or control loops (*rounded boxes*) which are connected through state variables provided by one reactor (*blue solid line*) with multiple possible clients (*green dashed lines*). (**b**) State information flow between the Pilot and the Vehicle reactors in an T-REX instance. Observations transit as a bottom-up flow and apply to the execution frontier τ. Goal information flow from the top to the bottom and relates to desired future values of state variables

Requests from a reactor can be *recalled* when it is determined that they are not part of its projected plan anymore. For example, if a reactor needs to replan due to an unexpected event, it will recall all requests produced that have yet to be executed. This recall is also sent to the reactor(s) that received the goal(s) to remove it/them from its current set of objectives.

Consider, for example, the T-REX agent instance we use on our AUV as shown in Fig. 3.20a. In this instance all the reactors (symbolized by colored boxes) rely on a publish/subscribe model on shared state variables. For instance the Pilot reactor needs the *externally* managed state information from Sensor_Data and Nav_cmd state variables in order to maintain the *internal* state variables Waypoint and Nav_policy. This relation applies in two ways:

- In order to identify its current *internal* state values the Pilot needs to know the current *external* state information it relies on. This information is provided in turn by the reactors that define these timelines as *internal*. In this specific case, the current Waypoint where the vehicle is heading can be evaluated based on the Nav_cmd executed along with useful Sensor_Data, both being provided by the Vehicle reactor.
- The future objectives a reactor has on its *internal* state variables will likely imply sub-objectives to be reported to the owner of its *external* state variables depending on their current values. Should the Pilot want to visit a specific

Waypoint—given its current position as provided by Sensor_Data—it can identify a sequence of Nav_cmd states that should eventually help it reach this location.

Given the above example, T-REX abstracts reactor interaction by constraining them to state variables. Each reactor is therefore agnostic where its *external* state information is managed as long as this information is available and maintained for managing *internal* state information. Such a formalism also constrains reactor interaction to state information exchange which comes in two forms as illustrated in Fig. 3.20b:

- *Observation on current state values*: This propagates *up* in the reactor dependency graph during the synchronization phase and occurs at the beginning of every tick. During this phase the agent provides updates to each reactor on their *external* state variable(s) to enable computation of their *internal* state variable values for this same tick. By doing so we propagate a consistent view of world state throughout the reactors as time advances. Further, observations can be both sensor measurements as well as state information, passed up from other reactors. The strict rules of timeline ownership enable a clear policy for conflict resolution: observations dominate expectations.
- *Goal request on future state*: This information propagates *down* in the reactor dependency graph. In order to satisfy its own *internal* objectives, a reactor may have a plan that relies on future values of one (or more) of its *external* state. When such a plan is identified the agent ensures that this part of the plan is provided as a goal to the owner of the given *external* state variable(s).

The choice of representation of state variables, observations, and goals is directly derived from EUROPA timelines and tokens. While such a rich representation allows exchange of state information in a formal yet flexible manner, its translation into planning frameworks that have an explicit representation of time (such as but not limited to EUROPA) is often trivial.

3.5.3 The Execution Cycle

Within T-REX planning/deliberation is a continuous cycle between all reactors. At the agent interface level, each reactor provides only two abstract calls that conceptualize execution:

- Synchronize takes the last set of *external* observations as an argument and returns the *internal* observations for a reactor or a failure report.
- Step that accepts new goals requested (if any) on the *internal* timeline of a reactor and will execute one step of deliberation. In the context of EUROPA, for example, one deliberation step corresponds to the resolution of a single flaw (Sect. 3.4.3) while ensuring that the resulting plan is not inconsistent. Backtracking is included in the step should it be necessary. The returned value

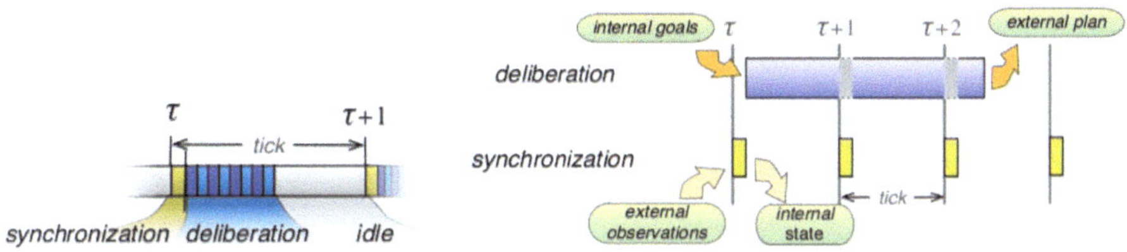

Fig. 3.21 Detailed and abstracted views of synchronization and deliberation within one and multiple tick cycles. (**a**) A view of synchronization and deliberation within a tick cycle. (**b**) A T-REX reactor's execution cycle: *Deliberation* is interrupted by *synchronization* at the beginning of every *tick* allowing integration of state information

by the `step` function is a boolean which indicates whether the reactor needs an extra step to deliberate or if this last step resulted in a complete plan within the reactor's look-ahead. Subsequently the part of the plan that covers the reactor's *external* timelines is dispatched as a goal.

Both these functions have different scope and temporal constraints. The `synchronize` call is considered as atomic and focuses on state inference for the current tick by identifying the current value of *internal* state variables from the latest updates on *external* state variables. This process propagates up in the reactor dependency graph at every tick to ensure that all reactors are aware of current world state. `step` relates to a single decision step of the planning process focused on the future evolution of state variables; it can take multiple steps spanning multiple ticks for a reactor to produce a plan that is both complete and valid. How the reactor identifies that its plan is complete and valid is its responsibility and depends on the way each reactor resolves its plan at each *step*. The general idea is that the reactor is able to find a valid strategy in order to satisfy as many of its *internal* goals as possible. For EUROPA, this means that the solver is unable to identify any further flaws to be resolved in the plan. Figure 3.21a illustrates the distinction between a tick, `step`, and synchronization within the tick.

When such a plan is identified, the reactors *external* timeline(s) are then dispatched to their respective reactor owners, which in turn can start their own deliberation `steps` to satisfy these new objectives. Consequently execution occurs as these goals and plans propagate down in the reactor hierarchy until they eventually become goals for the lowest level reactor which—instead of deliberating—may just transform this goal into a command for execution in hardware.

The manner in which the agent interleaves planning `steps` and synchronization for a single reactor is illustrated in Fig. 3.21b. Synchronization occurs at every tick while interrupting planning; this allows a reactor to identify its current state that can propagate through the reactor hierarchy in order to ensure a consistent view of the state of the world between all the reactors. These intertwined tasks together provide two natural information flows (bottom-up for state estimation and top-down for plan projection) that we expect to see in a control loop. As they are both inference-based processes it is natural to implement a reactor based on generic automated

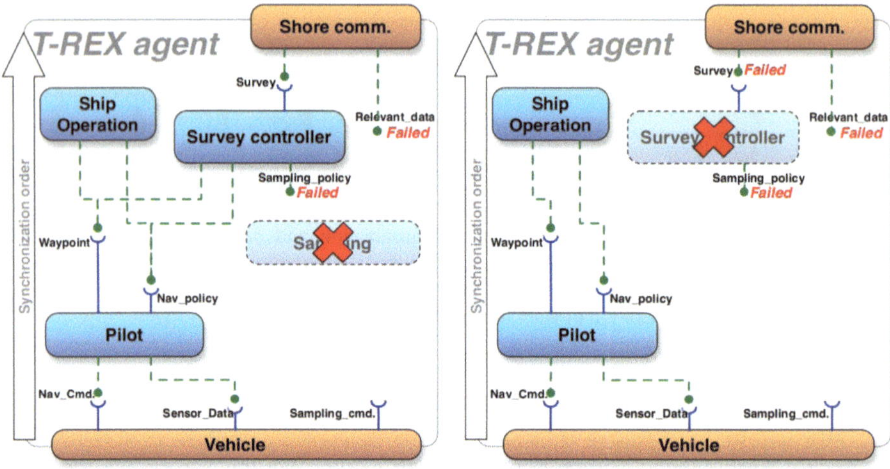

Fig. 3.22 Impact of the `Sampling` reactor failure in the AUV agent from Fig. 3.20. (**a**) When the `Sampling` reactor fails to synchronize, it is removed from the reactor dependency graph along with all its *internal* timelines while generating a `Failed` observation. (**b**) In the worst case this `Failed` observation may imply other reactors that depend on the `Sampling` reactor might fail to synchronize too leading to a bottom-up destruction of "faulty" reactors

planning. For example, in Fig. 3.20a, apart from Shore comm. and Vehicle acting as interfaces, every other reactor (in blue) is a different instance of a `EUROPA` reactor differentiated only by their model and `EUROPA` solver configurations.

3.5.3.1 Handling Reactor Failure in T-REX

The critical task of a `T-REX` reactor is to successfully identify its state through synchronization as the execution frontier advances. When a reactor fails to integrate its *external* state update and/or identify its *internal* state at any tick, it implies that it is unable to explain the state of the world. This in turn implies that any state information from this reactor is suspect, resulting in the exclusion of this reactor from the execution cycle and consequent destruction. As the agent destroys the faulty reactor, a special observation, `Failed`, is produced on all *internal* timelines. Reactors which declare these very state variables as *external* are informed of the failure of this timeline. All goals exchanged between reactors and these timelines are then considered *rejectable* by the reactor that manages them. As a result in such a situation, the failure to find a plan for these goals is *not* considered a critical failure of the reactor. It will only result in these goals being ignored until newer goals are received.

Take, for example, our AUV agent from Fig. 3.20a, and assume that the `Sampling` reactor failed to synchronize. In such a case this reactor is destroyed and both `Sampling_policy` and `Relevant_data` timelines generate the `Failed` observation as illustrated in Fig. 3.22a. The `Survey controller` will receive this `Failed` observation on its *external* `Sampling_policy` timeline which indicates that this state variable is no longer managed by its owner reactor. This leads to two *potential* outcomes based on how the reactor is implemented:

1. The `Survey controller` model could integrate the `Failed` observation. For example, it can identify that, subsequent use of the water sampler is no longer possible. The `T-REX` agent can continue to navigate since `Pilot` is still available. In such a case the agent structure remains as in Fig. 3.22a with no consequent impact.
2. The `Survey controller` model is unable to handle this unexpected failure resulting in its failure to synchronize. This results in the `Survey controller` also failing, with subsequent destruction and flagging via the `Failed` observation on its `survey` timeline as shown in Fig. 3.22b.

Even in the worst of the two cases presented above, we note that a reactor failure only propagates *up* in the reactor dependency graph. In our example, even if all the reactors that depend on `Sampling` are entrained by its failure, `T-REX` continues to keep the reactors `Vehicle`, `Pilot`, and `Ship operation` intact. This ensures maintenance of vehicle control via the `Pilot`, in addition to an abstract commanding mode for safe operation via the `Ship operation` reactor, illustrating how `T-REX` can handle system degradation despite the failure of one or more reactors, and doing so gracefully.

Consequently, by ensuring that lower-level reactors closest to the platform hardware are robustly designed, the `T-REX` framework ensures that system designers can focus their attention on critical components while factoring in deliberative control of their platforms.

3.5.4 The Deliberative Reactor

Within `T-REX`, the sole purpose of an instance of an `EUROPA` reactor is to deliberate. A typical reactor design takes its cues from `IDEA` [70, 73] where both planning and execution are based on a single unified model within a rich representation that provides support for temporal inference. In `T-REX`, the execution cycle consists of two deliberation processes which are implemented as two separate `EUROPA` solvers modifying the *same* PDB.

1. The *synchronization solver* is a specialized `EUROPA` solver that integrates new information in the plan. This signifies the evolution of *external* timelines and ensures that the reactor propagates changes to identify current state as well as to inform other reactors of any state change on its *internal* timelines. This solver is invoked at the beginning of every single tick.
2. The *deliberation solver* is used to either produce a new plan, to alter its current plan as new goals are provided, or when the synchronization solver identifies a conflict between current state and expectations from a previously generated plan. This process can span multiple ticks and can therefore be interrupted by the synchronization solver.

While the two processes are separate, they share the same plan internal to a reactor; consequently at every synchronization cycle the planning process is informed of the new world state and its impact on the plan. Conversely, when synchronization occurs after deliberation stops, its starting point is the last partial plan resulting from deliberation (see Fig. 3.21). This helps to target the synchronization problem by ensuring that the new *external* state updates are compatible with this plan. If so, synchronization propagates these events in the plan and can use it as an aid in resolving its *current state flaw*. It does so by using the plan information on the expected *internal* state evolution near the execution frontier.

If they are not compatible, then *synchronization* identifies a plan inconsistency. This results in the retraction of decisions made during previous *deliberation* steps. Such a relaxation can be viewed as a backtrack, allowing the reactor to identify that its current plan has failed to execute triggering further deliberation steps to find an alternate strategy.

Finally, when the deliberation process has identified a complete solution for its current goals, a part of its plan that applies to its *external* timelines is sent as a request to `T-REX` to handle their dispatch to corresponding reactors. Synchronization eventually receives *external* observations that confirms plan execution, in turn recording execution progress by having the part of the plan observed restricted to observation start times.

Synchronization offers an important challenge for embedding planning in a situated agent, typically an assumption to restrict the problem to "offline planning" defined in [9]:

> The planner is not concerned with any change that may occur in Σ[11] while it is planning; it plans for the given initial and goal states regardless of the current dynamics, if any.

This assumption, while helpful to reduce the scope and complexity of the planning problem, becomes problematic for planning within a dynamic environment. In a typical robot this implies that the only point where the planner can be informed about the current world state is prior to the commencement of a planning cycle by providing the planner an initial state and current objectives. During the planning, it is assumed that the agent can *maintain the world as perceived by the planner*. Lemai-Chenevier and Ingrand [106], Lemai-Chenevier [119] attempts to reduce the impact of planning by allowing local plan repair in the existing plan. In the eventuality that such repair is not feasible they require that their robot comes to a stop until (re-)planning is complete.

This assumption is problematic for situated agents in general and as Fig. 3.21b shows, for `T-REX`. Synchronization that tracks state evolution can occur several times during reactor deliberation. By making synchronization and deliberation share the same plan we provide a solution that relaxes this static world assumption. `IDEA` uses the same principle with all solvers in an agent sharing the same plan for communicating environmental change [72, 73]. Specifically their *Reactive*

[11] The world modeled by the plan domain.

planner—similar to our *synchronization* solver—is the solver targeting change at the execution frontier. Dias et al. [72] states:

> [...] as long as the planning horizon over which the deliberative planner is working never intersects the current execution time, deliberative planning can operate in parallel with reactive execution and does not affect the reactivity of the agent.

This suggests a deliberate nonoverlap between the *reactive planner* and *deliberative planner*'s horizons. In T-REX, and more specifically in our deliberative reactor, the planning horizon of *synchronization*—limited to the current tick—is a subset of the planning horizon of the deliberation steps—starting at the current tick and with an extent based on latency λ and look-ahead π of the reactor. In this manner we avoid creating a conflict between the deliberation and synchronization by controlling the scheduling of these two processes by the T-REX agent.

The following sections detail how synchronization and deliberation are implemented independently to show how their interaction through the plan helps relax this assumption. It does so by integrating planning in the reactor execution cycle even while allowing for more informed deliberation.

3.5.4.1 Synchronization and Identifying Internal State

We first show how a reactor can track and identify its state in the context where it does not have any compelling need to deliberate. In doing so we assume that we have a reactor that has no future goal. Under such a condition, T-REX requires each reactor to produce at every tick its *internal* state. For those reactors not owning *internal* timelines, T-REX makes sure that *external* observations are taken into account for every tick as in [26]. Such a requirement is enforced to ensure that all reactors share a consistent view of the world with advancing time.

External state updates are managed by introducing these observations in the plan structure as facts and ensuring that they remain compatible with the current plan maintained by the reactor. The challenging part of synchronization is to decide the *internal* state values for the current tick. In the shopping example from Sect. 3.4.4, for example, we have the following code snippet which helps identify when we can Buy a product:

```
Agent::Buy {
  eq(10, duration);
  // [...]
  contained_by(condition object.location.At currLocation);
  // [...]
}
```

If object.location.At is *external* and Agent is an *internal* timeline, a robot may encounter the situation where it is at the location where it can buy one (possibly among others) product. This results in a decision-making problem related to the state value of the Agent timeline; the robot can either *Buy* a specific product now, wait a little longer before doing so, or just simply not buy any product at this location.

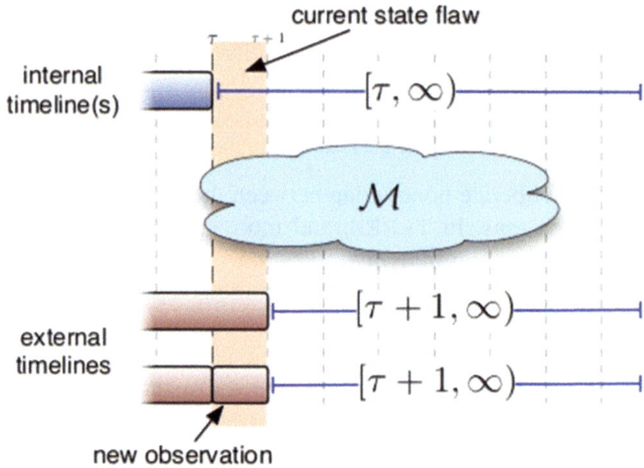

Fig. 3.23 Illustration of synchronization flaws in a reactor. The reactor receives new observations when they are produced by the owner(s) of its internal timelines. The open "holes" after the last token of each timeline represents the domain of all possible values for the end of this token. At every tick τ, the reactor needs to integrate the *external* state information it received and—based on its model $\mathcal{M}$—resolve its *internal* state that will then be provided in turn by the agent to other reactors

In the deliberative reactor, this decision-making requirement is a new type of flaw which requires identification of all values of *internal* state variables for the current tick. This flaw can be defined as

Definition 3.3. A *current state flaw* in Fig. 3.23 indicates that a reactor's *internal* timeline does not provide a fully defined state value for the execution frontier τ. Such a flaw can be the result of either:

- The absence of a token in this timeline which overlaps the execution frontier τ
- An open decision between terminating the last token for this timeline or extending its end in the future

The key idea is to ensure that at any single tick the *synchronization* solver will identify all the *internal* timelines for which the current state value is yet to be identified. We handle their resolution based on the following proposition:

Proposition 3.1. *A current state flaw can be resolved using one of the following choices which are evaluated in order:*

1. *Extend the previous state value to end after the execution frontier (i.e., restrict its end time to* $[\tau+1,\infty)$*).*
2. *Start the next active token in the timeline by restricting its start time to the single value* τ*.*
3. *Create and insert a new token in this timeline that will start at the current tick* τ *and attempt this for each possible token type for this timeline if necessary.*

Intuitively the first choice enforces an *inertial value assumption* where the reactor will tend to maintain its current state as long as the model and external observations allow for it. The second choice favors the advance of the reactor's current plan if any; this directs the *synchronization* decision towards assuming that its current plan is properly executed. Finally if none of these two choices provides a valid solution—i.e., it leads to an inconsistent state—the reactor tries all the possible state values remaining for the timeline until it finds one that results in plan consistency.

The synchronization process in turn can be described as:

1. Integrate the external state as provided by the owner of each external timeline into the PDB.
   - New observations are integrated by creating the corresponding token as a fact in the PDB.
   - If no observations have been received at this tick the last observation token restricts its `end` domain to the interval $[\tau+1, +\infty)$.
2. Propagate this information in the PDB using reactor model $\mathcal{M}$.
3. Resolve the current state value of each internal timeline based on Proposition 3.1.

Steps 2 and 3 above are handled by the `EUROPA` solver from Algorithm 1 which ensures that the *current state flaws* are selected last so as to follow the sequence given above.

As *synchronization* focuses on the execution frontier it also filters out all flaws that are outside of the current tick's scope. If the flaw correlates to a token that is *necessarily* ending before τ or starting after $\tau+1$ then the solver ignores it. Such a strategy helps to reduce the scope of synchronization exclusively to the current state of the reactor and in turn reduces the search space for the *synchronization* solver. To do so we need to assume that the current state value of an *internal* timeline does not depend on the future or more accurately that any choice made during this *synchronization* will not lead to a future inconsistency. Such an assumption impacts the set of possible domains the reactor can support while remaining algorithmically complete.

Take, for example, this rule from our `Shopping` model in Fig. 3.24. While this model is perfectly acceptable for planning it becomes problematic when execution is taken into account especially since it puts a strong emphasis between an evolving plan and an unknown future; `Going` to a location and the *expected* future outcome of ending at this location tie it strongly to the constraint in line 5 above. For instance, assume that the `Agent.location` is *external* to the reactor which receives the observation `Going(Home,SuperMarket)` which synchronization has resolved by producing the observation `Go(Home, SuperMarket)` in the `Agent` timeline. This would imply that the only possible outcome on the `Agent.location` timeline, is to be `At(SuperMarket)` in the directly foreseeable future.

In the real world the fact that the robot is `Going` to a certain location does not necessarily imply that it will succeed in ending up there. Due to an exogenous event, for example, the road to the `SuperMarket` could be closed, which could result in an alternate observation that the robot is `At(7th st. & Main)` contradicting

```
1 Agent::Go {
2   met_by(condition object.location.At origin);
3   eq(from, origin.loc);
4
5   equals(effect object.location.Going going);
6   eq(going.from, from);
7   eq(going.to, to);
8
9   meets(effect object.location.At destination);
10  eq(to, destination.loc);
11 }
```

Fig. 3.24 NDDL rule from the shopping agent example with impact on a future state

the existing plan. Further, it is important to understand the temporal direction of the *causal link* that is triggering the inconsistency. At the time the robot observes the contradiction of the `At` predicate, the rule that generated this conflict is connected to past observations `Going` and `Go`. These cannot be retracted easily without a potential impact on computational cost.

More generally tying a token to future outcomes in a model that is meant to be confronted by execution is problematic. It might be valid if one can guarantee that an outcome is predisposed to occur (e.g., the reactor managing the `Agent.location` timeline guarantees that the outcome of the `Going` will *always* result on ending at the target location). However, given the very nature of the planning problem and the ability to search outcomes and to replan, the model should be written so that a future projection can be manipulated.

One could therefore alter this part of the model to give the reactor more flexibility by relaxing the constraint in line 5 from a strict `equals` to a `contains` and to add the new constraint at line 11 to ensure that the last `Going` ends up in the location the agent actually wants the robot to be at. This first level of relaxation will therefore allow the agent to use multiple `Going` tokens should the first token fail to reach the objective, as shown below.

As the example in Fig. 3.25 shows, synchronization involves a deliberation process. The risk is such that the search for a solution is not a simple sequence of decisions where the resolution can be easily identified by the solver as any other option would lead immediately to plan inconsistency. Since such a process is required by the `T-REX` at *every* tick for *every* reactor the model needs to be designed carefully.

Synchronization is the only critical task of a reactor. Such a process is used in order for the agent to identify its current state and maintain it consistently throughout the agent's lifetime. For a reactor not being able to find a consistent solution during synchronization—implying that after evaluating all possible solutions provided by a model being unable to find a consistent plan—not only indicates that a reactor is unable to explain the state of the world but also more importantly it may jeopardize other reactors that rely on its *internal* state. For this reason a

```
Agent::Go {
  met_by(condition object.location.At origin);
  eq(from, origin.loc);

  contains(effect object.location.Going going);
  eq(going.from, from);
  eq(going.to, to);

  meets(effect object.location.At destination);
  eq(to, destination.loc);
  going meets destination;
}
```

Fig. 3.25 Model fragment to enforce a Going end condition

failure to synchronize immediately results in the T-REX agent killing a reactor and notifying any reactor that depends on its *internal* state variables; it does so by posting the special observation Failed. This special state is acceptable for all the *internal* and *external* timelines of a EUROPA reactor. By so doing we allow for graceful degradation where other reactors can readapt their plans on receiving this observation.

Finally, at the engine level, careful choices are made in both the focus of the search—limited only to tokens that can potentially overlap the current tick—and heuristics where the *current state flaw* is considered last during the search process. The latter limits the amount of backtracking for a solution—or at least how deep one needs to backtrack in a decision tree. Similarly the sequence in which we evaluate the possible resolution for such a flaw—described in Proposition 3.1—is selected according to the assumption, that any external observation that is compatible with an existing token in the current plan reflects the execution of the plan itself. In other words this means that our solver always prefers to use tokens *already existing* in the plan as a way to identify current internal state. By doing so synchronization exploits the structure of the plan produced during previous deliberation steps.

3.5.4.2 Deliberation: Planning for State Evolution

Deliberation inside a reactor is based on EUROPA capabilities described in Sect. 3.4.3. However this deliberation comes with some augmentation, namely the reactor's deliberation latency (λ) and its planning look-ahead (π) both expressed in ticks. In addition, the agent needs to interleave deliberation with not just synchronization but also with the deliberation of other reactors.[12] For this reason

[12]The current implementation of T-REX is running on a single process and it is the responsibility of the agent itself to emulate reactor multi-threading for deliberation and synchronization.

while the reactor relies on the EUROPA solver, it slices the execution of Algorithm 1 into atomic steps. This interruption occurs around the recursive call, line 8, while ensuring that the current partial plan is not inconsistent at any single step. This is managed by allowing the call to backtrack until a DecisionPoInt that is not exhausted (i.e., *decision.hasNext*() is `true` and nonempty) is found in the call stack.[13]

A critical aspect of this deliberation step is to ensure that at the exit of this call the *plan* is consistent. If the partial plan is found to be inconsistent during a deliberation step we either backtrack in the decision stack until a node of the search tree is found with an alternate solution or the plan is relaxed since there is no other alternative solution. We do this by removing all goals while keeping only recent observations.

Synchronization and the dispatching of new *goals* from T-REX can occur between two steps and alter the plan. These alterations exogenous to the deliberation process dynamically impact the set of flaws that need to be resolved. New flaws are created when a new goal is received from T-REX and conversely the recall of an objective could remove the corresponding flaws.[14] The impact of *synchronization* can be more subtle as its alteration of the plan at the execution frontier is more drastic. A possible list of alterations of the plan directly related to *synchronization* could be:

1. Previous flaws directly related to the execution frontier have been resolved by the *synchronization* solver.
2. The *synchronization* solver may have created new flaws that were outside of its scope but are now part of the *deliberation* solver's scope.
3. The last observation has proved that the current plan is invalid requiring the *synchronization* solver to relax the existing plan to identify its current *external* state. (This case is developed further in Sect. 3.5.4.3.)

When the *execution frontier* advances, the reactor planning scope moves further into the future. This is reflected by the *deliberation* solver which uses a new deliberation scope one tick further in the future. As Fig. 3.26 illustrates, such a scope allows filtering of tokens that are either necessarily ending before τ or necessarily starting after, focusing the planning problem on only those tokens that are within a "reasonable" future. This filtering window complements the T-REX *dispatch window* from Definition 3.2 which filters out, at the agent level, those goals that are before the reactor execution latency or too far into the future. The planning window refines this filtering further, as tokens can be excluded dynamically from deliberation as their insertion in the plan and scheduling with other generated tokens can alter their temporal scope to be outside of the planning window. This allows the planner to focus on its planning horizon and not report decisions on flaws that are beyond this scope. As time advances, these potential future flaws will eventually

[13]The call stack refers to the sequence of recursive function calls.

[14]However, if the flaws related to this goal were resolved, the removal of this goal may likely create new flaws.

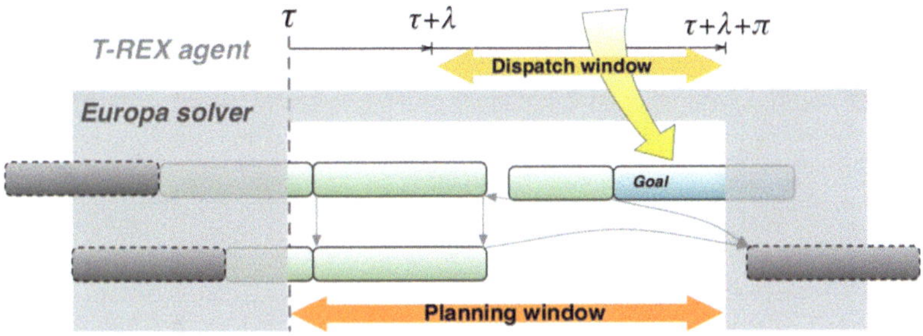

Fig. 3.26 The planning window of the deliberation solver and its relation to the *dispatch window*. The *dark-grey* tokens with *dashed* borders are excluded from deliberation as they do not overlap the interval $[\tau, \tau+\lambda+\pi]$. All the other tokens (*green*) along with the goals received from the agent (*blue*) which can overlap this window are evaluated

be part of the planning window of the reactor and will be then resolved along with potentially new goals. This, in turn, results in an apparent *continuous planning* [120] of the reactor. When there are no more flaws present in the current planning scope a partial plan solution is found. This has two effects:

1. This reactor does not need to deliberate until the next synchronization.
2. The token on the *external* timelines of this partial plan can then be posted. Eventually, at the end of the current tick, if this goal overlaps the owner reactor's planning window, it will be dispatched, resulting in further deliberation.

3.5.4.3 Interleaving Synchronization and Execution

A key aspect of the T-REX framework is interleaved planning and execution. Both *synchronization*—keeping track of execution—and *deliberation*—dealing with reactor planning—of one reactor share the same *single* PDB which they modify alternatively as the T-REX agent schedules their execution.

As shown in Sect. 3.5.4.1, it is synchronization which drives planning; so the comingling between these two processes is critical to maintaining state within a robot. Further, it is important that T-REX maintain one *plan* structure shared between synchronization and deliberation. This allows these two processes to not only be interleaved for agent responsiveness, but this also ensures a viable mechanism to communicate.

As also noted earlier, synchronization propagates *external* observations when a new tick occurs. Earlier, we isolated the case when a reactor has no future goal or no plan to enforce. By doing so we were able to develop the way we augment the EUROPA solver to propagate *external* observations to identify *internal* state. This ensured focus on synchronization as a state evaluation process that can be resolved within the EUROPA framework with few extensions. Similarly

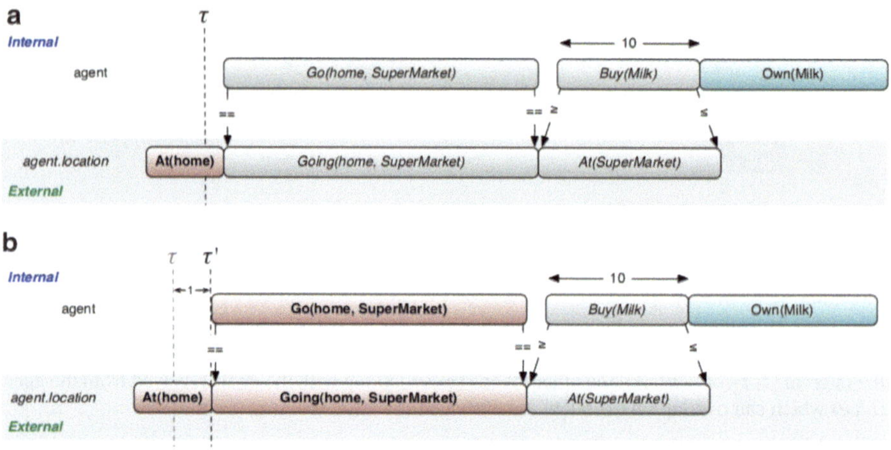

Fig. 3.27 The shopping example: Tokens in *red* indicate observations, tokens in *blue* the goal that this plan attempts to solve for, and tokens in *grey* the result of deliberation. *Arrows* indicate temporal constraints with == indicating "equality" and ≥ indicating "greater than equal to" in timepoints. (**a**) Result of synchronization after the *Going* observation was received by the reactor with the plan showed in Fig. 3.27a

deliberation was presented as an abstraction without discussing the disruptive nature of synchronization and how it impacts the planning process. That discussion occurs here in this section.

We first analyze the case when synchronization occurs after planning has completed, i.e., the last step of deliberation results in a plan that is considered complete for the current look-ahead. As synchronization starts with a plan, this deliberation will not only be a model-derived estimation uniquely based on *external* observations but will also include reactor intentions specified in the partial plan.

Consider, for example, the Shopping Agent model from Sect. 3.4.4, where the timeline `Agent.position` is *external* while `Agent` is implemented as an *internal* timeline. As the reactor adds a goal to `Own` milk and the current *external* timeline indicates that the robot was `At(home)`, it produces a plan as shown in Fig. 3.27a. The *external* part of the plan—namely the tokens `Going(home, superMarket)` and `At(SuperMarket)`—are installed as goals for the reactor managing this state variable internally.

Consider that at the next tick $\tau' = \tau + 1$ the reactor managing agent location is able to change the location state variable to `Going`. `T-REX` informs the reactor that owns the `Agent` timeline, of this observation which is added to its partial plan as a fact—starting exactly at τ' and lasting for an unknown duration which is at least one tick. As this new observation is integrated in the plan through synchronization, the solver can identify the similarity between this observation and the next token in the existing plan. Instead of initiating an exhaustive search of all possible implications, the synchronization process starts its execution with the assumption that these new observations are consistent with the currently maintained plan. As a consequence, at the end of synchronization, the plan depicted in Fig. 3.27b is likely. This is a result of propagation of information provided by the new observation in the existing plan while restricting the start time of `Going` and propagating temporal constraints, making token `Go` to start *exactly* at τ'. In doing so, the `EUROPA` solver's focus on the execution frontier (τ' in Fig. 3.27b) takes advantage of the existing plan to evaluate a solution that assumes that the observation is consistent with what was decided by the reactor. In the event of inconsistency, plan expectations do not match reality; then partial plan failure within the reactor is initiated. This is directed not only by heuristic choices made in the solver but also enforced by the heuristic order of the decision point resolution choices for the *current state flaw* as shown in Proposition 3.1. As noted, the first two choices are biased towards a conservative view with respect to the existing plan.

3.5.4.4 Dealing with Partial Plan Failures

The solver's heuristics first attempt to extend the previous observation for this *internal* timeline to avoid early commitment. If this fails `T-REX` attempts to start the next token in the plan to advance plan execution. Both choices attempt to resolve state with the least amount of plan perturbation by further constraining the start and end time of existing tokens generated by previous deliberations steps. By doing so, in a nominal situation, synchronization remains limited to restricting time points in the plan and in turn propagates their impact. Most of the computational effort of creating new tokens in the plan and scheduling them onto timelines are managed during deliberation steps. Consequently, most of the time synchronization is limited to very few nodes in the `EUROPA` search space often requiring no backtracking. Figure 3.27 shows how the existing plan can support the synchronization process. The observation `Going` occurring at τ' is fully compatible with the token already present in our plan. By updating the plan and propagating equality constraints, the resolution of the *current state flaw* for the *Agent* timeline becomes obvious as a solution, as the token `Go` from the plan presents itself.

Synchronization may become computationally more expensive when we need to identify a new token to be injected into the plan (the third choice in Proposition 3.1), as it needs to explore the outcome of inserting all possible token types the timeline can accept. This is the reason this choice is evaluated last in our flaw resolution strategy.

Synchronization also supports the tracking of plan execution. It does so by specifying token parameters such as time points that are propagated throughout the plan to identify plan validity. When invalid, identification occurs by synchronization failing to find a consistent solution. For instance, should the `Going` observation received be to the `hardwareStore`, this would break the partial plan in the reactor as the robot cannot possible be going simultaneously to two different locations; see Fig. 3.28a. As synchronization is not possible in such a situation the strategy followed is to relax all decisions made by previous plan steps. This is done by

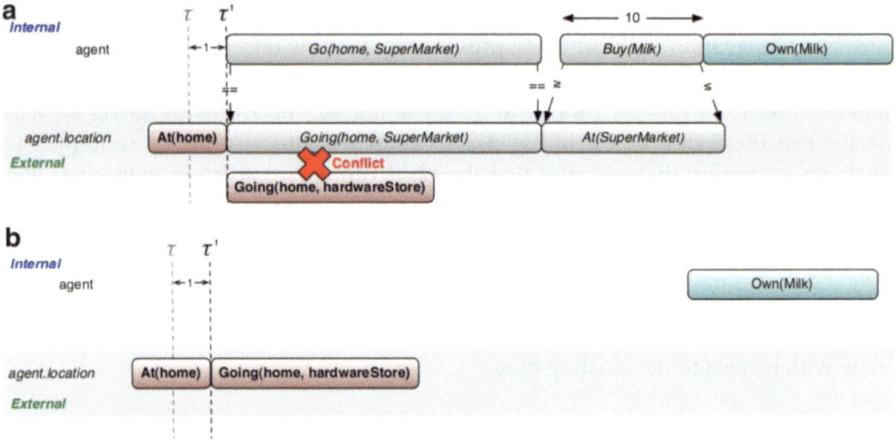

Fig. 3.28 Illustration of a conflict during synchronization and the resulting relaxed plan for recovery. (**a**) Conflicting observation. (**b**) Relaxed partial plan

keeping all past observations in both *external* and *internal* state variables as well as the existing *internal* goals (`Own(Milk)` in this example) received by the reactor. The resulting partial plan is shown in Fig. 3.28b. In this process the reactor informs the agent that all *external* goals it requested for the future are no longer valid, allowing timeline owners to divest maintaining such goals.

When the plan is no longer conflicted, synchronization can resume and find a solution as described above. Newly pending goals would be enforced at the next `step` to resume deliberation on this relaxed partial plan. By using synchronization the reactor is not only able to identify that the plan is consistent but also able to recover from it by offering a new incomplete partial plan to be resolved in the next deliberation `steps`. Synchronization, therefore, plays the role of what is typically termed as the *executive* in other architectures [17, 19, 47, 117, 121].

In `T-REX`, however, the interaction between planning and execution does not stop at this level. The overall scheduling of both *synchronization* and *planning* is such that the synchronization can occur in between any planning step. Not only is synchronization influenced by the outcome, but it also influences planning. Specifically when synchronization occurs before a complete plan is found, it will inject new facts (*external* observations and resulting *internal* state) in the PDB available for inference when the next planning `step` resumes. On this next step the `EUROPA` solver used for planning will have these new facts included in the partial plan which can introduce further new flaws in the plan while potentially having resolved previously existing flaws. The deliberation process is therefore perturbed externally by *synchronization* and consequently fully informed of the evolving world state as it is planning. This contrasts with other classical approaches, which avoid perturbing the planner during its search as it is not compatible with the "off-line planning" assumption, noted in Sect. 3.5.4.

Design choices made in the T-REX framework—both at the architectural level and the integration of the EUROPA framework for embedded deliberation—allow the implementation of a system that avoids this critical off-line planning limitation while enforcing a formal framework that allows for deliberation and model-based agent control. This tight integration of planning and execution allows the system to be more responsive to environmental change allowing informed decision making within a tighter reaction loop despite reactor latency.

3.6 Experimental Results

A significant driver of the applicability of our work was the growing need for AUV missions to be driven by sensor data. This trend is illustrated by the implementation of MBARI's *Gulper* water samplers [122], 2 liter syringe-style containers, 10 of which are encased within the mid-body section of our Dorado vehicle as shown in Fig. 3.29. However, the original design and intended use of operation continued to be dominated by a naive belief of collecting and returning water samples as traditional ship-borne methods often employ, namely using pre-scripted methods. With the installation of T-REX onboard the Dorado with its capability sense the environment and to appropriately synthesize plans and replan on the fly, more imaginative methods of sample acquisition were targeted. The Dorado uses two onboard computers: a main vehicle computer which is a 244 MHz PC/104 stack running the QNX real-time operating system, and a 367 MHz EPIC EPX-GX500 AMD Geode stack running Linux and T-REX. A socket-based communication protocol allows the exchange of goals and state updates between T-REX and the lower-level functional layer. Event logs from field experiments show that a peak load

Fig. 3.29 The Gulper water sampler [122] on MBARI's upper water-column Dorado AUV

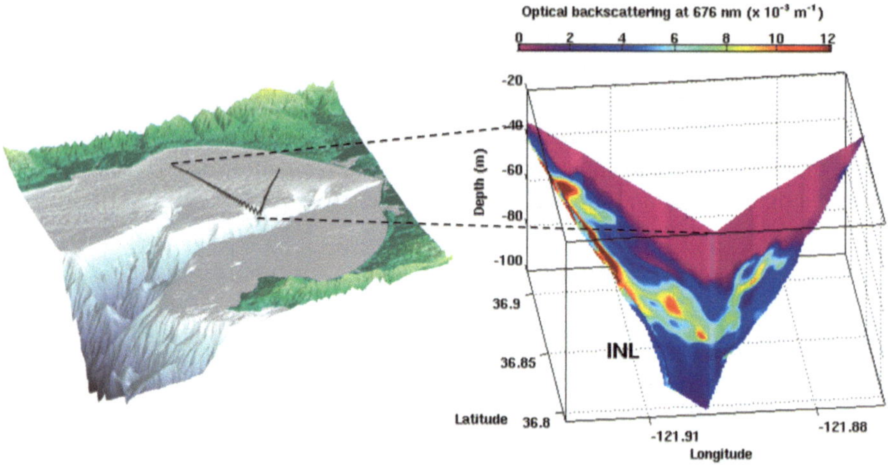

Fig. 3.30 An intermediate nepheloid layer (INL) detected by an AUV mission using a scripted mission plan in the Monterey Bay in August 2005. It shows the location of the AUV transect in the bay (*left*) and optical backscatter image at the 420 Nm frequency showing the INLs patchy nature (*right*). (Image courtesy: John Ryan, MBARI)

of up to 20 % occurs during replanning. Other desktop simulations [123] show that partitioning scales with problem size. Reactor design and the number of partitions, however, are still problem specific and determined empirically.

Our early work focused on observing and sampling within intermediate nepheloid layers (INLs)[15] [125]. The location and scale of INLs in the water column, which depend upon seasonal and episodic variability, are not predictable. In turn, they can be identified by their high turbidity and low chlorophyll fluorescence. Initial objectives focused on ways to enhance survey results by ensuring as large a coverage as possible so that INL fields were observed at tighter spaced transects and when no INL is detected, then transect spacing in the lawn mower can be incrementally increased. Follow on work directly targeted sample acquisition within INLs which are often sparse, patchy in large horizontal scales (in Kms) and small vertical scales (in meters). Figure 3.30 shows one example from an AUV transect from 2005. The intent of both observation and sample acquisition was to interrupt the nominal plan-execution cycle and inject an action to alter and re-factor the plan dynamically: in one instance altering the navigation, the other in actuating a sampler.

The implied assumption of installing the samplers on an AUV was that robotic methods to actuate sample acquisition (with no human control), would be competitive with human knowledge of where and when to sample a feature given available shipboard data. To support this assumption, it required new ways to couple sample acquisition with AUV navigation and control. While some simple and ad hoc approaches [126, 127] have shown promise, these techniques were tailored to

[15] INLs are fluid sheets of suspended particulate matter that originate from the sea floor [124].

solve a specific problem with assumptions on feature structure. Our approach, on the other hand, stemmed from the need to build a coarse-grained computational model of the INL feature where none existed and to use that model as a basis to close the *sense-sample-acquire* loop within the AUV. Fox et al. [128] document earlier work in using a well-known unsupervised method in classification, the *Kohonen network* or self-organizing maps (SOM) [129] to *reactively* acquire samples with plan alteration. In doing so we couple the immediacy of feature identification which is necessarily reactive, with the deliberation of an onboard temporal planner to trigger a sampler. This has two benefits. First, it goes to the core of having a deliberative approach to mission planning by balancing the immediate with that which is projected [130]. Doing so allows a more systematic approach in dealing with spatial factors in sampling large volumes. Second, it allows other sampling parameters which are not directly impacted by perception to be brought to bear; in our case, scientists would like to introduce constraints (spatial separation, feature bias, distribution of samples when tracking multiple features). These are critical not only for science but also for informed decision making which is challenging for pure reactive approaches.

While the early experiments in [128] coupled the classifier with the onboard planner, two further novel extensions to sampling were added. First, to enhance the classifier model by adding a temporal dimension with a hidden Markov model (HMM) [131, 132]. An HMM is a sequential state transition model in which the current state is not directly visible, but an observation, dependent on the state, is. Each state has a probability distribution over the possible observations. Therefore the sequence of observations gives some information about the sequence of traversed states in the HMM. Unlike identification models based on the simple classification scheme, HMMs use the dynamics of transition between modeled states to exploit sequential information of the sensor readings providing more robust formulation for environmental state identification. The temporal probabilistic transition then acts as a filter on top of the clustering model from [128], since it is not only derived from it, but it also informs the planner of a potential trigger state only when the models' transition occurs to an appropriate state of recognition of the INL feature [25]. Not only is deliberation aided by direct perception of the state (of the INL), but temporal transition functions also add a further level of validation between expected states in the INL model. This coupling of machine-learned model with a deliberative planner is novel in the ocean sciences and is a continuing target of investigation [133].

The second innovation involved dynamic resource allocation associated with the water samplers [130]. Using a priori sample utilities coupled with temporal and spatial information maintained by the deliberating planner in situ, the algorithm dynamically generates a policy on sample acquisition during the course of a mission. In so doing, utility functions can be used to balance sparcity or abundance of the feature encountered with the availability of samplers dynamically. Coupled with periodic subsampled science data returned by the vehicle when on the surface and goal-oriented commanding with `T-REX`, such a capability allows judicious use of the limited samplers by a scientist from her desktop.

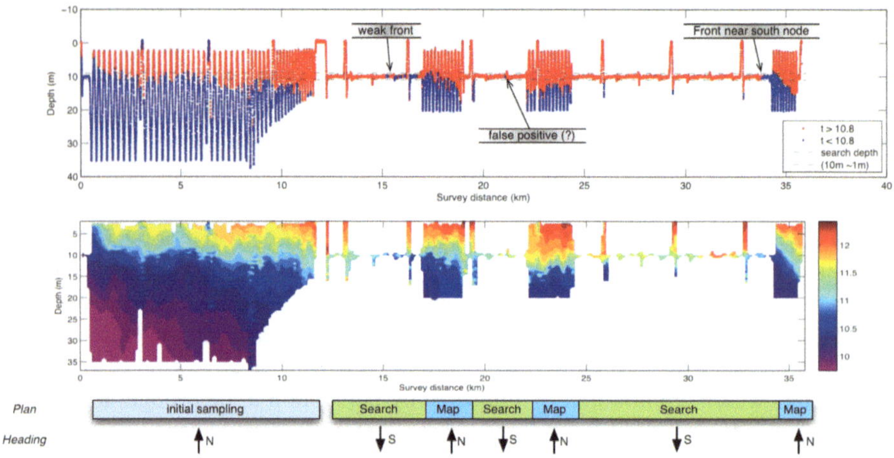

Fig. 3.31 Survey trackline of AUV following a thermal gradient. Image in the *middle* shows the temperature while the *bottom* figure shows different phases in the algorithm

3.6.1 Mixed-Initiative Frontal Tracking

We exploit the in situ decision-making capability of T-REX in order to find and continuously map a thermal gradient. The strategy encapsulated in algorithms onboard is described by the following three phases (Fig. 3.31):

Mapping: The vehicle performs a Yo–Yo transect between two designated waypoints and sends temperature sampled every 5 m to shore via satellite. The temperature field is reconstructed from interpolation of the transmitted data, and the scientist identifies the target front. Using a graphical user interface, she is then able to generate a high-level command which represents the front as a temperature threshold (t°) at a given depth (d). This goal is then appropriately packaged and sent to the AUV via satellite.

Search: After the mapping phase and on receipt of this message on the vehicle T-REX evaluates the goal, automatically synthesizes a set of waypoints by generating a plan and then dispatches the plan to execute by having the vehicle swim back and forth at depth d all the while reactively evaluating when a temperature threshold t° is crossed. In the process T-REX checks both the depth and temperature sensed by the AUV; when the depth is around d (± 1 meter to account for vehicle control uncertainty), it then compares the temperature read with t°. When the comparison value changes T-REX assumes that the front is identified and switches into the next phase.

Sampling: The vehicle undertakes a small transect around the identified front location while actuating water samplers, one at the center of the front with temperature threshold t°, and one on either side of the front. When this transect is completed, T-REX switches back to the Search phase to follow the frontal evolution. The survey is completed either when we reach the maximum mission duration or the vehicle returns to its initial waypoint during the search phase without crossing the threshold t°. The latter is indicative of the frontal zone having left the survey domain.

The novelty in this application was two-fold; in enabling the deliberation to decide when to trigger the phase changes and in showing viably how a human being coupled with a goal-oriented controller could enable joint problem solving. The approach, is general and can be adapted to any feature or scalar field.

3.6.2 *Lagrangian Surveys*

More recent experiments with T-REX controlling the Dorado AUV have centered on the CANON (controlled, agile, and novel observing network) program at MBARI [8]. The fundamental science goals of the program are related to detecting and tracking of spatio-temporal coastal ocean features. The dynamic features require a robotic platform to move with it, return water samples from across and within gradients and provide contextual observations to other drifting instruments. All of these observational strategies are tied to vehicle adaptation which is intrinsically tied to deliberation and challenging to accomplish otherwise.

Various adaptation methodologies in CANON are still being investigated; however what has become a critical goal for the program has been to provide contextual environmental information about water patches. The form and content of such "patches" varies; for instance initial attempts focused on harmful algal blooms (HABs). They have large spatial ($>50\,\text{km}^2$) and temporal (days to weeks) extent and are visible from space with visible coloring on the ocean surface. The societal impact of HABs has been in the introduction of toxins into the marine (and as a consequence human) food web, with [134, 135] estimating the consequent average economic impact in the United States to the tune of ~\$75 million from 1987 to 2000. However, the processes behind bloom initiation, evolution and collapse are poorly understood in part due to the complex interactions between microbial communities and the surrounding environment. As a result, our capacity to assess the range of potential future scenarios that might result from ocean temperature changes, acidification, or nutrient shifts is highly limited.

To augment our understanding of such microbial communities, the Dorado with T-REX has been utilized in a range of activities in a quasi-lagrangian mode. As the water mass is being tracked with a drifting instrument, the vehicle is expected to move around it gathering data to provide environmental context. While our model supports multiple patterns of observation (square, lawn-mower, transect(s)...) with any possible dimension factor, the majority of at-sea deployments have used a

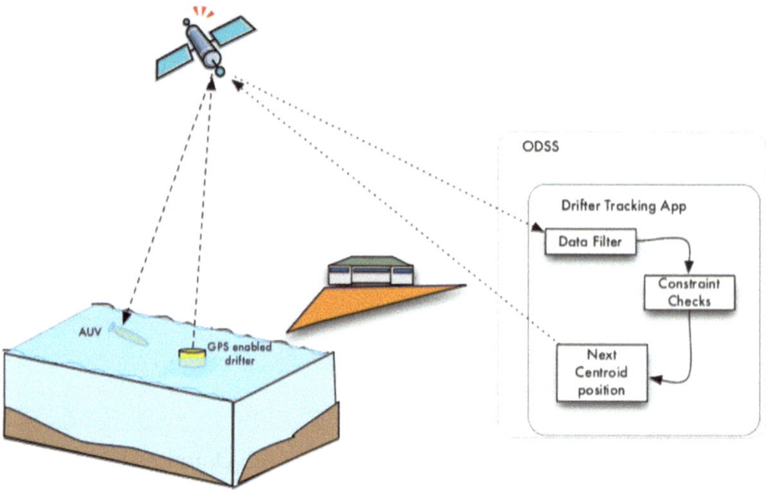

Fig. 3.32 Overall setup of CANON lagrangian experiments with a drifting instrument pack. Various combinations of computational methods can be included in the drifter tracking App which is a part of the oceanographic decision support system (ODSS) [136]

square pattern ($3 \times 3\,\text{km}^2$ or a $1 \times 1\,\text{km}^2$).[16] Plan synthesis occurs frequently, but within the context of a drifting point generated by a drifter. A drifter tracking application (DTA) is used on-shore with an oceanographic decision support system (ODSS; see below) [136], to process the drifter location from its onboard GPS and used as a target for the Dorado vehicle to provide a configurable and dynamic track-line enclosure to provide context to the drifting body. Fig. 3.32 shows the general setup of our experiments which can be tailored with the use of various computational methods within the DTA app. GPS data from the free-floating drifter is sent via satellite and processed onshore to determine a projection (speed and bearing) of the drifter. This data is sent to the AUV; at the completion of the current survey, the vehicle receives this data as a high-level goal which is decomposed and synthesized as a survey plan and subsequently executed. Currently, the vehicle completes the survey without any further dynamic updates from the drifter. In order to minimize the range of errors that the survey needs to encapsulate (data latency or surface or sub-surface currents for those drifters which have a drogue) the estimation is a projection in time and space of the drifters centroid sent with this goal from the shore-side app Fig. 3.33.

Typically this has meant that the vehicle computes a set of waypoints for the survey plan, moves to the start waypoint and then executes the survey even as the centroid is moving. Positional error of the vehicle and uncertainty in the speed and

[16]Often post-hoc reconstruction of the data set for visualization drives how much adaptation the vehicle can be allowed to undertake.

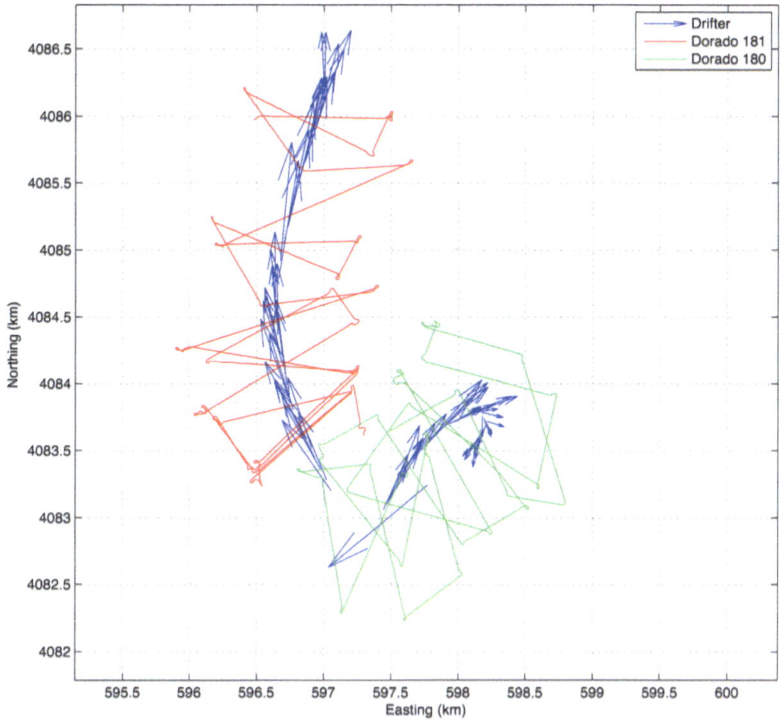

Fig. 3.33 A June 2011 drifter cruise shown in the Earth frame. Deviations in the AUV transect are based on the frame of reference in addition to positional errors [137]

bearing of the vehicle[17] exist; moreover, the observations of the survey to retain the environmental context require a change in reference frames so that the patch-center tagged by a GPS-tracked drifter remains within the survey perimeter at all times [137].

These theoretical results were validated in a five-day off-shore deployment carried out in September 2010, 160 kms off the California coast (Fig. 3.34a). A specialized drifter with a genomic sensor [139] hanging 20 m below, performing in-situ identification of micro-organisms was the target. The experiment was supported by crews on two support vessels, the R/V *Western Flyer* and the R/V *Zephyr*. The *Flyer* visited the drifter every 4 h to carry out a series of ship-based sampling experiments and lab analysis on water samples to ground-truth the drifter sensor data. The *Zephyr* was meanwhile focused on Lagrangian observation studies using the Dorado with the goal to monitor the nutrient budget at the perimeter of a

[17]In our September 2010 experiment [137] as the drifter moved over the Davidson Seamount [138] deviation in the California current, resulted in visible directional change that can be seen in Fig. 3.34b near the end of the mission.

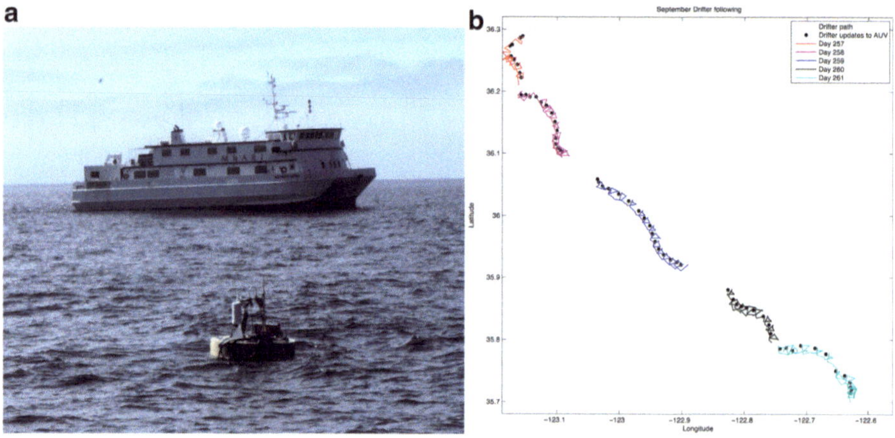

Fig. 3.34 The offshore CANON ESP-drifter following cruise [137]. (**a**) shows the ESP drifter foreground with the R/V *Western Flyer* with the image taken from the R/V *Zephyr*. (**b**) shows the 86 km transect of the Dorado following the ESP drifter about 100 Nm from California shores

1 km X 1 km water patch around the advecting drifter while the AUV performed a transformed box pattern around the drifter. A number of logistical issues were kept in mind while designing and executing the experiment. Each iteration began with the latest drifter update (position and velocity) received from the drifter through a satellite link which was transmitted to the vehicle for in-situ adaptation. `T-REX` used this input to compute the five waypoints necessary for an iteration of the box pattern, with the AUV traveling at constant velocity in the earth frame. Waypoints were computed once at the beginning of the survey with the AUV surfacing once within each survey, with each survey lasting $\sim$ 1–1.5 h. A total of 60 surveys were attempted over the course of 5 days. Figure 3.34b shows the overall transects over the course of the mission, with gaps in between days when the AUV was being recharged on deck Fig 3.33 shows the transect in Earth frame. Each deployment lasted $\sim$ 12 h. The black dots in the figure show the beginning of each iteration when drifter updates were received by the AUV.

Successive lagrangian experiments also followed a more innovative and robotically choreographed "ballet" between two AUVs, the Dorado and the LRAUV [140] in October 2010 as part of another series of CANON field experiments (Fig. 3.35). The slower vehicle with a fluorometer tracked a targeted chlorophyll patch seeded by a GPS-enabled drifter, sending periodic updates of the patch centroid to shore via the ODSS following a setup similar to Fig. 3.32, to the Dorado running `T-REX`.

More interesting experiments are being carried out as a basis to track more dynamic features such as ocean fronts [141] and tracking more capable drifting platforms. In particular experiments with a profiling CTD (Conductivity, Temperature, Depth) instrument, the `Wirewalker` [142] are resulting in new modalities of making Lagrangian and Eulerian observations. Figure 3.36 shows one such experiment in September 2011 with a GPS-enabled `Wirewalker` was tracked by the Dorado running `T-REX`. We are currently investigating spatio-temporal

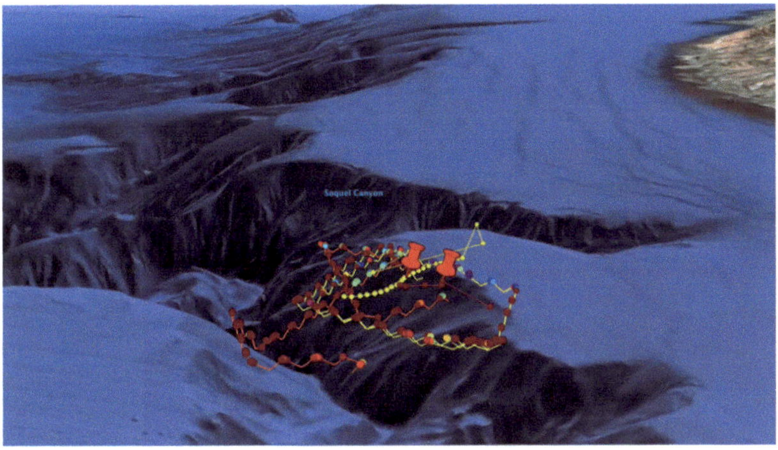

Fig. 3.35 An October 2010 robot "ballet" where a smaller slower AUV tracked a patch center, sending periodic updates to the Dorado using a setup shown in Fig. 3.32. The Google Earth post-hoc image shows *red track lines* of the `T-REX` commanded Dorado with the *yellow* transects of a separate AUV platform. *Yellow dots* show the track line of a GPS-enabled drifter used for seeding the experiment over the Monterey Canyon with Santa Cruz, California to the *right* of the figure

correlations [144] of CTD data obtained by the `Wirewalker` and tying it to the observations made in the contextual environment by Dorado's sensors [145].

3.6.3 Mixed-Initiative Autonomy

Large scale oceanographic field experiments akin to those targeted by `CANON` are usually logistically constrained driven largely by ship schedules. Once a feature set has been identified for scientific observation and science goals have been collated, appropriate robotic assets need to be assigned and deployed for observation and sampling. This asset allocation in turn depends on parameters like sensor payload, the speed of the robot in situ in comparison to the spatio-temporal scales of the evolution of the phenomena being sampled, the number of assets in a class, the kinds of measurements needed (e.g., synoptic or close-up), logistical considerations for deployments such as battery charge needed or distance/time to deploy, impact to long-range deployment plans, among others. All drive the strategy to balance short and long-term scientific and engineering needs. Our `CANON` deliberations during the 2010 field season, for example, were driven by all of the above factors [136]. Often these strategies were rebalanced during each decision-making cycle, especially in light of opportunistic science goals. For example, gliders running at 0.5 knots were often retargeted when surface support vessels noted the apparent dispersal of a bloom to subsurface layers when HABs were a `CANON` target. However, their retargeting was carefully monitored to balance the long-term desire of obtaining

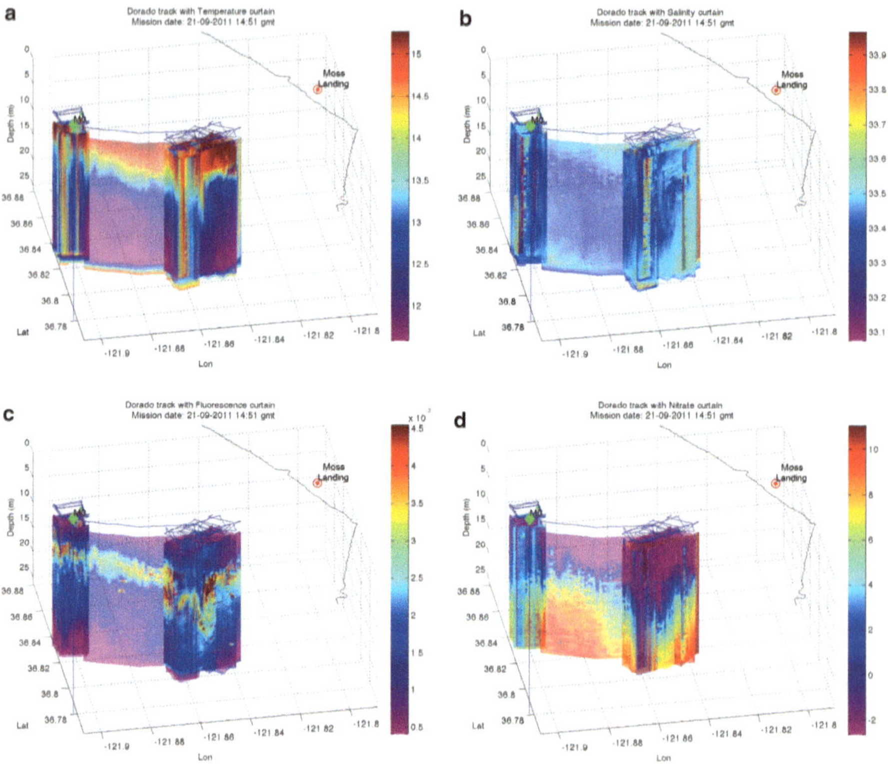

Fig. 3.36 Interpolated results of a September 2011 `T-REX` mission in the Monterey Bay with the profiling `Wirewalker` [142, 143] instrument. Plots show the instrument was entrained between two fronts. (**a**) Temperature. (**b**) Salinity. (**c**) Fluorescence line height. (**d**) Nitrate (Image courtesy: Rishi Graham, MBARI)

time series within fixed tracks and with the need to ensure these assets were not exposed to strong currents. Powered AUVs were targeted after advecting blooms, but the decision on their deployment locations and survey areas too close to the surf-zone were avoided. In patch-tracking experiments, the deployment of drifters close to shore and where depths of 20 m or less were encountered, was avoided even when contrary to scientific intent. The need to obtain high-quality data for post-hoc analysis was also driven by the response time of data returned by experiment partners (Figs. 3.37 and 3.38).

These distinctions make some vehicles more or less suitable to address certain science goals and suggest some degree of intelligent planning in the assignment strategy. Further, even if partial plans can be synthesized automatically, the need to capture qualitative "intent" and desires in the final outcome of an asset allocation strategy drives us towards a human-in-the-loop approach for asset allocation. The vast uncertainty and complex interaction between unknown quantities in the ocean environment dictate that any comprehensive experimental methodology must rely in some measure on human reasoning and adaptability. Furthermore, the multidisci-

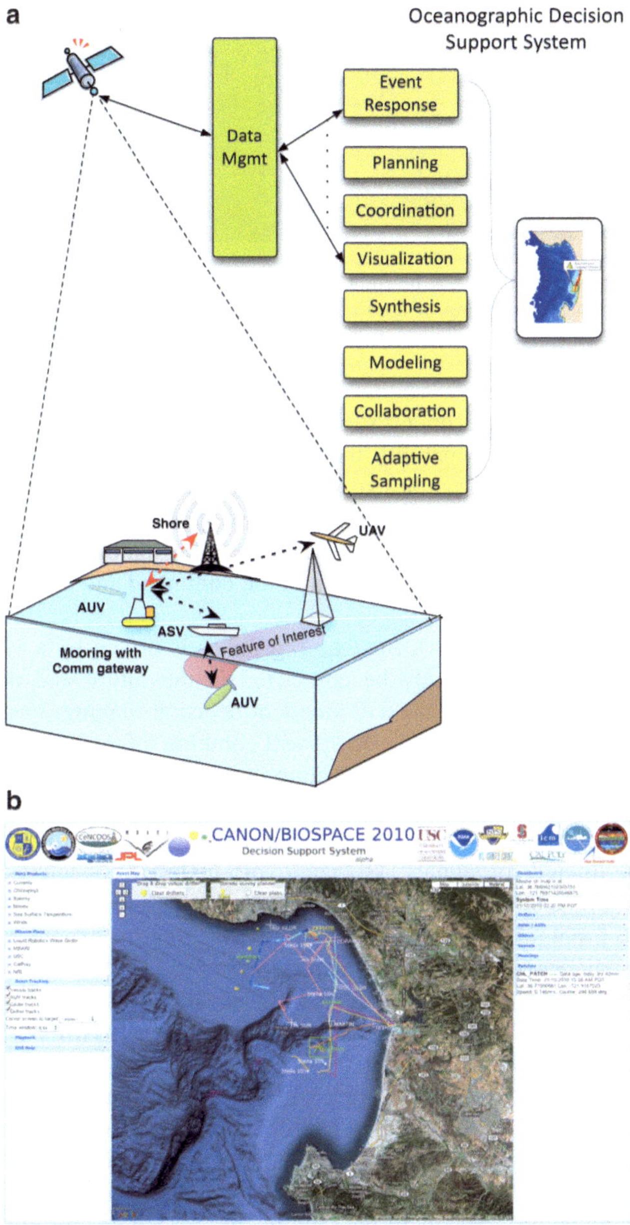

Fig. 3.37 The oceanographic decision support system (ODSS) is a single portal for planning, visualization, situational awareness, collaboration and archival of information before, during and after field experiments. The ODSS has been used since mid-year 2010 for CANON field experiments. (**a**) A conceptual view of the ODSS and how it fits into CANON field experiments. (**b**) Current implementation of ODSS for CANON

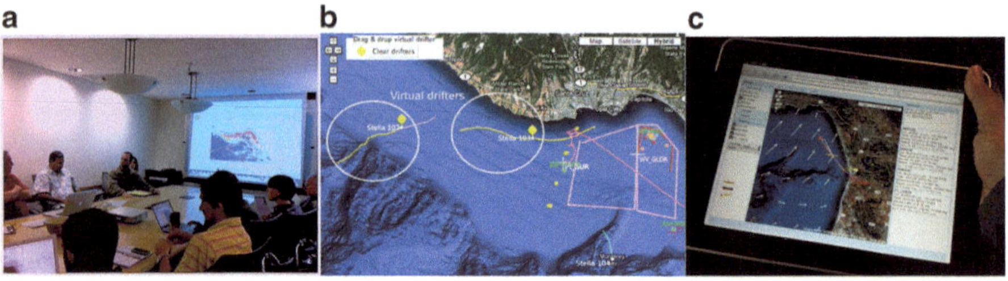

Fig. 3.38 (**a**) shows a 2010 CANON meeting with the ODSS in use. (**b**) shows a drifter tracking tool used during CANON planning; it used a physical model of advected surface currents from CeNCOOS [146] to plan AUV deployments. (**c**) shows ODSS use on an iPad

plinary nature of ocean experiments suggests collaboration between multiple human actors. Human cognitive capability then to capture these uncertainties, evaluate options, factor intent and synthesize data needed for asset allocation becomes increasingly important. Therefore, mixed-initiative approaches are viable as a means to apportion problem solving and inference between human and computational approaches. A viable example was shown in the command and control of the twin Mars rovers in [22, 23, 79].[18]

Additionally, our work on ODSS (see Fig. 3.37), is driven by a vision of how marine field robotics will likely be conducted in the future and the necessary computational infrastructure that will augment decision making, onshore as well as on-board. We believe ultimately ODSS will combine inferential, data gathering, archiving and analysis capabilities and more uniquely, provide an event response capability that could trigger the deployment of a robot and/or adapt its sampling strategy, all from a scientist's desktop on shore. Data returned from robotic assets at sea will be filtered for signals associated with feature(s) of interest along with information on its contextual environment. The association between the signal and a prospective event will be undertaken through tagging by experts and Machine Learning techniques used in recommender systems [147]. Adaptive sampling strategies will then balance decision making in situ with those onshore with computational models and human agents, the subject of a new NSF project in which the authors are participating.

Operationally, the ODSS has become a software system that the scientist consults during a CANON field campaign (see Fig. 3.38). S/he consults up-to-date model outputs, synoptic views of ocean color from remote sensing data and weather conditions for the next 24-h planning cycle and then queries the system to compute candidate sampling locations. In the near future, the ODSS will consult its situational awareness database, incorporate model outputs projecting these outputs into the space-time region of interest to generate candidate hypotheses with associated utility

[18]The MAPGEN system uses the same EUROPA planner used in T-REX. It continues to be used routinely to this day on mission-critical uplink process for the MER mission.

metrics. With the scientist's guidance the automated planner would then determine the most viable robotic option(s) to be dispatched (underwater, surface or aerial), given available resources, distance of travel, required payload and the environment. For instance, gliders could be chosen for longer-duration missions with smaller payload requirements in areas of moderate to weak ocean currents. The automated planner would balance these criteria to compute a viable trajectory plan for the selected vehicle. Vehicles with more controllability would be given a specified area to survey and the path generated in situ . Goals are communicated to the robot after analysis on shore for selective targeting of features by the conglomerate of vehicles in the ocean that are being tracked by the ODSS. To date, the ODSS has provided a simple yet functional tool for scientists to use in the daily complex decision making and the following experiment and survey design needs. It has helped train scientists to think of expanding their repertoire of tools for the scientific process and in the process it enabled a mixed team of engineers and scientists to interact at the functional level of the systems' capabilities.

3.7 Future Work

Our experience of using deliberative techniques in a highly interdisciplinary environment for targeted marine science applications has shown the need and applicability of such novel algorithms and methods. In addition to augmenting traditional AUV surveys, more advanced Lagrangian observations and mixed-initiative methods, we believe our colleagues at MBARI have understood the longer term implications of our work. Our current and consequently future efforts are therefore grounded in important science problems in an operational oceanographic setting.

The core of our problem domain for CANON has been towards Sampling. In large part given the spatial and temporal scales of the bio-geochemical processes we are interested in, in this project, it is imperative that a key direction is in formulating a multi-robot adaptive sampling problem. In this context while T-REX can be considered as a "multi-agent" system and could therefore imply its use in a multi-vehicle setting as a direct extension, sporadic communication with high-latencies in surface communication, not to mention low throughput acoustic communications imply that the extension of T-REX for distributed applications is non-trivial. Further, in order to efficiently execute coordinated observations of ocean processes with multiple autonomous sensing platforms, significant methodological as well as technical challenges must be addressed.

In this context we are working towards a systematic approach of using T-REX as a backend planner within the ODSS for CANON field experiments. The expectation is that ODSS users will initially be able to plan individual vehicle surveys which would be variants of straight-line transects. Lessons learned with human planners in the loop will be used towards planning larger surveys, also with simple fixed transects but using vehicle capabilities in ontologies similar to [148]. Doing so will

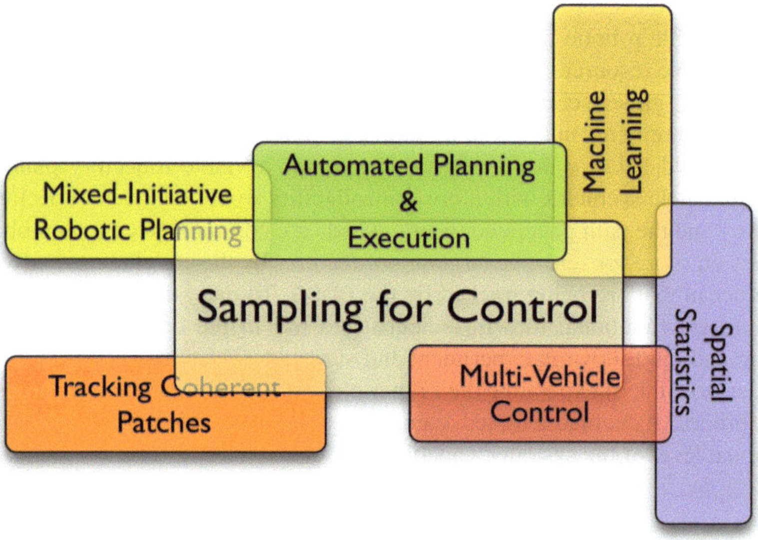

Fig. 3.39 Overlapping topics in our research in deliberation and autonomy with sampling the dominant theme

likely enable the right vehicle to target an ocean feature with appropriate sensors and sampling strategy. Apportioning the planning task between shore-side components with substantial contextual information and compute power, with robots having in situ reasoning capabilities similar to `T-REX`, is challenging in of itself and an open research problem. In pursuing this line of research, we expect to touch on a range of associated problems for decision-theoretic control of vehicles in the coastal ocean. Figure 3.39 shows what we believe is the scope and influence of the different topic areas we are currently focusing on.

Partitioned inference and decision making within `T-REX` also lends itself to graceful system degradation, making the overall approach effective towards increased mission robustness in the context of fault diagnosis, isolation and recovery (FDIR). There are two outstanding diagnosis and failure recovery challenges that we need to address; the first has to do with methodology related to model design of the instrument payload for a highly modular and configurable vehicle; this is local to a reactor. Doing so would enable (as a first step) using systematic FDIR algorithms as demonstrated in the `RAX` experiment [19, 149, 150] and more recently on AUVs [151–153]. The second challenge has to do with implementing a policy within the `T-REX` framework that explicitly deals with recovery since controller failure can be distributed across reactors. Since reactors are hierarchical in their relationship to one another, their well defined dependencies allow them to be removed from an agent systematically as noted in Sect. 3.5.3.1. Reactors more abstract in scope can be removed before those which are less abstract (and closer to the hardware). In the event a `Navigator` reactor, for example, has a fault, it can be removed safely with lower level functionalities (such as those in the executive). When the Navigator is removed, it will recursively remove more abstract reactors (such as a Mission Manager) that depend on it. Such a safety feature then allows the system designers to ensure that early on they can invest more in ensuring that reactors lower in the hierarchy are robust and potentially formally verified. It also means that there is a safe and sure way to take control of the vehicle at any level within the reactors hierarchy; as long as low level reactors are still active one will always been able to send goals/commands to them and execute what are traditionally known as runout sequences.

Feature recognition for Sampling is a general research thrust; however it has specific relevance in the context of T-REX and AUV control. As noted in Sect. 3.6 we have already demonstrated Machine Learning techniques for recognition of select features in the coastal ocean. A principal challenge that remains is that of obtaining expert labeled data for any supervised or semi-supervised learning methods. While supervised techniques for learning classifiers such as decision trees [154], instance-based classification [155], Bayesian classifiers [156] and artificial neural networks [157] are popular, given the vast amount of data available and poor understanding of correlation between physical and biogeochemical variables in the coastal ocean, our work on INLs [31, 125] and reliance on a oceanographic expert has shown that scaling to different problems remains problematic. This is a prime motivation to move towards more semi-supervised methods such as [133].

With the advent of the ODSS however, another direction our research is taking us is in Recommender Systems [147] as noted in Sect. 3.6. We are working to design and deploy software infrastructure that will learn from user tagged data. To do so, we will use tagging techniques similar to what users of commercial image sharing sites like Flickr and Picassa do, to learn the context of the data (including remote sensing). In addition, the ODSS will integrate statistical Machine Learning techniques with incoming asynchronous data stream obtained by sensors and platforms. Identification models will be based on existing training sets for INLs and plumes; models for other features of scientific interests driven by CANON requirements will be built in addition. Existing AUV archives for instance provide sufficient data for a learning base for a range of features of interest to scientists; in addition we will build software that will learn to incorporate incoming data streams to provide an existence proof of a feature of interest.

A number of issues remain to be explored improving T-REX and EUROPA integration particularly in the context of engineering models for execution.

Dispatchability of the *external* temporally flexible partial plans continues to be an open research problem. Substantial progress has been made [68, 69, 158, 159] by analyzing the plan structure of static offline temporal plans. However in the context of embedded plan execution, balancing model design while dealing with execution uncertainty requires further research. For instance it should be possible to distinguish part of the plan that contributes to a reactor goal, which often needs to be dispatched as early as possible with other tokens describing world evolution with a preference to dispatch as late as possible. Making this distinction would improve the quality of agent behavior while avoiding extra model complexity.

More abstractly at the architectural level, the ability to dynamically manipulate the reactor graph would be of importance while creating new reactors or modifying timelines. Doing so would provide more robust failure recovery mechanisms especially after a reactor has failed to synchronize. For example, an alternate reactor could take ownership of timelines left invalid by the destruction of another failed reactor ensuring a transparent implementation of component redundancy. While the current design broadly provides for such a capability, the T-REX API has yet to implement it while ensuring other design elements within are not compromised.

Finally and importantly, integration of reasoning about resources within T-REX and tying it to existing EUROPA capabilities needs to be demonstrated. Specifically T-REX needs to be able to update resource levels as incoming observations. Doing so would ensure plan adaptation occurs dynamically when planned and observed resource values diverge—for instance if the vehicle looses a bank of batteries during the course of a mission. Enhanced modeling capabilities and augmented domain-independent search within EUROPA would make it easier for T-REX users to create such domain models and to perform more proactive deliberation.

Along with all these potential extensions, we are also very cognizant that the learning curve associated with T-REX and EUROPA for a general user in the marine robotics community needs to be tackled. At present, the complex system that underlies T-REX and its underlying framework, even if well documented, can be intimidating for a naive first time user.

3.8 Conclusion

The Ocean Sciences are at a cusp, straddling traditional ship-based exploration with newer observatory-based methods. In the United States the NSF funded Ocean Observatories Initiative [160] and regional bodies such as CeNCOOS [146] are investing in new observational methods and technologies which aim to aid oceanographers by producing substantially higher-resolution data products. These will use moorings, cabled observatories and glider fleets generating near real-time data from one or more locations as a continuous stream. The tack chosen by CANON has been to complement such methods with the aim of developing methods for a *portable mobile* observatory which is dependent on autonomous platforms. In this context what we advocate in this chapter is an initial step in making robotic platforms more adaptive, scaling them towards fleets of vehicles (underwater and/or on the surface) and most importantly using them to solve the larger challenge of sampling dynamic spatio-temporal fields.

While our trajectory of research in the oceanographic domain follows that in spacecraft autonomy, from onboard [19] to mixed-initiative methods [23], the challenges are significantly different with the environment playing a far larger role. Moreover, prediction and projection of a future course of action for a robot given the level of uncertainty calls for representational and architectural methods which would enable adaptation at different levels of abstraction, from mission-planning to low level actuation and control.

Our efforts were initially targeted to full robotic autonomy; while this is an important goal that we continue to strive towards, it is increasingly clear that human-in-the-loop methods also have a sizable role to play in ocean exploration and by extension to the field of Marine Robotics. Not only is an embedded deliberative agent an important element of the effort towards smarter vehicles, but we are also working towards an implementation of `T-REX` which will perform multi-vehicle plan synthesis for the `ODSS`.

A key role that `T-REX` has played and will continue to play, is towards *event response* scenarios. In Sect. 3.6 we show the role Machine Learning techniques have played thus far in in-situ plan adaptation. This interaction between Planning and Machine Learning, we believe, will play a significant role going forward in exploration and discovery. Interpretation of incoming sensor data to deal with scientific surprise and re-adaptation for opportunistic science will be critical to dealing with the persistent problem of undersampling the ocean.[19]

Much of what we know of the ocean was derived by exploring in the blind. In the near future, we anticipate longer duration robotic vehicles with enough onboard computational capacity to run deliberative agents such as `T-REX`, make water-column measurements and reliably track and characterize dynamic features. While our own experiments have already demonstrated such capabilities, the exploration vs. exploitation tradeoff is better determined when vehicles can make choices over longer time periods (weeks and months) to study the impacts of bio-geochemical interaction driven by longer duration physical forcing and climate change.

Given our poor understanding of the coastal ocean in particular, mixed-initiative methods of command and control for Sampling with heterogenous assets is an important goal that is clearly at our doorstep. We believe deliberation is, but a small yet critical part of the solution to unfolding open research problems for AI and Marine Robotics.

Acknowledgments Rajan and Py are funded by a block grant from the David and Lucile Packard Foundation to MBARI and in part by NSF grant No. 1124975 and NOAA grant No. NA11NOS4780055. They thank their research collaborators John Ryan, Julio Harvey, Jim Bellingham and Robert Vrijenhoek at MBARI. The crew of the R/V *Zephyr* and the DMO support team led by Hans Thomas has been invaluable. The authors also thank other members of the EUROPA team over the years, including Conor McGann, Ari Jónsson, Jeremy Frank, Paul Morris, John Bresina, Michael Iatauro, Matthew Boyce and Tristan Smith. They are grateful to Martha del

[19]Walter Munk of the Scripps Institute of Oceanography has famously stated *"Most of the previous century could be called a century of undersampling"*—Testimony to the U.S. Commission On Ocean Policy, 18 April 2002.

Alto, David Korsmeyer, and Steven Zornetzer all at NASA Ames who not only made EUROPA available as open source but also encouraged its use outside the agency. Minh Bin Do (NASA Ames), William Yew Teck Tan (NUS, Singapore) and Rishi Graham (MBARI) provided valuable comments on early drafts; we are grateful for their help in making this manuscript more accessible. Finally, the authors are indebted to Nicola Muscettola (Google) for his ideas and early efforts in Constraint-based Temporal Planning while at NASA Ames which forms the basis of much of this work.

References

1. Brierley AS, Fernandes PG, Brandon MA, Armstrong F, Millard NW, McPhail SD, Stevenson P, Pebody M, Perrett J, Squires M, Bone DG, Griffiths G (2002) Antarctic Krill under sea ice: elevated abundance in a narrow band just south of ice edge. Science 295(5561):1890–1892
2. Ryan JP, Chavez FP, Bellingham JG (2005) Physical-biological coupling in Monterey Bay, California: Topographic influences on phytoplankton ecology. Mar Ecol Prog Ser 287:23–32
3. Thomas H, Caress D, Conlin D, Clague D, Paduan J, Butterfield D, Chadwick W, Tucker P (2006) Mapping AUV survey of axial seamount. EOS Trans AGU 87(52), Fall Meet. Suppl., Abstract V23B-0615, 2006
4. Yoerger D, Jakuba M, Bradley A, Bingham B (2007) Techniques for deep sea near bottom survey using an autonomous underwater vehicle. Int J Robot Res 26(1):41–54
5. Incze ML (2009) Optimized deployment of autonomous underwater vehicles for characterization of coastal waters. J Marine Syst 78(Supp 1):S415–S424
6. Rigby P, Pizarro O, Williams SB (2010) Toward adaptive benthic habitat mapping using gaussian process classification. J Field Robot 27(6):741–758
7. Rudnick D, Perry M (2003) ALPS: autonomous and lagrangian platforms and sensors, workshop report. http://www.geo-prose.com/ALPS, Tech. Rep
8. CANON: controlled, agile and novel observing network [Online]. Available: http://www.mbari.org/canon/
9. Ghallab M, Nau D, Traverso P (2004) Automated planning: theory and practice. Elsevier Science, San Francisco
10. Anderson CR, Siegel DA, Kudela RM, Brzezinskic MA (2009) Empirical models of toxigenic Pseudo-nitzschia blooms: Potential use as a remote detection tool in the Santa Barbara Channel. Harmful Algae 8:478–492
11. Brooks R (1986) A robust layered control system for a mobile robot. IEEE J Robot Autom RA-2:14–23
12. Brooks RA (1991) Intelligence without reason. In: Computers and thought, international joint conference on artificial intelligence. Morgan Kaufmann, San Francisco, pp 569–595
13. Brooks R (1991) Intelligence without representation. Artif Intell 47:139–159
14. Nilsson NJ (1984) Shakey the robot, Tech. Rep. Technical Note 323. Artificial Intelligence Center, SRI International, Menlo Park, CA, Apr 1984
15. Simmons R (1994) Structured control for autonomous robots. IEEE Trans Robot Autom 10:34–43
16. Haigh K, Veloso M (1998) Interleaving planning and robot execution for asynchronous user requests. Autonomous Robots 5:79–95
17. Alami R, Chatila R, Fleury S, Ghallab M, Ingrand F (1998) An architecture for autonomy. Int J Robot Res 17:315–337
18. Chien S, Knight R, Stechert A, Sherwood R, Rabideau G (2000) Using iterative repair to improve the responsiveness of planning and scheduling. In: Proceedings of the 5th international conference on artificial intelligence planning and scheduling (AIPS), Breckenridge, CO
19. Muscettola N, Nayak P, Pell B, Williams B (1998) Remote agent: to boldly go where no AI system has gone before. Artif Intell 103:5–48

20. Teichteil-Konigsbuch F, Fabiani P (2007) A multi-thread decisional architecture for real-time planning under uncertainty. In: Proceedings of the 3rd international workshop on planning and execution for real-world domains, international conference on automated planning and scheduling (ICAPS), Rhode Island
21. Rajan K, Bernard D, Dorais G, Gamble E, Kanefsky B, Kurien J, Millar W, Muscettola N, Nayak P, Rouquette N, Smith B, Taylor W, Tung Y (2000) Remote agent: an autonomous control system for the new millennium. In: Proceedings of prestigious applications of intelligent systems, European conference on artificial intelligence (ECAI), Berlin
22. Ai-Chang M, Bresina J, Charest L, Chase A, Hsu J, Jonsson A, Kanefsky B, Morris P, Rajan K, Yglesias J, Chafin B, Dias W, Maldague P (2004) MAPGEN: mixed initiative planning and scheduling for the Mars'03 MER mission. IEEE Intell Syst 19(1):8–12
23. Bresina J, Jonsson A, Morris P, Rajan K (2005) Activity planning for the mars exploration rovers. In: International conference on automated planning and scheduling (ICAPS), Monterey, California
24. McGann C, Py F, Rajan K, Thomas H, Henthorn R, McEwen R (2008) A deliberative architecture for AUV control. In: IEEE international conference on robotics and automation (ICRA), Pasadena, May 2008
25. McGann C, Py F, Rajan K, Ryan JP, Henthorn R (2008) Adaptive control for autonomous underwater vehicles. In: Association for the advancement of artificial intelligence, national conference (AAAI), Chicago, IL
26. Py F, Rajan K, McGann C (2010) A systematic agent framework for situated autonomous systems. In: International conference on autonomous agents and multiagent systems (AAMAS), Toronto, Canada, May 2010
27. WillowGarage (2008) http://www.willowgarage.com/pages/software/trex
28. Meeussen W, Wise M, Glaser S, Chitta S, McGann C, Mihelich P, Marder-Eppstein E, Muja M, Eruhimov V, Foote T, Hsu J, Rusu RB, Marthi B, Bradski G, Konolige K, Gerkey B, Berger E (2010) Autonomous door opening and plugging in with a personal robot. In: IEEE international conference on robotics and automation (ICRA), Anchorage, AK, May 2010, pp 729–736
29. McGann C, Berger E, Boren J, Chitta S, Gerkey B, Glaser S, Marder-Eppstein E, Marthi B, Meeussen W, Pratkanis T, Wise M (2009) Model-based, hierarchical control of a mobile manipulation platform. In: 4th workshop on planning and plan execution for real world systems, international conference on automated planning and scheduling (ICAPS)
30. Ceballos A, Bensalem S, Cesta A, de Silva L, Fratini S, Ingrand F, Ocon J, Orlandini A, Py F, Rajan K, Rasconi R, van Winnendael M (2011) A goal-oriented autonomous controller for space exploration. In: Proceedings of 11th symposium on advanced space technologies in robotics and automation, Noordwijk, the Netherlands, April 2011
31. McGann C, Py F, Rajan K, Ryan JP, Thomas H, Henthorn R, McEwen R (2008) Preliminary results for model-based adaptive control of an autonomous underwater vehicle. In: International symposium on experimental robotics (ISER), Athens
32. Bellingham J, Leonard J (1994) Task configuration with layered control. In: IARP 2nd workshop on mobile robots for subsea environments, May 1994
33. Benjamin M (2004) The interval programming model for multi-objective decision making. Computer Science and Artificial Intelligence Laboratory, MIT, Cambridge, MA (Tech. Rep., Jan 2004)
34. Zhang Y, McEwen RS, Ryan JP, Bellingham JG (2009) An adaptive triggering method for capturing peak samples in a thin phytoplankton layer by an autonomous underwater vehicle. In: Proceedings of marine technology society/IEEE oceans, Missisipi
35. Benjamin M, Grund M, Newman P (2006) Multi-objective optimization of sensor quality with efficient marine vehicle task execution. Robotics and Automation
36. Glenn S, Kohut J, McDonnell J, Seidel D, Aragon D, Haskins T, Handel E, Haldeman C, Heifetz I, Kerfoot J, Lemus E, Lictenwalder S, Ojanen L, Roarty H, Students AC, Jones

C, Webb D, Schofield O (2011) The trans-atlantic slocum glider expeditions: a catalyst for undergraduate participation in ocean science and technology. Marine Tech Soc 45:75–90
37. Carreras M, Ridao P, Garcia R, Battle J (2006) Behaviour control of UUV's. In: Roberts GN, Sutton R (eds) Advances in unmanned marine vehicles, IEE, ch 4. Institute of Engineering and Technology, London, UK
38. Boehm B (1986) A spiral model of software development and enhancement. Software Eng Notes 11:14–24
39. Bernard D, Dorais G, Gamble E, Kanefsky B, Kurien J, Millar W, Muscettola N, Nayak P, Rajan K, Rouquette N, Smith B, Taylor W, Tung Y (2000) Remote agent experiment: final report. NASA technical report, Pasadena, CA, USA
40. NDDL Reference (2011) [Online]. Available: http://code.google.com/p/europa-pso/wiki/NDDLReference
41. Feigenbaum E, Feldman J (1963) Computers and thought. McGraw-Hill, New York
42. Green C (1969) Application of theorem proving to problem solving. In: International joint conference on artificial intelligence (IJCAI), Washington, DC, May 1969
43. Fikes R, Nilsson N (1971) STRIPS: a new approach to the application of theorem proving to problem solving. Artif Intell 2:189–205
44. Muscettola N (1994) HSTS: integrating planning and scheduling. In: Fox M, Zweben M (eds) Intelligent scheduling. Morgan Kaufmann, San Francisco, pp 169–212
45. Jónsson A, Morris P, Muscettola N, Rajan K, Smith B (2000) Planning in interplanetary space: theory and practice. In: Artificial intelligence planning and scheduling (AIPS), Breckenridge, CO
46. Chien S, Sherwood R, Tran D, Cichy B, Rabideau G, Castano R, Davies A, Mandl D, Frye S, Trout B, Shulman S, Boyer D (2005) Using autonomy flight software to improve science return on earth observing one. J Aero Comput Inform Comm 2:196–216
47. Gat E (1998) On three-layer architectures. In: Kortenkamp D, Bonnasso R, Murphy R (eds) Artificial intelligence and mobile robots . MIT Press, Pasadena, pp 195–210
48. Ambros-Ingerson J, Steel S (1988) Integrating planning, execution and monitoring. In: Association for the advancement of artificial intelligence, national conference (AAAI), Jan 1988
49. Chien S, Knight R, Stechert A, Sherwood R, Rabideau G (1999) Integrated planning and execution for autonomous spacecraft. In: Proceedings of the IEEE aerospace conference, vol 1, pp 263–271, 1999
50. Knight R, Fisher F, Estlin T, Engelhardt B, Chien S (2001) Balancing deliberation and reaction, planning and execution for space robotic applications. In: Intelligent robots and systems (IROS), Maui, Hawaii, Jan 2001, pp 2131–2139
51. Dean TL, McDermott D (1987) Temporal data base management. Artif Intell 32:1–55
52. Dean TL, Boddy M (1988) Reasoning about partially ordered events. Artif Intell 36(3): 375–399
53. Boddy M (1993) Temporal reasoning for planning and scheduling. SIGART Bull 4:17–20
54. Muscettola N, Smith S, Cesta A, D'Aloisi D (1992) Coordinating space telescope operations in an integrated planning and scheduling architecture. IEEE Control Syst 12:28–37
55. Ghallab M, Laruelle H (1994) Representation and control in IxTeT, a temporal planner. In: Artificial intelligence planning and scheduling (AIPS), Chicago, IL, pp 61–67, 1994
56. Laborie P, Ghallab M (1995) Planning with sharable resource constraints. In: International joint conference on artificial intelligence (IJCAI), Montreal, Canada, pp 1643–1649, 1995
57. Cesta A, Oddi A (1996) Gaining efficiency and flexibility in the simple temporal problem. In: Proceedings of the 3rd international workshop on temporal representation and reasoning (TIME-96), Key West, FL, May 1996
58. Dechter R, Meiri I, Perl J (1991) Temporal constraint networks. Artif Intell 49(1–3):61–95
59. Mackworth AK (1977) Consistency in networks of relations. Artif Intell 8(1):99–118

60. Mackworth AK, Freuder E (1985) The complexity of some polynomial network consistency algorithms for constraint satisfaction problems. Artif Intell 25:65–74
61. Currie K, Tate A (1991) O-plan: the open planning architecture. Artif Intell 52:49–86
62. Vere S (1983) Planning in time: windows and durations for activities and goals. IEEE Trans Pattern Anal Machine Intell PAMI-5(3):246–267
63. Muscettola N, Pell B, Hansson O, Mohan S (1995) Automating mission scheduling for space-based observatories. In: Henry GW, Eaton JA (ed) Astronomical society of the pacific conference series, vol 79, 1995
64. Pell B, Gat E, Keesing R, Muscettola N, Smith B (1997) Robust periodic planning and execution for autonomous spacecraft. In: International joint conference on AI (IJCAI), 1997
65. Bernard DE, Dorais GA, Fry C, Gamble E Jr, Kanfesky B, Kurien J, Millar B, Muscettola N, Nayak P, Pell B, Rajan K, Rouquette N, Smith B, Williams B (1998) Design of the remote agent experiment for spacecraft autonomy. In: Proceedings of the IEEE aerospace conference, Snowmass, CO, 1998
66. Pell B, Gamble E Jr, Gat E, Keesing R, Kurien J, Millar B, Nayak P, Plaunt C, Williams B (1998) A hybrid procedural/deductive executive for autonomous spacecraft. In: Proceedings of autonomous agents, St. Paul, Minn, 1998
67. Williams BC, Nayak PP (1996) A model-based approach to reactive self-configuring systems. In: Proceedings of the national conference on artificial intelligence, pp 971–978, 1996
68. Muscettola N, Morris P, Tsamardinos I (1998) Reformulating temporal plans for efficient execution. In: Proceedings of 6th international conference on principles of knowledge representation and reasoning (KR), Trento, Italy, 1998
69. Morris P, Muscettola N (2000) Execution of temporal plans with uncertainty. In: Proceedings of the association for the advancement of AI, Austin, Tx, 2000
70. Muscettola N, Dorais G, Fry C, Levinson R, Plaunt C (2002) IDEA: planning at the core of autonomous reactive agents. In: Proceedings of 3rd international NASA workshop on planning and scheduling for space, Houston, Tx, Oct 2002
71. Finzi A, Ingrand FF, Muscettola N (2004) Model-based executive control through reactive planning for autonomous rovers. In: Intelligent robots and systems (IROS), 2004
72. Dias MB, Lemai S, Muscettola N (2003) A real-time rover executive based on model-based reactive planning. In: International symposium on artificial intelligence, robotics and automation in space (iSAIRAS), Nara, Japan, 2003
73. Aschwanden P, Baskaran V, Bernardini S, Fry C, Moreno M, Muscettola N, Plaunt C, Rijsman D, Tompkins P (2006) Model-unified planning and execution for distributed autonomous system control. In: Fall symposium on spacecraft autonomy, association for the advancement of artificial intelligence (AAAI), Washington, DC, 2006
74. Zilberstein S (1996) Using anytime algorithms in intelligent systems. AI Mag 17(3):73–83
75. Frank J, Jónsson A (2003) Constraint-based attribute and interval planning. Constraints 8(4):339–364
76. Barreiro J, Jones G, Schaffer S (2009) Peer-to-peer planning for space mission control. In: Proceedings of the IEEE aerospace conference, pp 1–9, Mar 2009
77. Wettergreen D, Cabrol N, Baskaran V, Calderon F, Heys S, Jonak D, Luders R, Pane D, Smith T, Teza J, Tompkins P, Villa D, Williams C, Wagner M (2005) Second experiments in the robotic investigation of life in the atacama desert of chile. In: Proceedings of the 8th international symposium on artificial intelligence, robotics and automation in space (iSAIRAS), Munich, Germany, Sep 2005
78. Bresina J, Jonsson A, Morris P, Rajan K (2003) Constraint maintenance with preferences and underlying flexible solution. In: Proceedings of online constraint solving: handling change and uncertainty CP2003 workshop, Kinsale Co. Cork, Ireland, Sep 2003
79. Bresina J, Jonsson A, Morris P, Rajan K (2005) Mixed-initiative planning in MAPGEN: capabilities and shortcomings. In: Workshop on mixed-initiative planning and scheduling, international conference on automated planning and scheduling (ICAPS), Monterey, California, 2005

80. Chien S, Sherwood R, Tran D, Castano R, Cichy B, Davies A, Rabideau G, Tang N, Burl M, Mandl D, Frye S, Hengemihle J, Agostino J, Bote R, Trout B, Shulman S, Ungar S, Gaasbeck JV, Boyer D, Griffin M, Burke H, Greeley R, Doggett T, Williams K, Baker V, Dohm J (2003) Autonomous science on the EO-1 mission. In: Proceedings of the international symposium on artificial intelligence, robotics, and automation in space (i-SAIRAS), Nara, Japan, May 2003
81. Interface and control systems [Online]. Available: http://www.interfacecontrol.com.
82. White DJ (1993) A survey of applications of Markov decision processes. J Oper Res Soc 44(11):1073–1096
83. Musliner DJ, Durfee E, Shin K (1995) World modeling for the dynamic construction of real-time control plans. Artif Intell 74(1):83–127
84. Degroote A, Lacroix S (2011) ROAR: resource oriented agent architecture for the autonomy of robots. In: IEEE international conference on robotics and automation (ICRA), Shanghai, 2011
85. Berthoz A (2000) The brain's sense of movement. In: Perspectives in cognitive neuroscience. Harvard University Press, Cambridge
86. Garey MR, Johnson DS (1979) Computers and intractability: a guide to the theory of NP-completeness. W. H. Freeman, New York
87. Cormen TH, Leiserson CE, Rivest RL, Stein C (2009) Introduction to algorithms. MIT Press, Massachusetts
88. Bruckera P, Drexlb A, Mohring R, Neumannd K, Pesche E (1999) Resource-constrained project scheduling: notation, classification, models and methods. European J Oper Res 112:341
89. Russel S, Norvig P (2003) Artificial intelligence: a modern approach, 2nd edn. Prentice Hall, New Jersey
90. Smith D, Frank J, Jónsson A (2000) Bridging the gap between planning and scheduling. Knowledge Eng Rev 15, 47–83
91. Marriott K, Stuckey P (1998) Programming with constraints. MIT Press, Massachusetts
92. Apt K (2003) Principles of constraint programming. Cambridge University Press, Cambridge
93. Bartak R (1999) Constraint programming, in pursuit of the holy grail. In: Proceedings of WDS99, Prague, 1999
94. Lustig IJ, Puget J (2001) Program does not equal program: constraint programming and its relationship to mathematical programming. Interfaces 31:29–53
95. Genesereth MR, Nilsson NJ (1987) Logical foundations of artificial intelligence. Morgan-Kaufman, CA
96. Hooker JN (2005) Unifying local and exhaustive search. In: Villasenor L, Martinez AI (eds) Advances in la Ciencia de la Computación. pp 237–243
97. Rossi F, Beek PV, Walsh T (2006) Handbook of constraint programming. Elsevier, New York
98. Puget J, Laconte M (1995) Beyond the glass box: constraints as objects. In: Proceedings 5th international logic programming conference (ILPS), 1995
99. Do MB, Kambhampati S (2001) SAPA: a domain independent heuristic metric temporal planner. In: Proceedings of the european conference on planning (ECP), Toledo, Spain, pp 109–121, 2001
100. van Beek P, Chen X (1999) CPLAN: a constraint programming approach to planning. In: Association for the advancement of artificial intelligence, national conference (AAAI), Orlando, FL, 1999
101. Vossen T, Ball M, Lotem A, Nau DS (1999) On the use of integer programming models in AI planning. In: International joint conference on artificial intelligence (IJCAI), Stockholm, Sweden, pp 304–309, 1999
102. Wolfman S, Weld DS (1999) The LPSAT engine and its application to resource planning. In: International joint conference on artificial intelligence (IJCAI), Stockholm, Sweden, pp 310–317, 1999

103. Joslin D, Pollack M (1995) Passive and active decision postponement in plan generation. In: Proceedings of the european conference on planning (ECP), Assisi, Italy, 1995
104. Joslin D, Pollack M (1996) Is "early commitment" in plan generation ever a good idea? In: Association for the advancement of artificial intelligence, national conference (AAAI), Portland, OR, pp 1188–1193, 1996
105. Ghallab M, Alami R, Chatila R (1987) Dealing with time in planning and execution monitoring. In: Bolles R, Roth B (eds) Robotics research 4. MIT Press, pp 431–443
106. Lemai-Chenevier S, Ingrand F (2004) Interleaving temporal planning and execution in robotics domains. In: Association for the advancement of artificial intelligence, national conference (AAAI), San Jose, CA, 2004
107. Cesta A, Cortellessa G, Fratini S, Oddi A (2009) Developing an end-to-end planning application from a timeline representation framework. In: Proceedings of the 21st conference on innovative applications of artificial intelligence (IAAI). AAAI, Pasadena, CA, 2009
108. EUROPA open source [Online]. Available: http://code.google.com/p/europa-pso/.
109. Smith D, Frank J, Cushing W (2008) The ANML language. In: Proceedings of the workshop on knowledge engineering for planning and scheduling (KEPS) at international conference on automated planning and scheduling (ICAPS), 2008
110. Rumbaugh J (1991) Object oriented modeling and design. Prentice Hall
111. Allen J (1984) Towards a general theory of action and time. Artif Intell 23(2):123154
112. Muscettola N (2004) Incremental maximum flows for fast envelope computation. In: International conference on automated planning and scheduling (ICAPS), Whistler, Canada, pp 260–269, 2004
113. Muscettola N (2006) Computing the envelope for stepwise-constant resource allocations. In: Principles and practice of constraint programming - CP 2002, Ithaca, NY, pp 109–119, 2006
114. Morris P, Bresina J, Barreiro J, Iatauro M, Smith T (2011) State-based scheduling via active resource solving. In: Space mission challenges for information technology (SMC-IT), 2011 IEEE 4th international conference on, pp 29–34, August 2011
115. Morris P, Bresina J, Barreiro J (2011) Stable grounded inference in flexible resource scheduling. In: Proceedings of the workshop on generalized planning, association for the advancement of AI, San Francisco, CA, 2011
116. Ingrand FF, Lacroix S, Lemai-Chenevier S, Py F (2007) Decisional autonomy of planetary rovers. J Field Robot 24(7):559–580
117. Nesnas IAD, Wright A, Bajrarcharya M, Simmons R, Eslin T (2003) CLARAty and challenges of developing interoperable robotic software. In: International conference on intelligent robots and systems, pp 2428–2423, Oct 2003
118. Rajan K, Py F (2012) T-REX: partitioned inference for AUV mission control. In: Roberts GN, Sutton R (eds) Further advances in unmanned marine vehicles. The Institution of Engineering and Technology (IET)
119. Lemai-Chenevier S (2004) IxTeT-EXEC: planning, plan repair and execution control with time and resource management. Ph.D. dissertation, Institut National Polytechnique de Toulouse, Toulouse, France, June 2004
120. Myers K (1999) CPEF: a continuous planning and execution framework. Artif Intell Mag 20:63–69
121. Williams B, Ingham MD, Chung S, Elliott P (2003) Model-based programming of intelligent embedded systems and robotic space explorers. Proc IEEE 91(1):212–237
122. Bird L, Sherman A, Ryan JP (2007) Development of an active, large volume, discrete seawater sampler for autonomous underwater vehicles. In: Proceedings of oceans MTS/IEEE conference, Vancouver, Canada, 2007
123. McGann C, Py F, Rajan K, Olaya A (2009) Integrated planning and execution for robotic exploration. In: International workshop on hybrid control of autonomous systems, in international joint conference on artificial intelligence (IJCAI), Pasadena, California, 2009

124. McPhee-Shaw E (2006) Boundary-interior exchange. reviewing the idea that internal-wave mixing enhances lateral dispersal near continental margins. Deep-Sea Res II 53:45–49
125. Ryan JP, Johnson S, Sherman A, Rajan K, Py F, Thomas H, Harvey J, Bird L, Paduan J, Vrijenhoek R (2010) Mobile autonomous process sampling within coastal ocean observing systems. Limnology & Oceanograhy: Meth, 8:394–402
126. Zhang Y, McEwen R, Ryan J, Bellingham J (2010) Design and tests of an adaptive triggering method for capturing peak samples in a thin phytoplankton layer by an autonomous underwater vehicle. IEEE J Oceanic Eng 35(4):785–796
127. Zhang Y, McEwen RS, Ryan JP, Bellingham JG, Thomas H, Thompson CH, Rienecker E (2011) A peak-capture algorithm used on an autonomous underwater vehicle in the 2010 gulf of Mexico oil spill response scientific survey. J Field Robot 28:484–496
128. Fox M, Long D, Py F, Rajan K, Ryan JP (2007) In situ analysis for intelligent control. In: Proceedings of IEEE/OES OCEANS conference, 2007
129. Kohonen T (2001) Self-organisation maps. Springer Series in Information Sciences, vol 30. Springer, Berlin
130. Olaya A, Py F, Das J, Rajan K (2012) An on-line utility based approach for sampling dynamic ocean fields. IEEE J Ocean Eng 37, 185–203
131. Rabiner LR (1986) An introduction to hidden Markov models. IEEE ASSP Mag 3:4–16
132. Rabiner LR (1989) A tutorial on hidden Markov models and selected applications in speech recognition. In: Proceedings of the IEEE, pp 257–286, 1989
133. Kumar S, Celorrio S, Py F, Khemani D, Rajan K (2011) Optimizing hidden Markov models for ocean feature detection. In: Proceedings 24th international Florida AI research society (FLAIRS) conference, Palm Beach, FL, 2011
134. Anderson D, Hoagland P, Kaoru Y, White A (2000) Economic impacts from harmful algal blooms (HABs) in the United States. In: Woods hole oceanographic institution technical report: WHOI 2000–2011, Tech Rep, 2000
135. Hoagland P, Scatasta S (2006) The economic effects of harmful algal blooms. In: Graneli E, Turner J (eds). Springer-Verlag
136. Das J, Maughan T, McCann M, Godin M, O'Reilly T, Messie M, Bahr F, Gomes K, Py F, Bellingham J, Sukhatme G, Rajan K (2011) Towards mixed-initiative, multi-robot field experiments: design, deployment, and lessons learned. In: Intelligent robots and systems (IROS), San Francisco, California, 2011
137. Das J, Py F, Maughan T, Messie M, O'Reilly T, Ryan J, Sukhatme GS, Rajan K (2012) Coordinated sampling of dynamic oceanographic features with AUVs and drifters. Int J Robot Res 31:626–646
138. Clague D, Lundsten L, Hein J, Paduan J, Davis A (2010) Spotlight 6: Davidson seamount. Oceanography 23, 126–127
139. Scholin C, Doucette G, Jensen S, Roman B, Pargett D, III RM, Preston C, Jones W, Feldman J, Everlove C, Harris A, Alvarado N, Massion E, Birch J, Greenfield D, Vrijenhoek R, Mikulski C, Jones K (2009) Remote detection of marine microbes, small invertebrates, harmful algae and biotoxins using the environmental sample processor (ESP). Oceanography 22:158–167
140. Bellingham J, Hobson B, Godin MA, Kieft B, Erikson J, McEwen R, Kecy C, Zhang Y, Hoover T, Mellinger E (2010) A small, long-range AUV with flexible speed and payload. In: Ocean sciences meeting, abstract MT15A-14, Portland, OR, Feb 2010
141. Gottlieb J, Graham R, Maughan T, Py F, Ryan J, Elkaim G, Rajan K (2012) An experimental momentum-based front detection method for autonomous underwater vehicles. In: IEEE international conference on robotics and automation, St. Paul, Minn, 2012
142. 142 Pinkel R, Goldin MA, Smith JA, Sun OM, Aja AA, Bui MN, Hughen T (2011) The wirewalker: a vertically profiling instrument carrier powered by ocean waves. J Atmos Ocean Tech 28(3):426–435
143. Rainville L, Pinkel R Wirewalker (2001) an autonomous wave-powered vertical profiler. J Atmos Ocean Tech 18:1048–1051
144. Gneiting T, Genton MG, Guttorp P (2006) Geostatistical space-time models, stationarity, separability, and full symmetry. In: Finkenstaedt B, Held L, Isham V (eds) Statistical methods

for spatio-temporal systems. Monographs in Statistics and Applied Probability. Chapman and Hall/CRC Press, Boca Raton, pp 151–175
145. Graham R, Py F, Das J, Lucas D, Rajan K (2012) Exploring space-time tradeoffs in autonomous sampling for marine robotics. In: International symposium on experimental robotics (ISER), Québec City, Canada, June 2012
146. CENCoos [Online]. Available: http://www.cencoos.org/. [Online] Available: http://www.cencoos.org/
147. Adomavicius G, Tuzhilin A (2005) Toward the next generation of recommender systems: a survey of the state-of-the-art and possible extensions. IEEE Trans Knowl Data Eng 17: 734–749
148. Patrón P, Lane DM, Petillot YR (2009) Interoperability of agent capabilities for autonomous knowledge acquisition and decision making in unmanned platforms. In: IEEE Oceans Europe, Bremen, 2009
149. Williams B, Nayak P (1996) Immobile robots: AI in the new millennium. AI Mag 17(3), 16–35
150. Williams BC, Nayak PP (1997) A reactive planner for a model-based executive. In: International joint conference on artificial intelligence (IJCAI), Nagoya, Japan, pp 1178–85, 1997
151. Wang M, Dearden R (2009) Detecting and learning unknown fault states in hybrid diagnosis. In: Proceedings of international workshop on principles of diagnosis (DX), Stockolm, Sweden, 2009
152. Ernits J, Dearden R, Pebody M (2010) Automatic fault detection and execution monitoring for AUV missions. In: IEEE ocean engineering society autonomous underwater vehicles, Monterey, CA, 2010
153. Dearden R, Ernits J (2011) Automated fault diagnosis for an autonomous underwater vehicle. IEEE J Oceanic Eng. http://www.cs.bham.ac.uk~rwd/research/publications.php
154. Quinlan JR (1993) C4.5: programs for machine learning. Morgan Kaufmann, San Mateo
155. Aha DW, Kibler D (1991) Instance-based learning algorithms. Machine Learning 6:37–66
156. Jensen FV (2002) Bayesian networks and decision graphs. In: Information science and statistics. Springer, New York
157. Zhang GP (2000) Neural networks for classification: a survey. IEEE Trans Syst, Man, and Cybernetics, Part C: Appl Rev 30(4):451–462
158. Vidal T, Fargier H (1997) Contingent durations in temporal CSPs: from consistency to controllabilities. In: Proceedings of IEEE TIME international workshop, Daytona Beach, FL, 1997
159. Tsamardinos I, Muscettola N, Morris P (1998) Fast transformation of temporal plans for efficient execution. In: Association for the advancement of artificial intelligence, national conference (AAAI), Madison, Wisconsin, 1998
160. OOI: ocean observing initiative [Online]. Available: http://www.oceanobservatories.org/

Chapter 4
Path Planning for Autonomous Underwater Vehicles

Liam Paull, Sajad Saeedi, and Howard Li

4.1 Introduction

This chapter addresses the task of motion or path planning for an autonomous underwater vehicle (AUV). Once a map of the environment is built, and the vehicle has been able to localize itself, the high-level task of path planning must be achieved in order for the platform to complete its mission objectives.

In its most general form, path planning consists of determining the sequence of robot configurations that will result in achieving the stated mission. Often, particularly in AUV applications, paths are represented by a set of waypoints to be hit in succession. In reality, these waypoints are used as an outer loop reference that assumes that some inner loop controller will be able to stabilize the vehicle and track the reference. Given that this outer loop reference is independent of the platform control, very generic algorithms can be developed that are agnostic to system hardware.

Waypoints can be generated one at a time based on current up-to-date information or can be fully preplanned ahead of time. For example, traditional seabed surveys use a preprogrammed series of waypoints that define the survey (usually a ladder or zigzag path). It is of particular importance that the AUV localization accuracy will drift without bound if the vehicle is submerged and there are no beacons, so the vehicle will surface periodically to obtain a GPS fix and thus bound localization errors.

An alternative to waypoint-defined paths is a more reactive behavior-based paradigm. For example, an AUV trajectory could be specified by maintaining an altitude over the seafloor or avoiding obstacles, or following a structure such as a dam.

L. Paull • S. Saeedi • H. Li (✉)
COBRA group Department of Electrical and Computer Engineering,
University of New Brunswick, Fredericton, NB, Canada, E3B 5A3
e-mail: liam.paull@unb.ca; sajad.saeedi.g@unb.ca; howard@unb.ca

M.L. Seto (ed.), *Marine Robot Autonomy*, DOI 10.1007/978-1-4614-5659-9_4,

In this style of approach, no model of the environment is built to aid in the decision-making process; therefore global goals are difficult to achieve.

As we will see, the task of path planning depends on many factors, such as the level of information about the environment, the desired quality of the planned path, and the mission objectives. If detailed environmental information is available beforehand, then path generation can be slow without consequence. However, if the AUV is required to preplan its mission *on the fly*, then an algorithm that can run in near real time will be required.

The majority of path planning research has been done for ground robotics. In this chapter we will summarize many of these existing path planning approaches and comment on how well suited they are to be applied to AUV planning.

In Sect. 4.2 a more detailed overview of the path planning problem will be given and the literature on AUV path planning will be summarized. In Sect. 4.3, the details of many path planning approaches are given, in Sect. 4.4 and Sect. 4.5 two case studies of actual AUV path planning algorithms are presented, and a final summary is given in Sect. 4.6.

4.2 Path Planning Overview

In order to formally define the path planning problem, we require a few fundamental terms to be defined:

Configuration Space (Q): The set of all configurations that the robot can achieve. Q can be subdivided into free configuration space, Q_{free}, and occupied configuration space, Q_{obs}. Any configuration in Q_{obs} will result in contact with an obstacle, and any configuration in Q_{free} will not.

Workspace or **Environment** (W): The world that the robot exists in.

Path: Parameterized curve through Q_{free}.

Degrees of Freedom: The minimum number of independent variables required to represent the robot configuration. To fully describe an AUV requires six degrees of freedom: $\{x, y, z, \phi, \theta, \psi\}$, where x, y, and z represent the position in 3D space and ϕ, θ, and ψ are the Euler angles. Sometimes planning for an AUV is done in 2D by assuming that $\theta = \phi = 0$ and that $z = z_{\text{ref}}$, in which case there are three DOF: $\{x, y, \psi\}$.

We can then define the task of path planning as finding a curve in the free configuration space, Q_{free}, that connects a start location, $q = q_s$, to a goal location, $q = q_g$. More specifically, as described in [1], the path can be expressed as a curve $\tau : [0,1] \to Q_{\text{free}}$ with $\tau(0) = q_s$ and $\tau(1) = q_g$.

Path planning tasks usually fall into one or more of the following four areas:

1. Navigation: Finding a collision-free path through an obstacle-laden environment
2. Coverage: Passing a sensor over every point in the environment

3. Localization: Using sensor data to determine the configuration of the robot within the environment
4. Mapping: Using a sensor to explore a previously unknown environment

The majority of the algorithms presented in this chapter apply to navigation. Coverage will also be briefly discussed because it is of particular importance to AUVs.

The following terms define properties of any path planning algorithm and can be used as a basis for comparison:

Optimality: An algorithm that optimizes (maximizes or minimizes) some objective.

Completeness: A plan is complete if it will always find a solution if one exists or determine that no solution exists in finite time. A path can also be considered **resolution complete** if it is complete subject to discretization. Alternately, an algorithm is said to be **probabilistically complete** if it is guaranteed to converge towards completeness.

Offline planning: All knowledge of the environment is known and the plan is completed before execution begins.

Online planning: The plan is incrementally constructed during execution.

Sensor-based planning: Sensor information is processed online and used for planning.

Deliberative: Sense $\rightarrow$ Plan $\rightarrow$ Act cycle. An entire representation of the environment is built on each iteration.

Reactive: Use sensory information to accomplish mission without representation of the entire environment.

Many algorithms will be presented in Sect. 4.3. These algorithms will be compared based on these criteria in Sect. 4.6.

4.2.1 Coverage Path Planning

Of the four types of path planning missions described in the previous section, the problem of coverage path planning (CPP) is of particular importance for AUVs. For example, for mine-countermeasure missions, bathymetric data gathering, or ocean sampling, the AUV must efficiently cover a desired area with a sensor.

CPP is a subproblem of the general path planning problem. Instead of generating a path that efficiently navigates from a start point to a goal point, the objective is now to generate a path that will result in every point in the workspace being covered with a sensor. The problem can be formulated as follows: consider a mobile robot with an onboard sensor with some 2D swath $\mathcal{A}$. A trajectory through the workspace W will result in a set of N swaths readings: $\{\mathcal{A}_1, \mathcal{A}_2, \ldots, \mathcal{A}_N\}$. The goal of CPP is to generate a path through the workspace such that $\bigcup_{i=1}^{N} \mathcal{A}_i \supseteq W$, meaning that at some point the sensor swath has passed over every point in the workspace.

In the context of CPP, some approaches can be considered as deterministic, which do not consider uncertainty and can provide provable completeness under good conditions [2]. Alternately, nondeterministic or probabilistic approaches are more robust to sensor and actuator noise but cannot guarantee completion. Simple CPP algorithms define structured paths that follow a predetermined shape like a lawn mower or a zigzag in order to ensure coverage. An alternate and more complex method is to generate a nonstructured trajectory from a search space that achieves some exploration objective [3].

As described in Choset's [4] survey of complete coverage methods, there are heuristic, random, and cell decomposition techniques (see Sect. 4.3.4). A heuristic defines a set of rules to follow that will result in the entire environment being covered. For example, Acar and Choset's complete coverage algorithm based on sensing critical points [5], and Wein's [6] method of building corridors based on maximizing some quality function. A key facet of these approaches is having obstacles to be able to generate the rules. Cell decomposition is used to divide up the environment into a manageable number of cells or areas that can be searched like a graph or tree. Once all cells have been covered, then the entire workspace has been covered. Decomposition can be approximate [7], semi-approximate, or exact [4]. The shape of the cells and type of decomposition can have a significant impact on the performance of the search algorithm. Other recently published results in the area of CPP include exact cell decomposition for vacuum cleaning [8] and unmanned aerial vehicles [9], 3D terrain coverage [10], and a trapezoidal decomposition for agricultural applications [11].

4.2.2 AUV Path Planning

Comparatively little research has been done on AUV path planning compared to ground vehicle or air vehicle planning. This section will summarize some of the more important and recent results in the field.

In most cases, an optimal path is found by some metric subject to holonomic or other constraints. For example, one of the first known papers to discuss path planning of AUVs was published by Warren in 1990 [12]. In this approach, potential fields (see Sect. 4.3.2) are used to avoid obstacles and local minima are avoided by considering the global path.

In [13], an optimal kinematic control scheme is proposed where the cost function to be minimized is the integral of a quadratic function of the velocity components.

In [14], a mixed integer linear programming method is used to find a path for adaptive sampling that maximizes the line integral of the uncertainty of sensor readings along the proposed path. This type of algorithm is used as an alternative to static buoys for collecting oceanic data such as temperature, salinity, etc. The metric for benefit in the objective function is a maximum sum of probabilities and paths planned are greedy.

In [15], mutual information is used as the benefit metric in the objective function, combined with a recursive greedy planner. The proposed grid decomposition results in very constricted paths, and the paths designed apply more readily to a glider-style AUV.

In [16], genetic algorithms (see Sect. 4.3.6) and dynamic programming are used to generate paths for multiple AUVs.

In some cases, such as highly cluttered harbors, obstacle avoidance is an important consideration. In [17], a path is planned using a fast-marching-based method. In [18], a sequential-quadratic programming approach is used to avoid obstacles detected with a multi-beam sonar.

A significant consideration in start-to-goal path planning in an underwater environment is accounting for ocean or estuarine currents, for example, refer to [19–24].

In [25], a coverage algorithm for MCM with a side-looking sonar (SLS) is proposed that uses cell decomposition and exploits the supposed fact that mines are normally placed in lines.

In [26], a boustrophedon decomposition [27] is combined with the generalized Voronoi diagram (GVD) (see Sect. 4.3.3) to derive paths for coverage of a highly unstructured or non-convex environment. This algorithm presumes that absolute knowledge of the environment is known *a priori* and all planning is done offline.

Algorithms have been presented that optimize the design of structured paths for CPP. For example, in [28], the Cramer-Rao lower bound is used to optimize the track spacing for a ship hull inspection mission. In [29], a coverage algorithm for MCM with a SLS is proposed that optimizes the spacing between parallel tracks. The metric for optimality is maximizing the mean probability of detection of the environment. The dependence of probability of detection on seabed type and range is described.

In [30], an algorithm is presented that uses an online sensor-driven approach to develop unstructured path efficiently.

4.3 Popular Path Planning Approaches

This section will present descriptions, pseudo-code and results for many common path planning approaches.

Path planning approaches come in many flavors. The first type that we will encounter are simple heuristic, reactive methods that are able to traverse a workspace by following simple rules. These types of approaches are extremely limited and will always be prone to becoming trapped by unforeseen circumstances, but they require no prior knowledge of the environment.

If knowledge is available, then the options for planners increase dramatically. One type of algorithm uses the locations of obstacles and the goal point to generate a potential function over the entire workspace. The robot then follows the direction

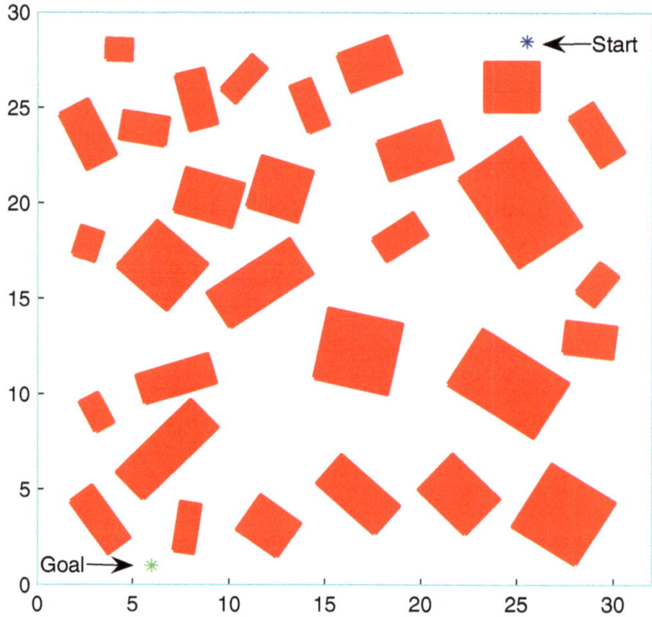

Fig. 4.1 Simple environment that will be used for comparison

of the negative gradient of the potential function at each location to navigate towards the goal.

The dimensionality of the free space can also be reduced to form something analogous to a roadmap. A roadmap must follow the rules that it is possible to get from any point in the free space onto the roadmap in some simple way, and that the roadmap itself is connected. Various types of such roadmaps will be presented.

Another common approach is to decompose the environment into a collection of smaller pieces, which are usually called cells.

In the case of roadmaps and cell decomposition, the result is that the workspace is transformed into an abstract representation that can be treated as a *graph* with *nodes* and that are connected by *edges*. Once the graph is built, it must be searched. Algorithms for traversal or search of graphs abound in the literature, but a few of the more common ones will be discussed.

More recently, advanced methods inspired by biological processes have also been proposed to solve a variety of problems including searching.

In order to compare and contrast the various algorithms, the simple 2D environment shown in Fig. 4.1 will be used. Red randomly placed blocks represent obstacles and the start and goal are as indicated.

4.3.1 Heuristic-Based Reactive Algorithms

The bug algorithms are simple heuristic methods that help a robot navigate from a start point to a goal in an obstacle-laden environment.

Bug1 In Bug1 [31], it is assumed that the robot has only a contact sensor and knowledge of where the goal is. The robot begins by moving towards the goal. If an obstacle is encountered, it is circumnavigated to find the point on the obstacle that is closest to the goal. The robot then travels to that point and continues until the goal is reached. If the heading line intersects with a previously visited obstacle, it is inferred that there is no path to the goal. It has been proven that this method is complete, meaning that the robot is guaranteed to find the goal or report that there is no possible path to the goal in finite time. The details of the approach are presented in Algorithm 1.

Algorithm 1: Bug1 algorithm

```
Require: Map
Ensure: Path
while Not arrived at the goal do
    Take a step in the direction of the goal
    if Robot encounters an obstacle then
        if Obstacle is previously unvisited then
            Circumnavigate the obstacle until encountering the initial contact point
            Move to the point on the obstacle that is closest to the goal
            Mark obstacle visited
        else
            Path does not exist
        end if
    end if
end while
```

The sample path generated with the Bug1 algorithm is shown in Fig. 4.2.

Bug2 In Bug2 [31], an imaginary line is drawn between the start point and the goal point. The robot begins by following this line. If an obstacle is encountered, it begins to circumnavigate the obstacle until it intersects the line and then it continues. Bug2 is generally faster than Bug1 but has the disadvantage that it is not guaranteed to find the goal in finite time, as it can be trapped by certain oddly shaped obstacles. If the robot ever encounters an already visited point in the map, then the algorithm fails.

The details of the approach are presented in Algorithm 2.

The path generated is shown in Fig. 4.3.

There is also a variation of Bug2 called TangentBug. In TangentBug [32], it is assumed that the robot is equipped with a range sensor. The robot begins by traveling

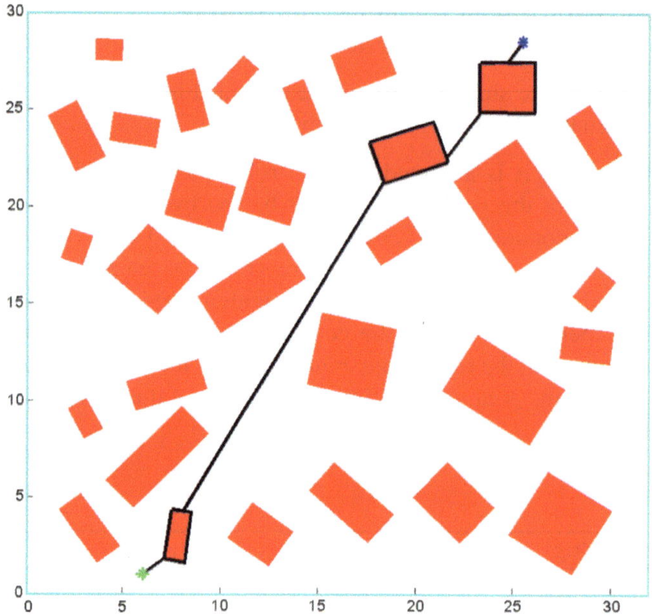

Fig. 4.2 Path planned using Bug1 algorithm

Algorithm 2: Bug2 algorithm

Require: Map
Ensure: Path
Generate line l that intersects start and goal
while Not arrived at the goal **do**
    Take a step along l
    if Obstacle encountered **then**
        Move around the obstacle until hitting l
        if Robot revisits an already visited point **then**
            Search fails
        end if
    end if
end while

---

towards the target point. If an obstacle is detected by the range sensor then the robot alters its trajectory to the point on the range sensor that will avoid the obstacle and bring it towards the goal. The resulting path is tangential to the obstacles. This method vastly outperforms Bug1 and Bug2 but requires the use of an additional sensor. In all bug algorithms, the direction to circumnavigate the obstacles must be chosen prior to the search. Usually this choice is made randomly but can have a significant effect on the length of the outputted path.

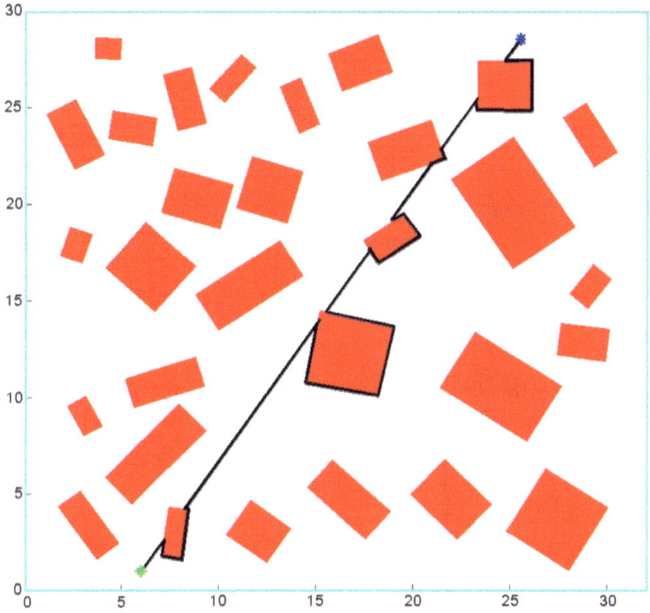

Fig. 4.3 Path planned using Bug2 algorithm

4.3.2 Potential Function Approaches

Available knowledge of the environment such as goal and obstacle locations can lead to the development of a potential function over the free configuration space. Paths are generated by following the negative gradient of the potential function. The approach tends to be computationally simple because no searching is required to determine the path; instead the direction of the negative gradient of the potential field is automatically chosen.

Potential Fields In the potential fields method, the potential function is generated by considering the goal as an attractive force and the obstacles as repulsive. The potential function at any point in the space is the sum of the positive and negative forces [33]:

$$U(q) = U_{\text{att}}(q) + U_{\text{rep}}(q),$$
$$\nabla U(q) = \nabla U_{\text{att}}(q) + \nabla U_{\text{rep}}(q), \tag{4.1}$$

where the positive force could be defined as

$$U_{\text{att}}(q) = \frac{1}{2}\xi d^2(q, q_{\text{goal}}),$$
$$\nabla U_{\text{att}}(q) = \xi(q - q_{\text{goal}}), \tag{4.2}$$

where $d(q_i, q_j)$ is a function that defines the Euclidean distance between two configurations and ξ is a scalar multiplier and is the attractive force towards the goal, and the negative force could be defined as

$$U_{\text{rep}_i}(q) = \begin{cases} \frac{1}{2}\eta(\frac{1}{d_i(q)} - \frac{1}{Q_i^*})^2, & d_i(q) \leq Q_i^*, \\ 0, & d_i(q) > Q_i^*, \end{cases} \tag{4.3}$$

$$\nabla U_{\text{rep}_i} = \begin{cases} \eta(\frac{1}{Q_i^*} - \frac{1}{d_i(q)})\frac{q-c_i}{d_i^3(q)}, & d_i(q) \leq Q_i^*, \\ 0, & d_i(q) > Q_i^*, \end{cases} \tag{4.4}$$

$$\nabla U_{\text{rep}} = \sum_{i=1}^{n} \nabla U_{\text{rep}_i}, \tag{4.5}$$

where $d_i(q)$ is the minimum distance to obstacle i, c_i is the closest point to q on obstacle i; therefore $\nabla d_i(q) = \frac{q-c_i}{d_i(q)}$. Q_i^* is the domain of influence for obstacle i, and η is a scalar multiplier. The path is generated by following the direction of the negative gradient $(-\nabla(U(q))$ which corresponds to the direction that is progressing towards the goal the fastest.

In this form, complete knowledge of the location of obstacles in the environment is required assuming Q_i is large enough. However, online versions of the potential fields method have been suggested [34]. For example, if the robot is equipped with a range sensor, then the Q_i values could be adjusted to the range of the sensor. The main drawback of this method is that it is possible for the robot to enter locations that are not the goal where the sum of the vectors is zero. These are referred to as *local minima*.

The potential field method does not require expensive computation before the search for a path can actually start. The potential field approach searches a smaller graph. There are various techniques to construct good potential fields and efficiently escape their local minima [35]. The potential field approach is both fast and capable of handling many degrees of freedom (Fig. 4.4).

The algorithm for the method is shown in Algorithm 3. In the first loop, the attractive and repulsive force vectors and gradients are calculated. The gradient descent is applied in the while loop where at each time step we move some distance Δ in the direction of the negative gradient.

The path generated by the potential fields algorithm is shown in Fig. 4.5. The number of obstacles has been reduced so that local minima are avoided in this case.

Algorithm 3: Potential fields

Require: Map, Q, start q_{start}, goal q_{goal}
Ensure: Path, P

for all locations q in Q **do**
  $U_{att}(q) \leftarrow \frac{1}{2}\xi|q - q_{goal}|^2$
  for all obstacles o_i, $i = 1..n$ **do**
    if q is in o **then**
      $U_{rep}(q) \leftarrow \infty$
    else
      $d \leftarrow$ minDistToPolygon(q, o_i)
      $c_i \leftarrow$ closestPointOnObstacle(q, o_i)
      if $d < Q_i^*$ **then**
        $U_{rep_i}(q) \leftarrow \eta(\frac{1}{Q_i^*} - \frac{1}{d_i(q)})\frac{q-c_i}{d_i^3(q)}$
      else
        $U_{rep_i}(q) \leftarrow 0$
      end if
    end if
  end for
end for
while $q \neq q_{goal}$ **do**
  add q to P
  $\nabla U(q) \leftarrow \nabla U_{att}(q) + \sum_{i=1}^{n} \nabla U_{rep_i}(q)$
  $q \leftarrow q - \Delta \nabla U(q)$
end while

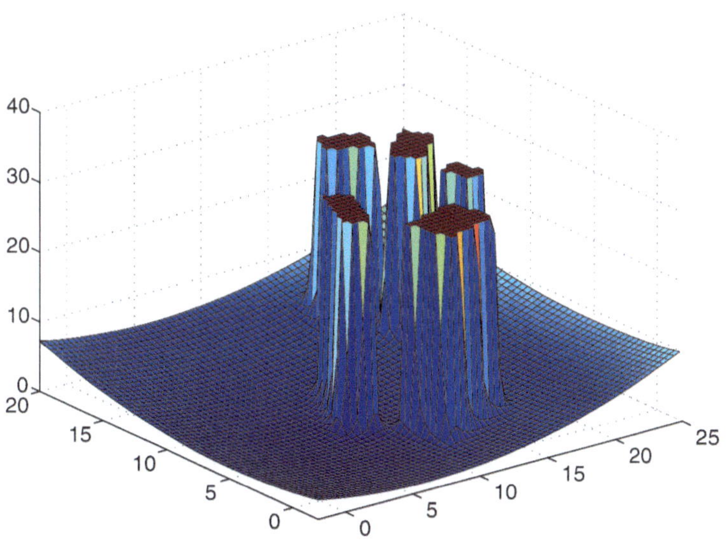

Fig. 4.4 The potential function

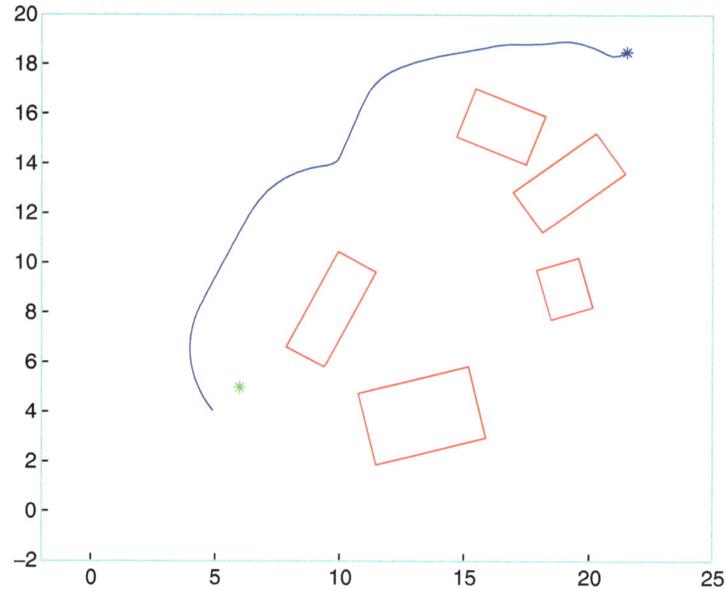

Fig. 4.5 Path generated from the potential field algorithm

Wavefront Planner The wavefront planner [36] is similar to the potential fields method but differs in several important ways. The potentials at each point in the free space are generated by simulating a wave moving from the goal location to the start. The details are given in Algorithm 4. The free space is represented by a grid, all of which is initialized as unvisited. To initialize, the goal cell is marked as visited and given a value of 2. The algorithm then iteratively finds unvisited neighbors of visited cells and assigns them values of one greater than the visited cell. The result will be a "wavefront" radiating outwards from the goal cell as new cells are assigned increasing values.

Once the wavefront reaches the start cell, the potential function is complete and a path can be planned using gradient descent. This is accomplished by beginning at the starting location and iteratively choosing the neighbor with the minimum value.

The wavefront planner has the advantage over the potential fields method in that no local minima are generated. However, it is computationally more expensive and requires prior knowledge of the locations of all of the obstacles (offline algorithm).

Figure 4.6 shows an example implementation of the algorithm. In the figure, the lighter areas of free space are at a higher potential. The values of the potential function are shown in the color bar on the right. The obstacles shown in black are considered infinite potential. The start point is shown in green and the end point is shown in red. In this case, the free space is discretized into cells and only four neighbors are considered in the gradient descent step.

Algorithm 4: Wavefront algorithm

Require: Map Q, start cell q_{start}, goal cell q_{goal}
Ensure: Path P

mark all free space as unvisited
$index \leftarrow 2$
$q_{goal} \leftarrow index$
mark q_{goal} visited
while q_{start} not visited **do**
    set all unvisited cells neighboring a cell with value *index* to $index + 1$
    mark all cells with value $index + 1$ as visited
    $index \leftarrow index + 1$
end while
$q \leftarrow q_{start}$
add q to P
while q_{goal} not in P **do**
    $q \leftarrow$ neighbor of q with minimum value
    add q to P
end while

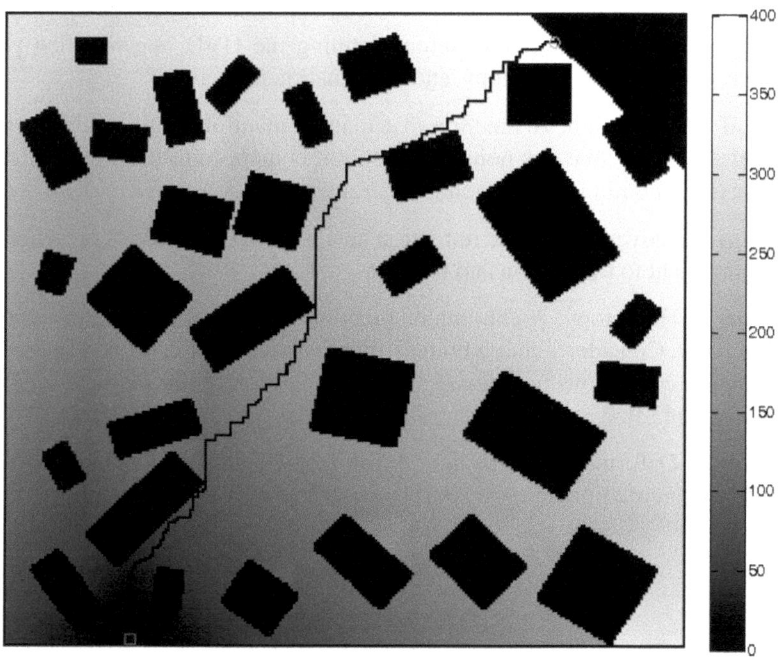

Fig. 4.6 Path planned using the wavefront algorithm

4.3.3 Roadmaps

A roadmap [37] is generated as a subset of the configuration space that satisfies three main rules:

1. It is possible to get from any start point in the configuration space to the roadmap.
2. it is possible to get from any point on the roadmap to any other point on the roadmap.
3. It is possible to get from the roadmap to any goal point in the configuration space.

Different spaces that have been proven to be roadmaps include the visibility graph [38], which is generated by connecting every vertex of obstacles that can be connected with a straight line, and the Voronoi diagram [39], which is the set of points that are equidistant from at least two obstacles.

Once a roadmap is generated, it can be represented as a graph of edges and nodes, which can be searched with classical search methods, such as A*, breadth-first search (BFS), and many others. Generating the roadmap requires prior knowledge of the positions of all of the obstacles. However, online methods where the roadmap is built as the robot traverses the environment do exist and are referred to as "sampling-based methods."

Generalized Voronoi Diagram Before defining the GVD, we will first present definitions of retraction, homotopy, and deformation retraction.

Definition 1. Retraction: A function that maps a manifold onto a subset of itself and has the property that any point in the subset is mapped to itself. Specifically, a retraction is any continuous function $f: X \rightarrow A$ with $A \subset X$ and $f(a) = a \, \forall a \in A$ [33].

An intrinsic advantage of the retraction as a topological world representation is that it is invariant to translation and rotation.

Definition 2. Homotopy: A continuous function that smoothly deforms one function to another. Consider f and g both as functions with domain U and range V. A homotopy is a continuous function $H : U \times [0,1] \rightarrow V$ such that $H(x,0) = f(x)$ and $H(x,1) = g(x)$ [33].

Definition 3. Deformation Retraction: A homotopy that smoothly maps the identity function into a retraction. $H : X \times [0,1] \rightarrow X$ with [33]

$$H(x,0) = x \tag{4.6}$$

$$H(x,1) \in A \tag{4.7}$$

$$H(a,t) = a \text{with} a \in A \text{and} t \in [0,1]. \tag{4.8}$$

The GVD is the set of all points in the free space of a map that are equidistant from at least the two closest obstacles. The GVD is a retraction because all points on the GVD will be mapped to themselves and all points in free space will be mapped onto the GVD. Therefore, a function that smoothly maps the free space onto the GVD is a deformation retraction. This formulation leaves us with two important properties of the GVD that will be exploited:

1. The GVD is connected because the set of free space is connected and connectivity is maintained under a deformation retraction.
2. The GVD of a map is unique and is invariant to transformations and rotations because it is a retraction.

The GVD can be interpreted as a topological representation of the map structure that contains the key information intrinsic to the map but in a much more compact, one-dimensional form.

An easy way to generate the GVD is to assign a circle with a small radius to each grid of the map, q_i, then increase the radius of the circle until it comes into contact with at least one obstacle. If there is only one contact point, then q_i does not belong to GVD; otherwise, it does. Algorithm 5 gives the detailed approach.

Algorithm 5: Generate GVD

Require: Map Q
Ensure: GVD

GVD $\leftarrow \emptyset$
for all $q_i \in Q_{\text{free}}$ **do**
    $r \leftarrow 0$
    $C \leftarrow$ circle with center q_i and radius r
    $cp \leftarrow \emptyset$
    while $cp = \emptyset$ **do**
        $r \leftarrow r + \Delta r$
        $cp \leftarrow C \bigcap Q_{obs}$
    end while
    if number of elements in cp is greater than 1 **then**
        Add q_i to GVD
    end if
end for

An example of the GVD is shown in Fig. 4.7.

Visibility Graph The visibility graph is an alternate roadmap representation of the free space.

In a visibility map, all obstacle vertices are nodes in the graph. Edges connect nodes that are in line of sight of each other in the free space. The full algorithm for generating the visibility graph is given in Algorithm 6. This algorithm requires that obstacles be polygonal. The set $\{v_i\}_{i=1}^{V}$ represents all obstacle vertices where V is the total number of vertices in all obstacles. The closest configurations to each possible combination of vertices is selected and a straight line in the configuration space is made that connects the two, denoted by Q_{line}. If all configurations on this line do not intersect with any obstacles then an edge between these two nodes is added to the graph.

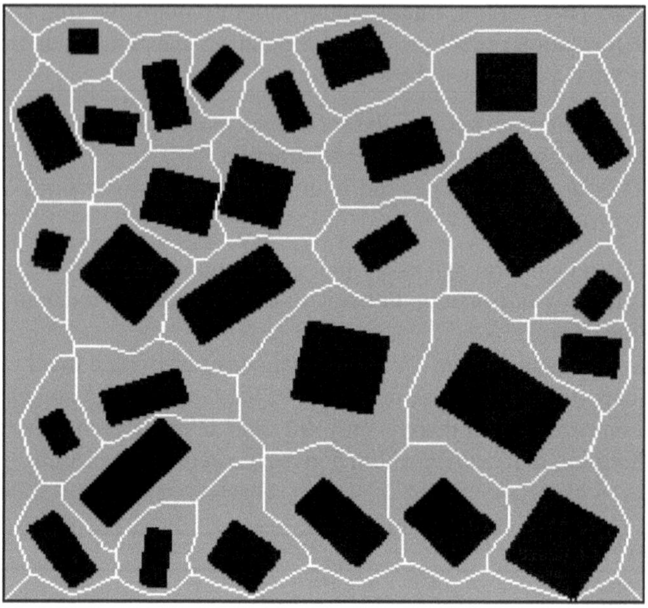

Fig. 4.7 A GVD

This algorithm produces a graph with a high level of connectivity. The graph can be pruned by considering the following two definitions:

- *Supporting lines* are tangent to two obstacles with both of them on the same side of the line.
- *Separating lines* are tangent to two obstacles with each on opposite sides of the line.

It can be proven [33] that the best solution will be made of supporting and separating lines. Therefore all other edges can be removed.

A visibility graph is shown in Fig. 4.8. In reality it can be useful to implement a safety margin around all obstacles so that the paths do not closely approach obstacles.

The visibility has the advantage over the GVD roadmap that it is very simple to generate. However, the resulting graph in general has more edges and nodes and therefore will be harder to search. In addition, the path generated from following the roadmap is guaranteed to come very close to obstacles, and the algorithm requires that obstacles be represented as polygonal.

Probability Roadmaps In some cases, generating the roadmap using either the GVD or the visibility graph is either too difficult or results in a graph with too many nodes and edges that it cannot be searched efficiently. This can be the case if there are many obstacles or particularly if the dimensionality of the configuration space is high. For example, an underwater robotic arm with multiple links that is used for

Algorithm 6: Visibility graph

Require: Map, Q
Ensure: Visibility graph VG
Add each obstacle vertex v_i to VG as node n_i
for $i = 1 : V$ **do**
  for $j = 1 : V$ and $j \neq i$ **do**
    $q_i \leftarrow$ closest valid configuration to vertex v_i
    $q_j \leftarrow$ closest valid configuration to vertex v_j
    $Q_{line} \leftarrow$ set of configurations connecting q_i to q_j by a straight line in configuration space
    if $Q_{line} \bigcap Q_{obs} = \emptyset$ **then**
      Add edge between n_i and n_j in VG
    end if
  end for
end for

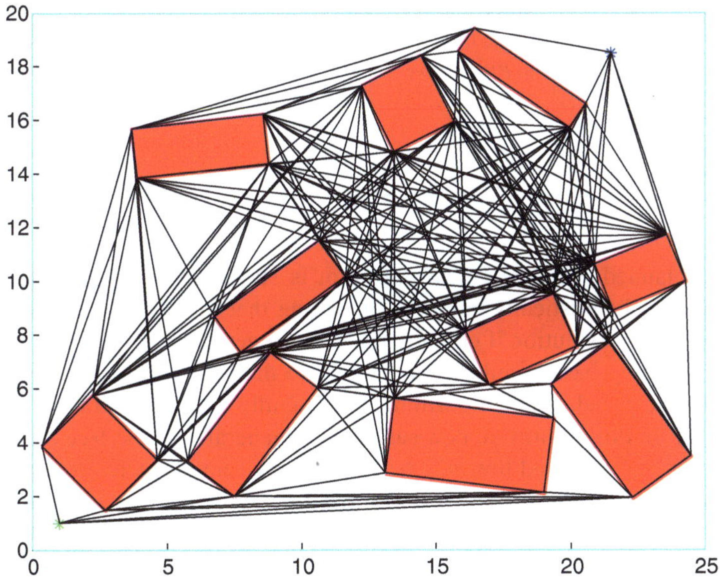

Fig. 4.8 A visibility graph

docking or precise tasks. To account for these problems, sampling-based methods have been proposed. While these algorithms do not possess the nice qualities such as being complete or optimal, they have been shown to be effective at solving real life problems efficiently.

There are many different types of sampling-based algorithms, but essentially they all aim to build a roadmap by randomly generating samples in the free configuration space. The first sampling-based algorithm was the probabilistic roadmap (PRM) planner [40]. According to this planner, the entire free configuration space is covered with random samples which are then connected in some way to generate a roadmap. The PRM planner is designed such that the roadmap building and the query phases of operation are separate; the entire roadmap is built before a path is found from start to goal. The challenging aspects of this type of algorithm are (1) determining whether a randomly generated sample lies in the free configuration space and (2) determining whether the edge between two nodes remains in the free configuration space. A simple implementation of the PRM planner is given in Algorithm 7.

Algorithm 7: Probabilistic roadmap

Require: Map, Q
Ensure: Probabilistic Roadmap PRM
Add q_{start} and q_{goal} to PRM
while No path exists from q_{start} to q_{goal} in PRM **do**
  Randomly generate q from Q
  if $q \in Q_{free}$ **then**
    Add q to PRM
    for all node $n \in PRM$ with $n \neq q$ **do**
      if Good path exists from n to q **then**
        Add edge in PRM between n and q
      end if
    end for
  end if
end while

A path planned with the PRM algorithm is shown in Fig. 4.9. Note that this algorithm is probabilistically complete, meaning that it is guaranteed to converge towards an optimal solution if enough nodes are generated.

Other sampling-based methods find a path to the goal as the samples are being generated. One popular such algorithm is the rapidly exploring random tree (RRT) planner [41]. In this algorithm, new samples are generated in the free configuration space randomly but biased towards the goal. Next, if no obstacles are met, a small step is taken towards the randomly generated point, and the process continues until the goal is found. As mentioned, these methods are particularly useful for finding acceptable solutions in the case when the configuration space dimensionality is very high.

4.3.4 Cell Decomposition

Cell decomposition refers to any method that partitions the configuration space into a set of smaller cells [4]. Once the decomposition has been done, a connectivity graph can be built, where the cells are the nodes, and cells that share a common edge are connected in the graph. This method is very effective for planning paths

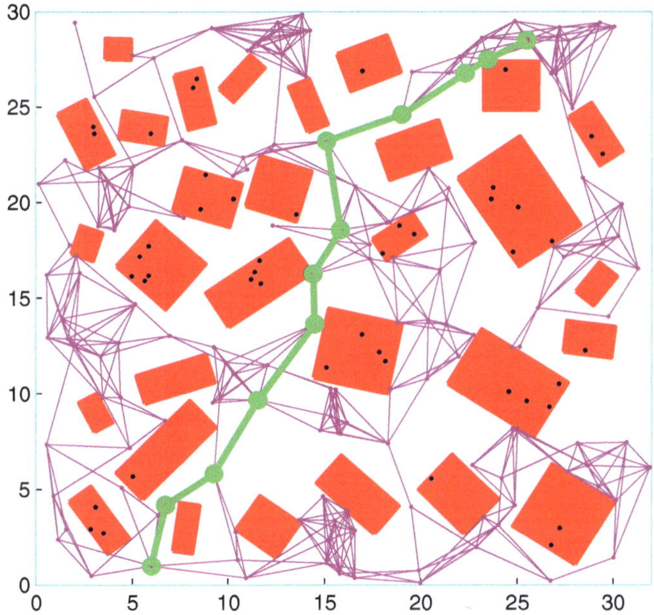

Fig. 4.9 Path planned using the probability roadmap algorithm

that achieve complete coverage as the entire space will be covered once every node in the graph has been visited, but can also be used for start-to-goal path planning.

The shape and style of the cell decomposition can have a significant impact on the effectiveness of the planned path. Cell decompositions can be classified as either exact, semi-approximate, or approximate [4]. In [7], an approximate rectangloid decomposition is used to help a ground vehicle navigate from a start point to a goal configuration while hitting as many targets as possible along the way. This approach has considerable benefits, such as that obstacle avoidance is built in, and that targets can be searched using the information gain metric. However, it does assume that information about the presence of obstacles and targets is known beforehand. In addition, the searching of the connectivity graph can take considerable time.

A simple cell decomposition for a 2D environment with polygonal obstacles is the Boustrophedon [42] or Morse [43] cell decomposition. The decomposition is generated by detecting *critical points* which are vertices of the polygonal obstacles that cause the connectivity of the free configuration space to change. At each critical point, a *critical slice* is drawn that passes through the critical point but does not pass through any other obstacles or environment boundaries. The resulting partitioning of the free space is the cell decomposition. This method is an improvement over the trapezoidal decomposition which extends lines from each obstacle vertex because it is applicable to environments with non-polygonal obstacles and it generates fewer cells making searching easier and faster. A simplistic description of the algorithm is given in Algorithm 8.

Algorithm 8: Cell decomposition

Require: Map, Q
Ensure: Cell Decomposition
```
for left-most point to right-most point define vertical slice s do
    if s is tangential to an obstacle then
        add s to set of critical slices
    end if
    Add all cells to graph as nodes.
    for all critical slices do
        add edges in the graph connecting bordering cells
    end for
end for
```

Once the cell decomposition is made, a graph is generated where each cell represents a node in the tree and adjacent cells are connected with an edge. In the context of the Boustrophedon decomposition, this graph is referred to as the *Reeb graph* [5]. As mentioned, this type of decomposition is most useful for achieving coverage, but start-to-goal planning can be done by reducing each cell to one or a few points; in Fig. 4.10 each cell has been reduced to its centroid. The centroids of cells in the figure are denoted by black dots. The blue lines represent the critical slices. Note that every critical slice is tangent to an obstacle but cannot penetrate through an obstacle. The obstacles are shown in red. The path found by connecting centroids is shown as the yellow line. Note that in this simplistic case, a suboptimal path is generated because of the reduction of the cells to only their centroids.

4.3.5 Search Algorithms

Once the free space has been transformed into a graph using a technique such as roadmaps or cell decomposition, then the graph can be searched using one of any number of algorithms. The graph can more formally be defined as $G = (n, E)$ where n are nodes and E are edges that connect two nodes.

In this section, we will present some of the most popular methods of searching graphs.

Breadth-First Search In BFS, the closest nodes are exhaustively searched with no heuristic involved. The search is guaranteed to find the optimal solution; however in general it is extremely slow and is only useful for small problems.

The algorithm is described in detail in Algorithm 9. The first-in-first-out (FIFO) queue L maintains the set of nodes that are to be expanded. The FIFO nature of L results in nodes closest to the root nodes always being expanded first.

A more detailed tree is shown in Fig. 4.11. It can be seen that BFS does indeed find the optimal search. However, every node is expanded.

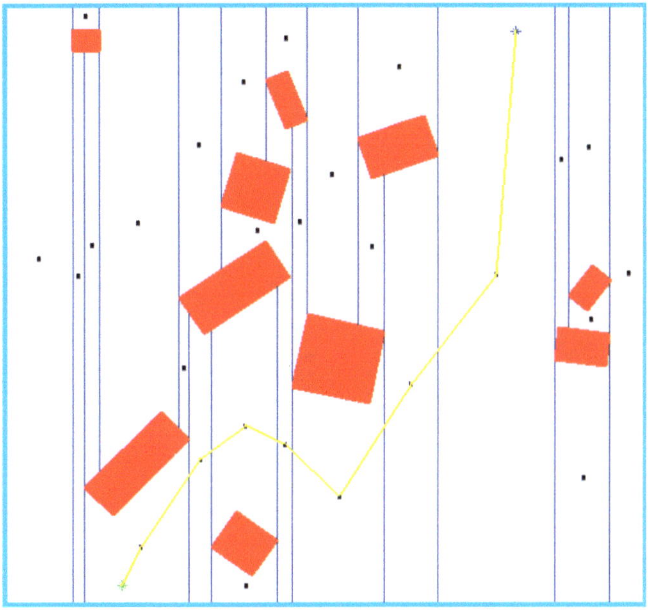

Fig. 4.10 A path planned using the Morse cell decomposition

A* search A* is an extension of Dijkstra's search algorithm. It is known for its better time performance using a distance-cost heuristics. Algorithm 10 explains the details.

Algorithm 9: BFS

Require: Graph, G, with root n_s
Ensure: Path P
$L \leftarrow n_s$
n_s.visited $\leftarrow$ true
while $L \neq \emptyset$ **do**
  $n \leftarrow$ POP(L)
  if n is the goal **then**
    $P \leftarrow n$
    while $n_i \neq n_s$ **do**
      $P \leftarrow \{n_i.\text{parent}, P\}$
      $n_i \leftarrow n_i$.parent
    end while
    Search complete, return P.
  end if
  for all unvisited neighbors $\hat{n}$ of n **do**
    $\hat{n}$.visited $\leftarrow$ true
    $\hat{n}$.parent $\leftarrow n$
    PUSH $\hat{n}$ onto L
  end for
end while

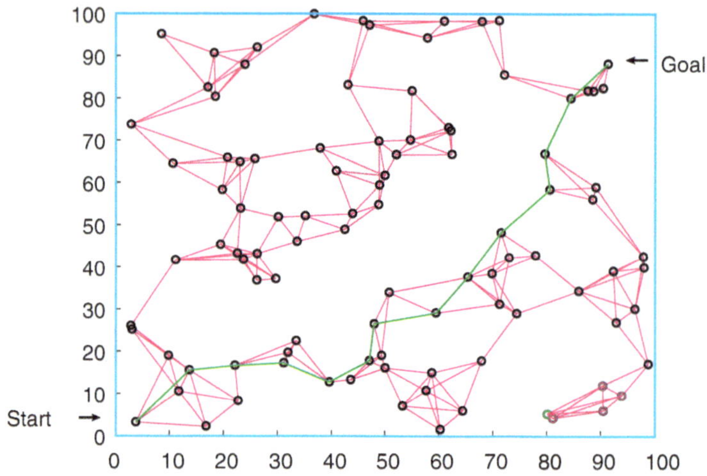

Fig. 4.11 Path planned using the BFS algorithm

Two sets are maintained, an open set and a closed set. The closed set contains all nodes that have been processed. The open set contains nodes that have been visited and are waiting to be expanded. An essential element for A* search is the definition of some heuristic cost that relates each node to the goal node. Typically, this cost can be calculated using the Manhattan or Euclidean distance, for example. Each node contains a true cost to reach that node through the graph from the start, represented by g, and a heuristic cost that represents the distance that the node is away from the goal node, represented by h. The node in *OPEN* with the minimum sum of these two costs, represented by $f = g + h$, is chosen as the next node to expand. When a node is expanded, all neighboring nodes not already in *CLOSED* are considered. If the neighbor is not already in *OPEN*, then its costs are calculated and it is added to *OPEN*. In addition to the costs, we must also keep track of which node was expanded to reach this neighbor. This is denoted as *parent* and can be seen as a pointer that will allow us to recover the actual path. A slightly tricky situation can occur when a neighboring node is already in *OPEN*. This can arise if there are two paths through the graph to one node. In this case, the two paths are compared, and the one with the smaller true cost g (they will both have the same heuristic cost) is chosen to be the node in *OPEN*.

As mentioned, once the goal is reached the path can be recovered by following the *parent* pointers backwards from the goal node to the start node [44].

A figure showing a search tree with a solution is shown in Fig. 4.12. The grey edges and nodes were not expanded. Like the BFS, the A* is capable of finding the optimal solution; however it does so without exhaustively searching.

Algorithm 10: A* algorithm

```
Require: Start node, n_s, Goal node , n_g, graph G
Ensure: Path P
n_s.g ← 0
n_s.h ← heuristicCost(n_s, n_g)
n_s.f ← n_s.g + n_s.h
OPEN ← n_s
CLOSED ← ∅
while OPEN ≠ ∅ do
    n ← node with minimum f value in OPEN
    remove n from OPEN
    add n to CLOSED
    for all neighbors n_i of n do
        if n_i ∈ CLOSED then
            skip
        end if
        g ← n.g + dist(n, n_i)
        if n_i ∈ OPEN then
            if n_i.g > g then
                n_i.g ← g
                n_i.f ← n_i.g + n_i.h
                n_i.parent = n
            end if
        else
            n_i.g ← g
            n_i.h ← heuristicCost(n_i, n_g)
            n_i.f ← n_i.g + n_i.h
            n_i.parent = n
            Add n_i to OPEN
        end if
        if n_i = n_g then
            P ← n_g
            while n_i ≠ n_s do
                P ← {n_i.parent, P}
                n_i ← n_i.parent
            end while
            Search complete, return P.
        end if
    end for
end while
```

4.3.6 Genetic Algorithms

Genetic algorithms (GA) are inspired by Darwin's theory of evolution. All living organisms consist of cells. In each cell there is the same set of chromosomes. Chromosomes are strings of DNA and serve as a model for the whole organism. A chromosome consists of genes, which are blocks of DNA. Each gene encodes a particular protein. During reproduction, crossover (or recombination) occurs first. Genes from parents form in some way a whole new chromosome. A complete set of

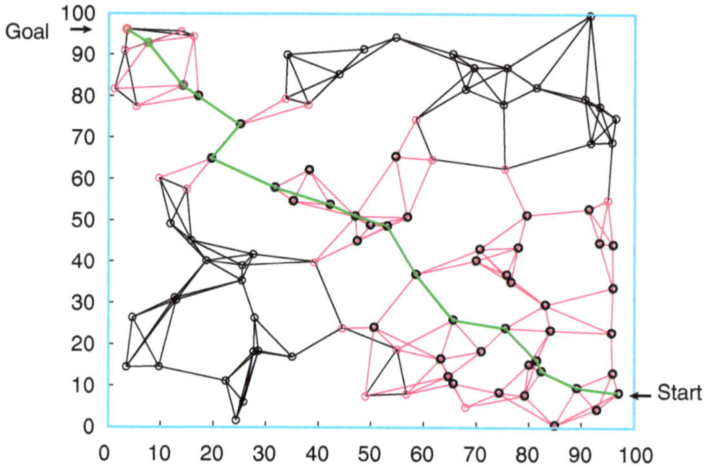

Fig. 4.12 Path planned using the A* algorithm

genetic material (all chromosomes) is called a genome. Evolution is a remarkable problem-solving machine [45]. First proposed by John Holland in 1975 [46], GAs are an attractive class of computational models that implement the natural evolution theory to solve problems in a wide variety of domains.

The generic algorithm is presented in Algorithm 11. The population consists of a set of chromosomes where each chromosome represents a possible solution in the search space. During each iteration, chromosomes are selected based on their "fitness" and recombined to create a new offspring until an entirely new generation is created. Each member of the new generation can also be optionally "mutated" to maintain diversity of the chromosomes. Another option (not shown in the algorithm) is to "clone" a certain number of the best chromosomes and copy them from one generation to the next so that the best solutions are not lost. This is referred to as "elitism."

The main tasks for implementing the GA are how to encode the chromosomes to represent potential solutions and how to define the fitness function.

More details of the approach including an implementation and paths planned for multiple AUVs will be presented in Sect. 4.4.

4.3.7 *Artificial Neural Networks*

Another popular biologically inspired planning algorithm is the artificial neural network [47]. Inspired by the shunting model of neurons, a neural network architecture can be developed whose dynamic neural activity landscape (as shown in Fig. 4.13) represents the dynamically varying workspace.

Algorithm 11: Genetic algorithm

Require: Start node, n_s, Goal node, n_g, graph G
Ensure: Path P

```
Randomly initialize population POP
i ← 0
while Maximum fitness ⇐ threshold or maximum number of iterations not exceeded do
    i ← i + 1
    Evaluate fitness of each chromosome
    for j from 1 to population size do
        Randomly select two chromosomes where probability of selection is based on fitness, c_1^{i-1}, c_2^{i-1}
        c_j^i ← crossover(c_1^{i-1}, c_2^{i-1})
        c_j^i ← mutate(c_j^i) (Optional)
    end for
end while
P ← chromosome with maximum fitness
```

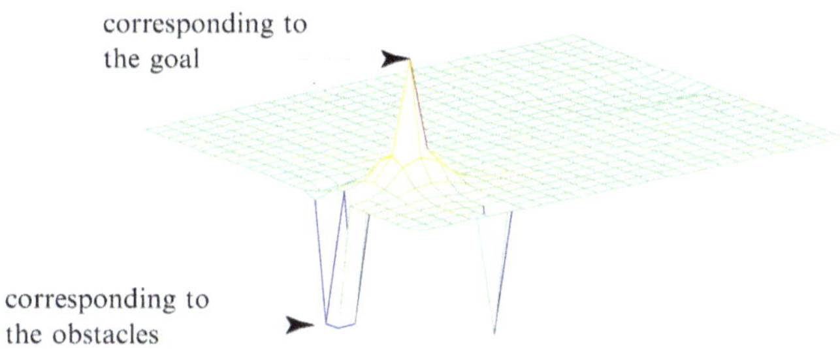

Fig. 4.13 Landscape of neural activities

The dynamics of the ith neuron x_i is characterized by a shunting equation,

$$\frac{\mathrm{d}x_i}{\mathrm{d}t} = -Ax_i + (B - x_i)([I_i]^+ + \Sigma_{j=1}^{m} w_{ij}[x_j]^+) \\ -(D + x_i)[I_i]^-, \tag{4.9}$$

where

- Parameters A, B, and D represent the passive decay rate, the upper and lower bounds of the neural activity, respectively.
- Variable x_i is the neural activity of the ith neuron, which has a bounded continuous value $x_i \in [-D, B]$.
- t is a virtual time index that only depends on the occurrence of events.
- The excitatory input, $[I_i]^+ + \Sigma_{k=1}^{m} w_{ij}[x_j]^+$, results from the target and the lateral connections among neurons.
- The external input I_i to the ith neuron is defined as $I_i = E$, if there is a target; $I_i = -E$, if there is an obstacle; $I_i = 0$, otherwise, where E is a positive constant.

- The inhibitory input $[I_i]^-$ results from the obstacles.
- The connection weight w_{ij} from the jth neuron to the ith neuron is defined as $w_{ij} = f(|d_{ij}|)$, where $d_{ij} = |q_i - q_j|$ represents the Euclidean distance between q_i and q_j in Q.
- Function $f(a)$ is a monotonically decreasing function, for example, a function defined as $f(a) = \mu/a$, if $0 < a < r_0$; $f(a) = 0$, otherwise, where μ and r_0 are positive constants.
- The neuron has only local connections in a small region $(0, r_0)$, which is called its receptive field $\mathcal{R}_{ij}$.
- Parameter m is the number of neurons located within $\mathcal{R}_{ij}$.

We may implement a topologically organized neural network which is expressed in a configuration space Q. The location of the ith neuron at the grid in Q, denoted by a vector $q_i \in \mathbb{R}^F$, where F is the degrees of freedom, uniquely represents a state of a robot in Q. Each neuron has local lateral connections to its neighboring neurons that constitute a subset in Q, which is called the receptive field of the ith neuron in neurophysiology. The dynamics of the ith neuron is characterized by the shunting equation.

For a given present configuration in Q, denoted by q_p, the next state q_n is obtained by

$$q_n \Leftarrow x_{q_n} = \max\{x_i, i = 1, 2, \ldots, k\}, \tag{4.10}$$

where i is the number of the neighboring neurons including itself, i.e., all the possible next locations. The present location adaptively changes according to the varying workspace. x_i represents the neural activity of the landscape at configuration q_i.

Algorithm 12: Artificial neural network algorithm

Require: Map Q, start cell q_{start}, goal cell q_{goal}
Ensure: Path P

```
q_n ← q_start
while q_n ≠ q_goal do
    for all configurations q_i in Q do
        x_i ← neural activity of the landscape at configuration q_i
        dx_i/dt ← −A x_i + (B − x_i)([I_i]^+ + Σ_{j=1}^m w_ij [x_j]^+) − (D + x_i)[I_i]^−
    end for
    x_i ← all neighbor cells of q_p
    n ← argmax_i {x_i, i = 1, 2, ..., k}
    Add q_n to P
end while
```

By properly defining the external inputs from the workspace and internal neural connections, the target and obstacles are guaranteed to stay at the peak and the valley of the activity landscape of the neural network, respectively. The target can globally attract the robot in the whole workspace through the neural activity propagation, while the obstacles have only local effect to avoid conflicts. The path planning is realized through the dynamic activity landscape of the neural network.

4.4 Case Study A: Multiple AUV Planning with Genetic Algorithms

In this first case study, multiple AUV paths are planned using the GA approach. Details of the GA formulation are presented as well as results showing that paths are generated for the AUVs that maintain a safe distance from the seabed while exploring new areas and ultimately reaching a goal location.

The possible paths for multiple AUVs can be represented by strings of control points that are used to generate Bezier curves as will be described in detail. Thus, the GA can be applied to obtain the optimal paths for all the AUVs to achieve the best solution for the path planning problem for multiple AUVs.

In order to find the optimal paths, some specific values should be found, which are the best among all the possible solutions. It is time-consuming to find a specific solution in a big search space. For the path planning problem of multiple AUVs, all the possible paths are solutions in the search space. All the possible paths are evaluated by the evaluation function corresponding to them. The best set of paths, which is one point among all the feasible values, is one point in the search space. Finding a set of paths is then equal to looking for some extreme value in the search space. Algorithm 11 is adapted to suit this particular problem leading to the following procedure:

Step 1: Define the population size.
Step 2: Initialize generation 0 of the population, encode multiple waypoints in each set of paths to form chromosomes, and initialize evaluation values.
Step 3: Calculate curves of each path by using chromosomes within one generation.
Step 3.1: Use the control points on each path to obtain the evaluation value.
Step 3.2: Evaluate
Step 3.2.1: If the evaluation value is good enough, go to step 5.
Step 3.2.2: If not all the individuals in the population are evaluated and the evaluation value is not good enough, go to step 3.
Step 3.2.3: If all the individuals in the population are evaluated and the evaluation value is not good enough, go to step 4.
Step 4: Reproduce
Step 4.1: Sort by the evaluation value.
Step 4.2: Select good individuals.
Step 4.3: Crossover.
Step 4.4: Mutate and obtain new offspring, go to step 3, and use the new generation.
Step 5: Store the optimal set of paths selected by the GA and calculate curves of each path.

4.4.1 Representation

The basis of GAs is that a population of solutions can form a set of chromosomes, which are strings encoding all the possible solutions. Strings can be binary or have many possible alternatives (genes) at each position. For this problem, the chromosome is designed as a binary string. Chromosomes are used to represent values of possible sets of paths. With a set of initial chromosomes (here we choose a population of 25 individuals), GA manipulates the most promising chromosomes in the set and creates the new population of improved chromosomes. The jth gene of the ith individual in the population of one generation is defined as $chromosome(i, j)$.

Each path consists a string of waypoints, while each path curve is represented by Bezier curves defined as [48]

$$x(t) = \Sigma_{i=0}^{m} C_m^i t^i (1-t)^{m-i} x_i, \tag{4.11}$$

$$y(t) = \Sigma_{i=0}^{m} C_m^i t^i (1-t)^{m-i} y_i, \tag{4.12}$$

$$z(t) = \Sigma_{i=0}^{m} C_m^i t^i (1-t)^{m-i} z_i, \tag{4.13}$$

$$C_m^i = \frac{m!}{i!(m-i)!}, \tag{4.14}$$

where t is the parameter of the curve whose values vary uniformly between [0, 1]. (x_i, y_i, z_i) are the coordinates of the control points which define the path coordinates $(x(t), y(t), z(t))$. m is the total number of control points in the chromosome. The first and the last control points are known since they are the starting point and the ending point. A small number of control points will consume less time.

4.4.2 Fitness Function

The chromosomes are evaluated according to an evaluation function. It is important to select a good evaluation method; otherwise, the best chromosomes can be lost in the noise.

The fitness function is defined based on the total distance traveled by the AUVs, the clearance, the turning angles, the areas covered by multiple AUVs, and the number of repetitive routes of multiple AUVs.

The fitness function is modeled as

$$f_k = \frac{1}{\Sigma_{j=1}^{5} a_j F_j}, \tag{4.15}$$

where f_k is the fitness function for one AUV, k is the AUV index number, F_1 is the total distance traveled by the AUVs, F_2 is the clearance, F_3 is the turning angles, F_4

is the area covered by multiple AUVs, and F_5 is the number of repetitive routes of multiple AUVs.

F_1 can be calculated as

$$F_1 = \Sigma_{i=1}^{m}[(x_{i+1} - x_i)^2 + (y_{i+1} - y_i)^2 + (z_{i+1} - z_i)^2]^{\frac{1}{2}}, \tag{4.16}$$

where m is the number of control points on the path and x, y, and z are coordinates on the curve of the planned path.

F_2 can be calculated as

$$F_2 = \Sigma_{i=1}^{m} p_i, \tag{4.17}$$

where p_i is the penalty value related to the clearance of the points on the curve.

$$p_i = \begin{cases} 1 & \text{if} z_{\text{curve}} - z_{\text{seabed}} < \text{limit}, \\ 0 & \text{otherwise}. \end{cases}$$

If a point on the curve z_{curve} is too close to the seabed z_{seabed}, the penalty value is 1.

F_3 can be calculated as

$$F_3 = \Sigma_{i=1}^{m} p_i, \tag{4.18}$$

where p_i is the penalty value related to the turning angle of the AUVs.

$$p_i = \begin{cases} 1 & \text{ifturningangle} > \text{limit}, \\ 0 & \text{otherwise}. \end{cases}$$

If the turning angle is too sharp, the penalty value is 1.

F_4 can be calculated as

$$F_4 = -\Sigma_{i=1}^{m} r_i, \tag{4.19}$$

where r_i is the reward value related to new locations visited by the AUVs.

$$r_i = \begin{cases} 1 & \text{ifnewlocation}, \\ 0 & \text{otherwise}. \end{cases}$$

The reason that we need the reward value is because that we need the AUVs to discover more unknown areas in the region.

F_5 can be calculated as

$$F_5 = \Sigma_{i=1}^{m} p_i, \tag{4.20}$$

where p_i is the penalty value related to the repetitive locations visited by AUVs.

$$p_i = \begin{cases} 1 & \text{iflocationsalreadyvisited}, \\ 0 & \text{otherwise}. \end{cases}$$

Finally, the total fitness function is defined as

$$f = \Sigma_{k=1}^{K} f_k, \tag{4.21}$$

where K is the total number of AUVs.

4.4.3 GA Operators

Following the fitness evaluation and selection, a new generation, a new population of chromosomes, is generated by applying a set of genetic operators to the original population. The basic genetic operations are selection, crossover, and mutation which are inspired by the biological evolution process.

4.4.3.1 Selection Mechanism

The selection mechanism is responsible for selecting the parent chromosome from the population and forming the mating pool. The selection mechanism emulates the survival of the fittest mechanism in nature. It is expected that a fitter chromosome receives a higher number of offsprings and thus has a higher chance of surviving on the subsequent evolution while the weaker chromosomes will eventually die. In this work we use the roulette wheel selection method [49] which is one of the most common selection mechanisms. Basically it works as follows: each chromosome in the population is associated with a sector on a virtual wheel. According to the fitness value of the chromosome, the sector will have a larger area when the corresponding chromosome has a better fitness value while a lower fitness value will lead to a smaller sector. In order to select the more promising individuals, the fitness value fitness(i) for each chromosome chromosome(i) $(i = 0, \ldots, \text{pop\_size} - 1)$ is calculated.

Then the total fitness of the population is given by

$$F = \sum_{i=0}^{\text{pop\_size}-1} \text{fitness}(i), \tag{4.22}$$

and the probability of a selection is given as

$$P_i = \frac{\text{fitness}(i)}{F}. \tag{4.23}$$

After that, the cumulative probability Q_i for each chromosome chromosome(i) $(i = 0, \ldots, \text{pop\_size} - 1)$ is calculated by

$$Q_i = \sum_{j=0}^{i} P_j. \tag{4.24}$$

The selection process is based on spinning the roulette wheel pop_size − 1 times. Each time we select a single chromosome for the new population in the following way:

Step 1: Generate a random number r within the range $[0,1]$.
Step 2: If $r < Q_1$, then select the first chromosome chromosome(0). Otherwise select the ith chromosome such that $Q_{i-1} \leq r < Q_i$ while $1 \leq i \leq \text{pop\_size} - 1$.

Obviously, some chromosomes will be selected more than once. The best chromosomes get more copies and the worst ones die off.

4.4.3.2 Crossover

The crossover process recombines genetic information of two parent chromosomes to produce offspring. Thus the offspring inherits some of the good characteristics from their parents. If two chromosomes are *chromosome* $1 = (1\ 1\ 1\ 1\ 1\ 1\ 1)$ and *chromosome* $2 = (0\ 0\ 0\ 0\ 0\ 0\ 0)$, one-point crossover at the fourth point produces two new chromosomes, *new chromosome* $1 = (1\ 1\ 1\ 0\ 0\ 0\ 0)$ and *new chromosome* $2 = (0\,0\,0\,1\,1\,1\,1)$.

4.4.3.3 Mutation

Finally, the mutation process selects a random position of a random string and negates the bit value. For instance, if mutation is applied to the eighth bit of string *new chromosome* 1, the transformed string becomes $(1\ 1\ 1\,0\,0\,0\,1)$.

The new population would contain solutions that are more promising than the previous generation. This process continues until an acceptable set of paths for multiple AUVs is found.

4.4.4 Parameter Settings

The decision on the population size and its initial values has a significant impact on the performance of the genetic algorithm. An extremely small population size could lead to local optima or early convergence while an extreme large population would exhaust the processing power and the memory. Appropriate initial values would guarantee the randomness in the population and result in a higher chance that the global optima are found. The following parameter settings are used: Population size is set to 25; probability of mutation is set to 15 %. In the simulation, we have seven control points in total, which includes the starting point and the ending point. For all the AUVs, they start and finish at the same location.

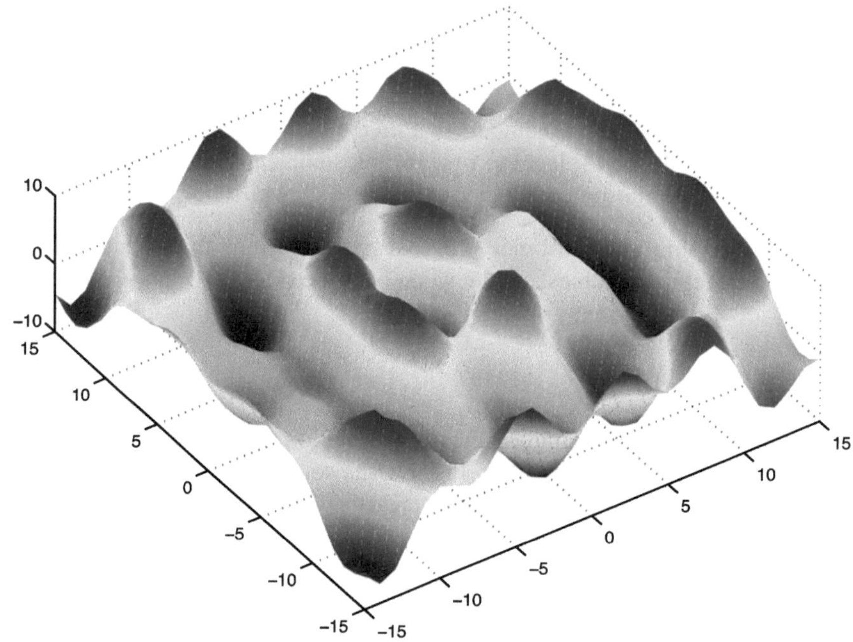

Fig. 4.14 The trigonometric function based seabed

4.4.5 Results

For simplicity without losing generality, the seabed is modeled as a trigonometric function-based surface, as shown in Fig. 4.14. The trigonometric function-based seabed has a continuous geometry in every required point due to continuous function descriptions, such as sine and cosine function. Figure 4.15 shows the paths of multiple AUVs generated using the proposed algorithm. The paths of four AUVs are optimized based on the five criteria we have proposed.

4.5 Case Study B : Sensor-Driven Online Path Planning

Sensor-driven methods have become more popular because they are capable of handling uncertainties in the environment. Most of the previously mentioned methods require that complete knowledge of the environment is available before planning the path. Even if good information is available, it is possible that this information can be incorrect or incomplete. In modern robots, maps are built simultaneously as worlds are explored. As a result, a robot should be able to modify an existing plan or make a new plan from scratch based on information obtained *in situ*. A popular way of handling this problem recently has been to use information

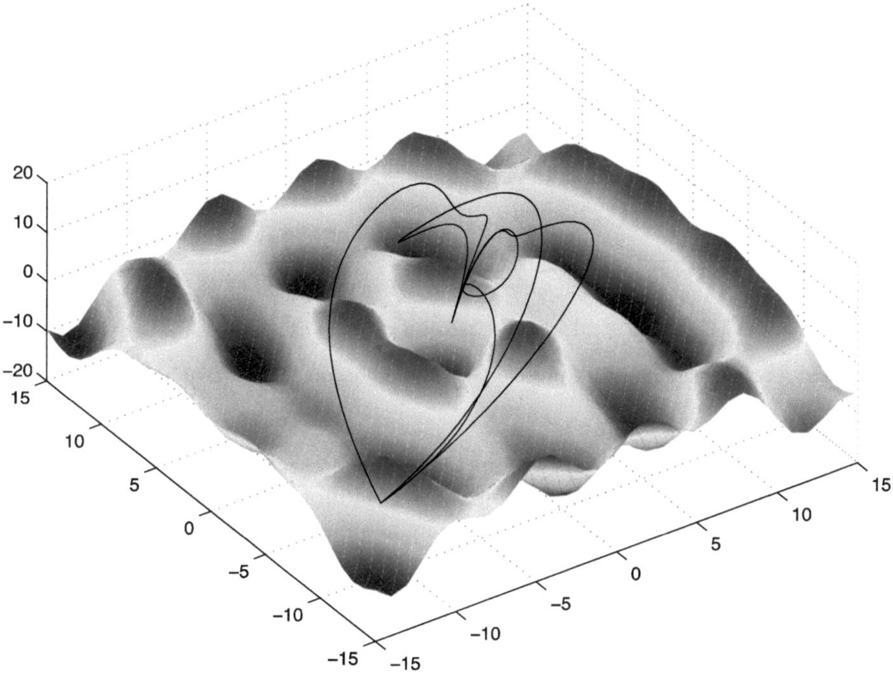

Fig. 4.15 The planned paths for multiple AUVs

theory to quantify the benefits of a finite set of proposed actions. The rest of this section will provide a specific example of sensor-driven path planning with information gain. We will begin with an overview of the information gain theory and how it is applied. The problem as stated is for the AUV to cover an area of seabed with mines notwithstanding the variable environment and the challenging underwater environment.

Define the workspace to be covered as a discretized grid of N cells: $W = \bigcup_{i=1}^{N} w_i$. The *confidence* at each grid location is the probability that if a mine is present, it will be correctly detected. Before the survey, the confidence at all cells is 0.5: $c^{w_i} = 0.5$, $i = 1..N$.

Define a track for AUV number k by the start and end 2D positions: $\mathcal{C}^k$: $(x_1^k, y_1^k, \psi_1^k) \rightarrow (x_2^k, y_2^k, \psi_2^k)$. In order for the track to be valid, we require that motion be roughly rectilinear. The traversal along this track will result in an area of seabed being covered with the sidescan sonar sensor (SSS). Define this area as $A \subset W$. Assume A is comprised of N_A of the grid cells in W and define those cells by a_j: $A = \bigcup_{j=1}^{N_A} a_j$ with $a_j \in W, \forall j = 1..N_A$. For those cells that were in the coverage area, their confidence is updated according to $c^{a_j} = \mathcal{P}(y)$ where y is the lateral range of

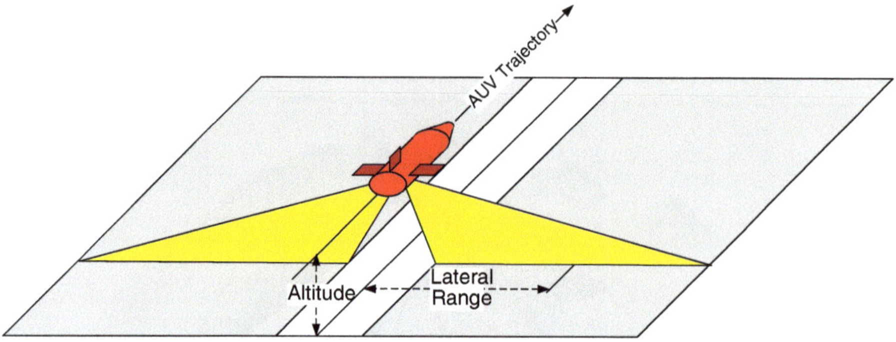

Fig. 4.16 An example of the AUV trajectory and corresponding area covered by its SSS

cell a_j from track $\mathcal{C}^k$ and $\mathcal{P}(\cdot)$ is a function that represents the sensor performance based on many factors such as environmental conditions.

Therefore, the objective is to define a way to generate the tracks to guarantee an average confidence greater than some desired threshold:

$$\frac{1}{N}\sum_{i=1}^{N} c^{w_i} > c_{\text{thresh}}. \tag{4.25}$$

4.5.1 Sidescan Sonar Sensor

Most underwater MCM missions are conducted with a side-looking sensor: either a synthetic aperture sonar (SAS) or a SSS. Historically, this task was executed using towed arrays from a vessel. However, the reduction in size of onboard sensors, the increased computational capacity of mirco-computers, and the advances in battery technology have allowed these missions to be executed by AUVs with onboard sonar arrays. In general this has been carried out by having the AUV follow a predefined structured survey pattern such as a "lawn mower" or a "zigzag" pattern [60].

The SSS uses the returns from emitted high-frequency sound to generate an image of the seabed. An object sitting on the seabed will cast a sonar shadow that can be analyzed to determine if the shape is suggestive of a mine. The onboard SSS gathers data as the AUV moves forward in rectilinear motion and leaves a narrow channel of unscanned seabed directly beneath it. An AUV path and corresponding SSS coverage swath are shown in Fig. 4.16. SSS returns are combined with onboard navigation data to provide geo-referenced mosaics of the seabed. A geo-referenced and mosaiced image from a Klein 5500 SSS is shown in Fig. 4.17. The target is highlighted. Often targets will be detected using automated systems [50], which have recently been implemented onboard AUVs operating in real time [51]. When the sonar makes sharp turns, areas on the outside of the turn are missed completely

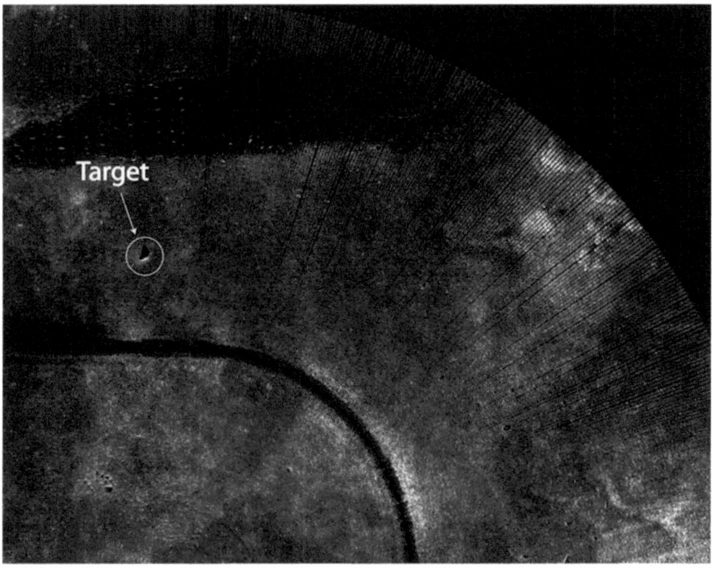

Fig. 4.17 Sample geo-referenced SSS data

due to the finite ping rate of the sonar and areas on the inside of the turn can become completely distorted. In both cases, it becomes very difficult for automatic target recognition (ATR) systems that rely on template matching to identify targets in these areas [52]. The angle of incidence of the sonar beam with the seabed has a significant effect on the resolution of the shadow cast by an object and therefore the probability of successful mine detection and classification. The extensible performance and evaluation suite for sonar (ESPRESSO) is a tool developed by the NATO underwater research centre (NURC) to evaluate the sonar performance characteristics for a specific set of environmental conditions [53]. The program generates a $\mathcal{P}(y)$ lateral range curve that indicates the probability that a target at a specified lateral range from a sonar's track will be detected. Parameter values that affect the generation of the $\mathcal{P}(y)$ curve include environmental factors such as seabed type, target properties, sonar attributes, and platform variables such as speed, depth, and navigation error. The performance curves follow the general shape of a Rayleigh distribution. Figure 4.18 shows the $\mathcal{P}(y)$ curves generated by ESPRESSO for three different seabed types, cobble, sand, and clay, all at a depth of 10 m for the Klein 5,500 sonar.

4.5.2 Information Gain

In the proposed method, information theory will be used to quantify utility.

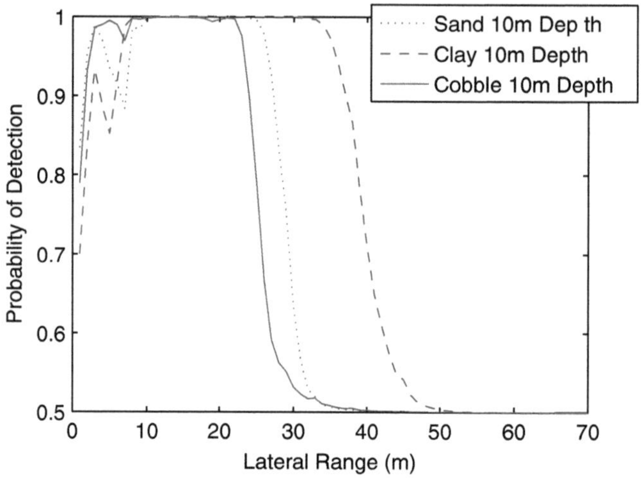

Fig. 4.18 $\mathcal{P}(y)$ curves for three different seabed conditions

The Shannon entropy of a discrete random variable (RV) X, with a probability density of $f_X(x)$, is defined as

$$H(X) = -E\{\log f_X(x)\} = -\int f_X(x)\log f_X(x)\mathrm{d}x \tag{4.26}$$

and represents a measure of the compactness of the distribution [54,55].

Conditional entropy can also be defined. Consider two RVs X and Z. The *conditional entropy* of X given Z is defined as

$$\begin{aligned} H(X|Z) &= -E\{\log P(X|Z)\} \\ &= -\int P(X|Z)\log P(X|Z)\mathrm{d}x. \end{aligned} \tag{4.27}$$

Now we can define the *mutual information* or *entropy reduction* as

$$I(X,Z) = H(X) - H(X|Z). \tag{4.28}$$

The mutual information defines a scalar quantity that represents the amount of information about X contained in Z or, put another way, how much the entropy of X will be reduced by the information provided by Z. Typically, X would represent some state vector and Z would represent some sensor measurement. *A priori*, we don't have direct access to the measurement because it hasn't been made yet; therefore we cannot evaluate the mutual information directly. As a result, it is useful to define the expected entropy reduction (EER), which represents the expected reduction in entropy that will come about by making measurement Z.

$$\bar{I}(X|Z) = H(X) - \bar{H}(X|Z). \tag{4.29}$$

To evaluate $\bar{H}(X|Z)$ we take the expectation over the measurement Z:

$$\begin{aligned}\bar{H}(X|Z) &= E_z\{H(X|Z)\} \\ &= -\int P(Z)\int P(X|Z)\log P(X|Z)\mathrm{d}x\mathrm{d}z.\end{aligned} \tag{4.30}$$

$P(Z)$ is the probability of obtaining measurement Z.

The essential aspect of this definition is that it specifies a way of combining sensor measurements *additively*. Consider some control action at time t to be $U(t)$. If the ratio of the control frequency to the sensor frequency is n then each control action, $U(t)$, will result in a set of n independent measurements $\{Z_1, Z_2, \ldots, Z_n\}$. The total expected *information gain* of $U(t)$ can be expressed as:

$$B(U(t)) = \sum_{k=1}^{n} \bar{I}(X|Z_k). \tag{4.31}$$

4.5.3 *The Information Gain Behavior*

To define the information gain objective function, information gained must be formulated as a function of desired heading ψ_d. This is achieved by defining a track starting at the AUVs current location, (x, y), and traveling a fixed distance, r, at every potential heading ψ_d. The measurements that will be made can be predicted and then (4.31) can be used to evaluate the expected information gained from traveling along the given track.

Define the variable $M_{ij} \in \{0,1\}$ to represent the actual presence or absence of a target at the point (i, j) in the discretized workspace, W. Then, consider the variable $\bar{M}_{ij} \in \{0,1\}$ to be our belief about the presence or absence of a mine at location (i, j). The confidence at location (i, j) represents the probability that if a mine exists, then it will be detected (refer to Fig. 4.18). Therefore, we can define a binary RV T_{ij} such that

$$\begin{aligned}P(T_{ij} = 1) &= P(\bar{M}_{ij} = M_{ij}) = c_{ij}, \\ P(T_{ij} = 0) &= P(\bar{M}_{ij} \neq M_{ij}) = 1 - c_{ij}.\end{aligned} \tag{4.32}$$

Then the entropy of T_{ij} can be represented as:

$$H(T_{ij}) = -c_{ij}\log(c_{ij}) - (1 - c_{ij})\log(1 - c_{ij}). \tag{4.33}$$

From (4.33) it follows that

$$\lim_{c_{ij} \to 1} H(T_{ij}) = 0. \tag{4.34}$$

This implies that maximizing the confidence over the environment minimizes the entropy of T_{ij}. As a result, the information gain objective function can be defined in terms of gaining information about T_{ij}.

The Probability of Mine Detection From the above formulation it is possible to derive the probability that a mine actually exists given that we detected one at location (i,j). If the number of mines in a given area A can be known or estimated as N_{mines}, then the probability that a mine exists at any given location, (i,j), can be approximated by

$$p_{\text{mine}}^{ij} \approx \frac{N_{\text{mines}}}{A}, \tag{4.35}$$

then, from Bayes' formula

$$P(M_{ij}=1|\bar{M}_{ij}=1) = \frac{c_{ij}p_{\text{mine}}^{ij}}{2c_{ij}p_{\text{mine}}^{ij} - c_{ij} - p_{\text{mine}}^{ij} + 1}, \tag{4.36}$$

defines the probability that a mine actually is present given that we think one is present. This equation can be used to validate the assumptions used in this chapter such as the validity of the sensor models and the target recognition systems.

Let the proposed path to be evaluated be represented by $\mathcal{C}$. The path begins at the AUV's current location (x,y) and moves a distance r at heading ψ_d:

$$\mathcal{C}(0) = (x,y), \tag{4.37}$$

$$\mathcal{C}(1) = (x + r\cos(\psi_d), y + r\sin(\psi_d)). \tag{4.38}$$

Let the proposed action, $U(t)$ from (4.30), be defined by the proposed track. Since r, x, and y are assumed constant, the information gain resulting from following the proposed track can be defined as a function of the desired heading, ψ_d. It should be noted that these tracks could not be followed in reality due to dynamic constraints of the robotics platform. However, this framework can be used to evaluate the expected benefit of potential desired headings. These desired headings are used as a reference input to an inner loop controller that defines the control plane values. The confidence over the environment is updated based on the actual heading, not the desired heading. As a result, the actual trajectory will be smoother with less variations in actual heading.

Assume that following this track will result in a series of n SSS returns that represent measurements $Z = \{Z_1, Z_2, \ldots, Z_n\}$. Then, (4.30) can be used to compute the expected entropy at location (i,j) in the workspace as a result of any individual measurement $Z_k, k = 1 \ldots n$ as [54]:

$$\begin{aligned} \bar{H}(T_{ij}|Z_k) &= E_{z_k}\{H(T_{ij}|Z_k)\} \\ &= -\sum_{Z_k} P(Z_k)[-c'_{ij}\log c'_{ij} - (1-c'_{ij})\log(1-c'_{ij})], \end{aligned} \tag{4.39}$$

where c_{ij} is the prior confidence at location (i,j) and c'_{ij} is the confidence at location (i,j) after measurement Z_k. c'_{ij} is generated by combining the current measurement with all of the previous measurements at that location. Past research has shown that it can be beneficial to observe potential targets from different aspect angles, as this will improve the probability of detection [56,57]. Past approaches to exploiting this fact have preplanned paths that will guarantee observation of an object of interest from optimal angles [58,59].

Evaluation of (4.39) requires knowledge of the distribution of Z_k. This distribution represents the probability of obtaining the set of environmental conditions that would produce a given lateral range curve $\mathcal{P}(y)_{ij}^{[Z_k]}$. If there is no knowledge of environmental conditions beforehand, this distribution can be initialized as uniform across all environment parameters. As the AUV traverses the workspace, unknown environmental conditions such as seabed type can be determined and the distributions can be updated based on the new knowledge. The updates to distributions are done based on the measurements of environmental conditions that can be measured *in situ* and will depend on the accuracy of the sensor measurements.

The distribution $P(T_{ij}|Z_k)$ is determined by the lateral characteristic curve given the environmental conditions $\mathcal{P}(y)_{ij}^{[Z_k]}$. The perpendicular distance of point (i,j) to the path $\mathcal{C}$ can be found using a simple orthogonal projection, and is the lateral range used to sample the curve $\mathcal{P}(y)_{ij}^{[Z_k]}$.

The new confidence from $\mathcal{P}(y)_{ij}^{[Z_k]}$ should be combined with the existing confidence at (i,j), c_{ij} to produce the new confidence at that location c'_{ij}.

Then the EER at location (i,j) caused by measurement Z_k then follows from (4.29) as

$$\bar{I}(T_{ij}|Z_k) = H(T_{ij}) - \bar{H}(T_{ij}|Z_k). \tag{4.40}$$

Define the line that is perpendicular to $\mathcal{C}$ and aligns with SSS reading Z_k as $\mathcal{C}^{\perp}$. The EER over the entire workspace, W, brought about by a measurement Z_k is then the sum of the EER along the line $\mathcal{C}^{\perp}$.

$$\bar{I}(W|Z_k) = \sum_{(i,j)\mathrm{on}\mathcal{C}^{\perp}} \bar{I}(T_{ij}|Z_k). \tag{4.41}$$

Given that there is no overlap between subsequent sonar pings from a SSS, the total expected information gain brought about by moving along the path $\mathcal{C}$ can be expressed as

$$B(\psi_d) = \sum_{k=1}^{n} \bar{I}(W|Z_k). \tag{4.42}$$

As the $\mathcal{P}(y)$ curve does not have a closed form representation, gradient-based optimizations are not possible. For this reason, IvP is suitable. The accuracy of the piecewise linear representation of the underlying function can be tuned by altering the desired number of linear pieces used to approximate the function.

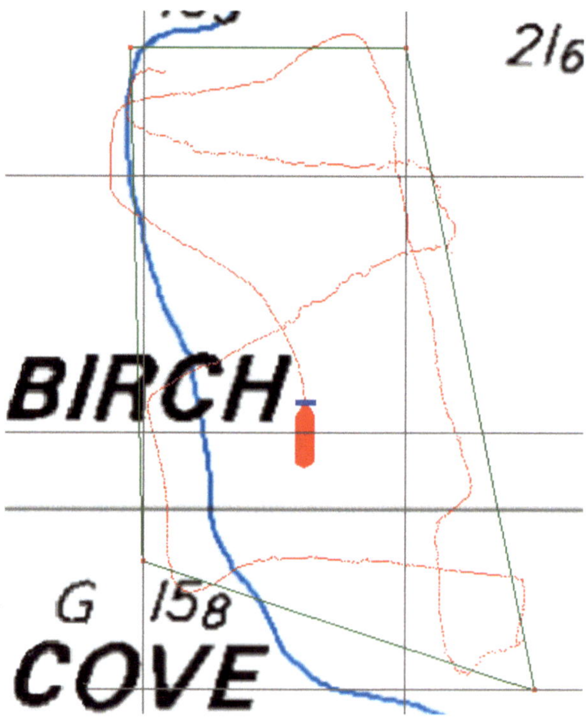

Fig. 4.19 The path to cover a simple environment with a SSS using developed planner in Bedford Basin, Nova Scotia, Canada. *Red line* represents the AUV path, *green line* represents the environment boundary. Other lines are bathymetry, etc.

It should also be noted that the seabed environment that is being sensed is assumed static. As a result, information is never lost in the prediction stage; information is only gained by sensing the environment.

4.5.4 In-Water Trials

Tests were performed on OceanServer's IVER2 AUV in Bedford Basin, Nova Scotia, Canada. The vehicle is equipped with an inertial navigation system (INS), a Doppler velocity log (DVL), a GPS, and a dual frequency 330/880 Hz Yellowfin sidescan sonar.

The AUV was able to successfully cover two environments within the limited operating region. A plot of a sample path taken in a simple convex environment and the corresponding final confidence map are shown in Figs. 4.19 and 4.20. The runs were stopped when confidence values reached 95%.

A comparison lawn mower mission was also performed on the simpler environment. The path is shown in Fig. 4.21.

When comparing the two paths from Figs. 4.19 and 4.21, it is interesting to note that, although the desired tracks for the lawn mower are straight lines, the actual path oscillates across these paths. This is largely due to the inability of the frontseat

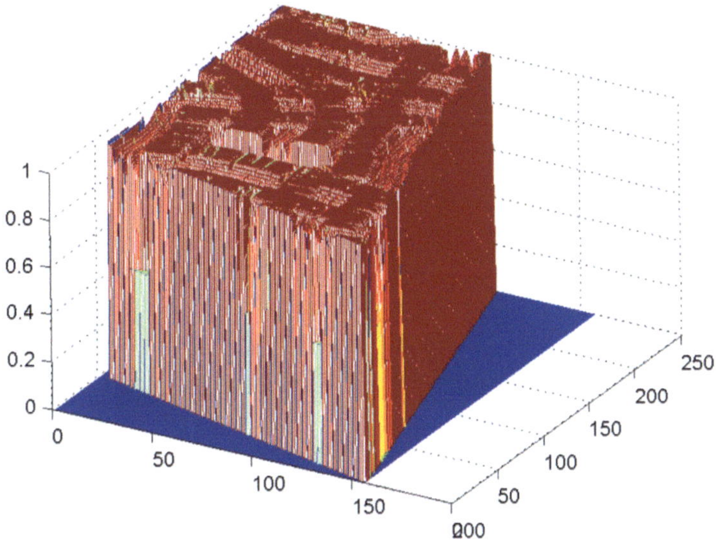

Fig. 4.20 The confidence map resulting from the path shown in Fig. 4.19. *Darker* areas represent areas of higher confidence

Fig. 4.21 A lawn mower path that achieves coverage of the simple environment

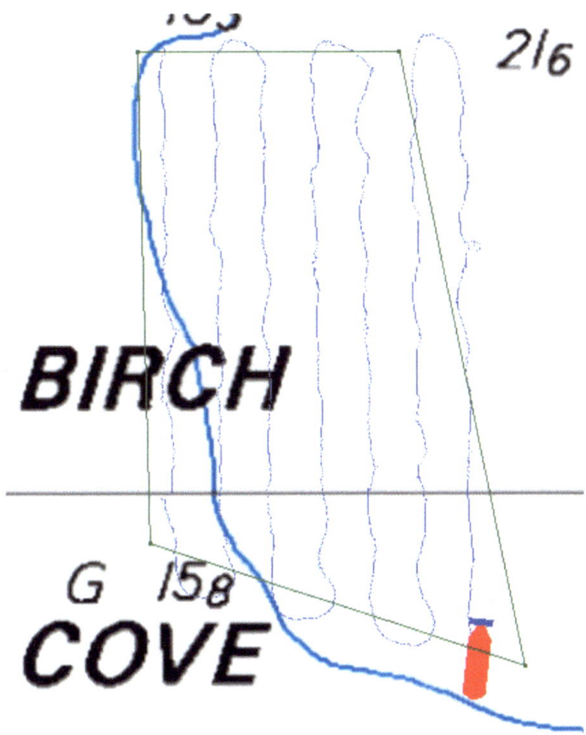

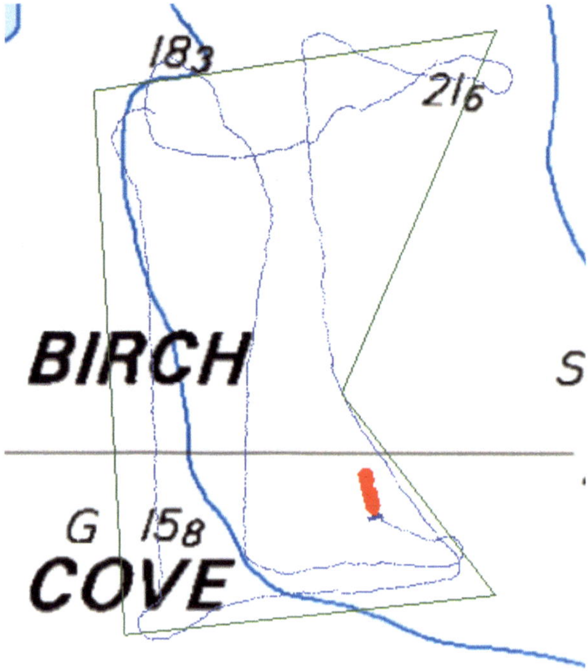

Fig. 4.22 The path to cover a more complex non-convex using developed planner in Bedford Basin, Nova Scotia, Canada

Table 4.1 Sample path lengths for paths planned during hardware trials

	Path length (m)	Workspace area (m^2)
Proposed planner (Fig. 4.19)	1,203	28,000
Lawn mower (Fig. 4.21)	1,580	
Proposed planner (Fig. 4.22)	1,661	41,250

Tests were stopped when an average confidence of 95 % was obtained

controller to stabilize the heading in the presence of currents. It should be noted that the currents on the day when this trial was conducted were extremely small, on the order of 0.2 knots at most. Because the proposed planner is designed to maintain headings, the outputted sonar data will be of higher quality and targets will be more easily identified.

A more complex non-convex environment test was also conducted with results shown in Fig. 4.22 and the corresponding confidence map shown in Fig. 4.23.

The path lengths for each of the three trials are shown in Table 4.1.

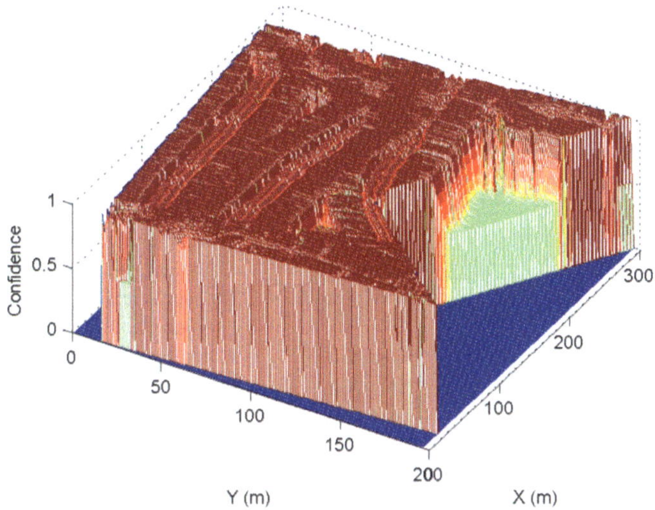

Fig. 4.23 The confidence map resulting from the path shown in Fig. 4.22. *Darker* areas represent areas of higher confidence

4.6 Summary

In this chapter, we have introduced AUV path planning concepts, given detailed algorithms for many of the more popular approaches, and considered two in-depth case studies on real AUV path planning methods.

The simplest methods are termed "Bug" algorithms and are able to find the goal by following basic rules. Other more complex methods such as wavefront and potential fields define a potential over the free space and then use a gradient descent approach to move from the start to the goal. The concept of a roadmap is also defined and some examples are provided. The roadmap represents a searchable construction that has provable guarantees about how the free space is covered. In the case that the configuration space is too high, it is often more realistic to sample the free space in some random way to generate a search space. This is referred to as a PRM. The free space can also be subdivided into regions that are termed cells. In the case of a cell decomposition or a roadmap, the free space representation is converted to a graph with nodes and edges that can be searched using any number of search methods. Two are provided: A* and BFS. Meta-heuristic methods such as artificial neural networks and genetic algorithms can also be used to find acceptable solutions in the case when the search space becomes extremely large.

It should be noted that underwater path planning is often less concerned with obstacle avoidance and more concerned with target observation or data acquisition. Nevertheless, the basic path planning concepts presented here can be easily adapted to suite any mission objective.

Table 4.2 Summary of path planning approaches

Method	Optimality	Completeness	Reactive / deliberative
Bug1	Nonoptimal	Complete	Reactive
Bug2	Nonoptimal	Not complete	Reactive
Wavefront	Optimal	Complete	Deliberative
Potential fields	Nonoptimal	Not complete	Deliberative
A* Search	Optimal	Complete	Deliberative
Breadth-first search	Optimal	Complete	Deliberative
Probability roadmaps	Nonoptimal	Probabilistically complete	Deliberative
Generalized Voronoi diagram	Nonoptimal	Complete	Deliberative
Visibility graph	Nonoptimal	Complete	Deliberative
Genetic algorithms	Optimal	Not complete	Deliberative
Neural networks	Nonoptimal	Not complete	Reactive
Cell decomposition	Nonoptimal	Resolution complete	Deliberative
Sensor-driven	Nonoptimal	Probabilistically complete	Deliberative

To summarize, Table 4.2 categorizes all of the presented algorithms based on the terms defined.

Acknowledgements This work was supported by Defence Research and Development Canada and Natural Sciences and Engineering Research Council (NSERC) of Canada.

References

1. LaValle S (2011) Motion planning. IEEE Robot Autom Mag 18(1):79–89
2. Gasparri A, B.K., Sukhatme G (2008) A framework for multi-robot node coverage in sensor networks. Ann Math Artif Intell 52:281–305
3. Gonzalez E, Gerlein E (2009) Bsa-cm: a multi-robot coverage algorithm. In: Web intelligence and intelligent agent technologies, 2009. WI-IAT '09. IEEE/WIC/ACM international joint conferences, vol 2, pp 383–386, sept 2009
4. Choset H (2001) Coverage for robotics - a survey of recent results. Ann Math Artif Intell 31:113–126
5. Acar EU, Choset H (2002) Sensor-based coverage of unknown environments: incremental construction of morse decompositions. Int J Robot Res 21:345–367
6. Ron Wein JvdB, Halperin D (2008) Planning high-quality paths and corridors amidst obstacles. Int J Robot Res 27:1213–1231
7. Cai C, Ferrari S (2009) Information-driven sensor path planning by approximate cell decomposition. IEEE Trans Syst, Man, and Cybernetics - Part B: Cybernetics 39(3):672–689
8. Baek S, Lee TK, Se-Young OH, Ju K (2011) Integrated on-line localization, mapping and coverage algorithm of unknown environments for robotic vacuum cleaners based on minimal sensing. Adv Robot 25(13–14):1651–1673
9. Li Y, Chen H, Er MJ, Wang XM (2011) Coverage path planning for uavs based on enhanced exact cellular decomposition method. Mechatronics 21:876–885

10. Jin J, Tang L (2011) Coverage path planning on three-dimensional terrain for arable farming. J Field Robot 28(3):424–440
11. Oksanen T, Visala A (2009) Coverage path planning algorithms for agricultural field machines. J Field Robot 26(8):651–668
12. Warren C (1990) A technique for autonomous underwater vehicle route planning. IEEE J Oceanic Eng 15(3):199–204
13. Biggs J, Holderbaum W (2009) Optimal kinematic control of an autonomous underwater vehicle. IEEE Trans Automatic Contr 54(7):1623–1626
14. Yilmaz N, Evangelinos C, Lermusiaux P, Patrikalakis N (2008) Path planning of autonomous underwater vehicles for adaptive sampling using mixed integer linear programming. IEEE J Oceanic Eng 33(4):522–537
15. Binney J, Krause A, Sukhatme G (2010) Informative path planning for an autonomous underwater vehicle. In: IEEE international conference on robotics and automation (ICRA), May 2010, pp 4791–4796
16. Cheng CT, Fallahi K, Leung H, Tse C (2010) An auvs path planner using genetic algorithms with a deterministic crossover operator. In: IEEE international conference on robotics and automation (ICRA), May 2010, pp 2995–3000
17. Petres C, Pailhas Y, Patron P, Petillot Y, Evans J, Lane D (2007) Path planning for autonomous underwater vehicles. IEEE Trans Robot 23(2):331–341
18. Petillot Y, Tena Ruiz I, Lane D (2001) Underwater vehicle obstacle avoidance and path planning using a multi-beam forward looking sonar. IEEE J Oceanic Eng 26(2):240–251
19. Blanco M, Wilson P (2010) Autonomous underwater vehicle minimum-time navigation in a current field. In: OCEANS 2010, pp 1–4, Sept 2010
20. Kruger D, Stolkin R, Blum A, Briganti J (2007) Optimal auv path planning for extended missions in complex, fast-flowing estuarine environments. In: IEEE international conference on robotics and automation 2007, pp 4265–4270, April 2007
21. Smith RN, Chao Y, Li PP, Caron DA, Jones BH, Sukhatme GS (2010) Planning and implementing trajectories for autonomous underwater vehicles to track evolving ocean processes based on predictions from a regional ocean model. Int J Robot Res 29(12):1475–1497
22. Soulignac M (2011) Feasible and optimal path planning in strong current fields. IEEE Trans Robot 27(1):89–98
23. Alvarez A, Caiti A, Onken R (2004) Evolutionary path planning for autonomous underwater vehicles in a variable ocean. IEEE J Oceanic Eng 29(2):418–429
24. Thompson D, Chien S, Chao Y, Li P, Cahill B, Levin J, Schofield O, Balasuriya A, Petillo S, Arrott M, Meisinger M (2010) Spatiotemporal path planning in strong, dynamic, uncertain currents. In: IEEE international conference on robotics and automation (ICRA) 2010, pp 4778–4783, May 2010
25. Stack J, Smith C (2003) Combining random and data-driven coverage planning for underwater mine detection. In: Proceedings of OCEANS 2003, vol 5, pp 2463–2468, Sept 2003
26. Fang C, Anstee S (2010) Coverage path planning for harbour seabed surveys using an autonomous underwater vehicle. In: OCEANS 2010 IEEE - Sydney, pp 1–8, May 2010
27. Choset H (2000) Coverage of known spaces: the boustrophedon cellular decomposition. Autonomous Robots 9:247–253
28. Kim A, Eustice R (2009) Toward auv survey design for optimal coverage and localization using the cramer rao lower bound. In: OCEANS 2009, pp 1–7, Oct 2009
29. Williams D (2010) On optimal auv track-spacing for underwater mine detection. In: IEEE international conference on robotics and automation (ICRA) 2010, pp 4755–4762, May 2010
30. Paull L, Saeedi S, Li H, Myers V (2010) An information gain based adaptive path planning method for an autonomous underwater vehicle using sidescan sonar. In: IEEE conference on automation science and engineering (CASE), pp 835–840, 2010
31. Lumelsky V, Stepanov A (1987) Path planning strategies for point mobile automaton moving amidst unknown obstacles of arbitrary shape. Algorithmica 2:403–430
32. I Kamon ER, Rivlin E (1998) Tangentbug: A range-sensor based navigation algorithm. Int J Robot Res 17(9):934–953

33. Choset H, Burgard W, Hutchinson S, Kantor G, Kavraki LE, Lynch K, Thrun S (2005) Principles of robot motion: theory, algorithms, and implementation. MIT Press, Cambridge
34. Khatib O (1986) Real-time obstacle avoidance for manipulators and mobile robots. Int J Robot Res 5:90–98
35. Koditschek DE, Rimon E (1990) Robot navigation functions on manifolds with boundary. Adv Appl Math 11:412–442
36. Barraquand J, Langlois B, Latombe C (1992) Numerical potential field techniques for robot path planning. IEEE Trans Man and Cybernetics 22:224–241
37. Latombe J (1991) Robot motion planning. Kluwer Academic Publishers, Norwell
38. Lozano-Perez T, Wesley M (1979) An algorithm for planning collision-free paths among polyhedral obstacles. Commun ACM 22(10):560–570
39. Aurenhammer F (1991) Voronoi diagrams a survey of a fundamental geometric structure. ACM Comput Surveys 23:345–405
40. Kavraki LE, Svestka P, Latombe J-C, Overmars MH (1996) Probabilistic roadmaps for path planning in high-dimensional configuration spaces. IEEE Trans Robot Autom 12(4):566–580
41. Kuffner JJ, LaValle SM (2000) Rrt-connect: an efficient approach to single-query path planning. In: IEEE international conference on robotics and automation, pp 995–1001, 2000
42. Choset H (2000) Coverage of known spaces: the boustrophedon cellular decomposition. Autonomous Robots 9:247–253
43. Acar EU, Choset H, Rizzi AA, Atkar PN, Hull D (2002) Morse decompositions for coverage tasks. Int J Robotic Res 21(4):331–344
44. Szymczyk M, LaMothe A (2002) Mac game programming. Muska & Lipman/Premier-Trade
45. Srinivas S, Patnaik LM (1994) Genetic algorithms: a survey. IEEE Comput 27(6):17–26
46. Holland J (1945) Adaptation in natural and artificial systems. University of Michigan Press, Ann Arbor
47. Li H, Yang S, Seto M (2009) Neural network based path planning for a multi-robot system with moving obstacles. IEEE Trans Syst, Man and Cybernetics, Part C - Appl Rev 39
48. Pehlivanoglu YV (2007) Path planning for autonomous uav via vibrational genetic algorithm. Aircraft Eng Aerospace Tech: Int J 79(8):532–539
49. Davis L (1991) Handbook on genetic algorithms. Van Nostrand Reinhold, New York
50. Zerr B, Fawcett J, Hopkin D (2009) Adaptive algorithm for sea mine classification. In: 3rd international conference & exhibition on underwater acoustic measurements: technologies & results, pp 319–326, 2009
51. Williams D, Groen J, Fox W (2011) A fast detection algorithm for autonomous mine countermeasures. In: NATO underwater research center, Nurc-fr-2011-006, Oct 2011
52. Chapple P (2002) Automated detection and classification in high-resolution sonar imagery for autonomous underwater vehicle operations. Dsto-gd-0537, Maritime operations division DSTO defence science and technology organisation, Dec 2002
53. Davies G, Signell E (2006) Espresso scientific user guide. In: NATO underwater research centre, Nurc-sp-2006-003, 2006
54. Grocholsky B (2002) Information-theoretic control of multiple sensor platforms. Ph.D. thesis, Australian Centre for Field Robotics, University of Sydney
55. Papoulis A, Pillai SU (2002) Probability, random variables and stochastic processes, 4th edn. McGraw Hill, New York
56. Fawcett J, Myers V, Hopkin D, Crawford A, Couillard M, Zerr B (2010) Multiaspect classification of sidescan sonar images: Four different approaches to fusing single-aspect information. IEEE J Oceanic Eng 35(4):863–876
57. Fawcett JA, Crawford A, Hopkin D, Myers V, Couillard M, Zerr B (2008) Multi-aspect computer-aided classification of the citadel trial side-scan sonar images. Drdc atlantic tm 2008-029, Defence R&D Canada - Atlantic, 2008
58. Bourque FA, Nguyen B (2011) Optimal sensor configurations for rectangular target detection. In: 9th IEEE international conference on control and automation (ICCA) 2011, pp 455–460, Dec 2011

59. Bays M, Shende A, Stilwell D, Redfield S (2011) A solution to the multiple aspect coverage problem. In: IEEE international conference on robotics and automation (ICRA) 2011, pp 1531–1537, May 2011
60. Nguyen B, Hopkin D, Yip H (2008) Autonomous underwater vehicles a transformation of mine counter-measure operations. Defense and Security Analysis 243:247–266

Chapter 5
An Ontology-Based Approach to Fault Tolerant Mission Execution for Autonomous Platforms

David Lane, Keith Brown, Yvan Petillot, Emilio Miguelanez, and Pedro Patron

5.1 Introduction

5.1.1 Background

Autonomous underwater vehicles (AUVs) have become a standard tool for data gathering and intervention in security, offshore and marine science applications. In these environments, mission effectiveness directly depends on vehicle's operability. Operability underlies the vehicle's ultimate availability. Two main vehicle characteristics can improve operability: reliability depends on the internal hardware components of the vehicle and survivability, a concept that is closely related with vehicle failures due to external factors or damages.

In recent years, efforts to increase the AUV's ability to keep operating have focused on reducing the susceptibility and vulnerability of the platform. For example, recent approaches in rules of collision [1] and wave propagation techniques for obstacle avoidance and escape scenarios [2] have focused on reducing susceptibility by adapting the vehicle's trajectory plan.

However, when active and passive measures fail to protect the vehicle, or unexpected hardware failures occur, the focus of the mission should shift to "reconfigure" itself to use alternative combinations of the remaining resources. The underwater domain has scarce communication bandwidth and tight response constraints to keep the operator in the loop. In such a challenging environment,

D. Lane (✉) • K. Brown • Y. Petillot
Ocean Systems Laboratory, School of Engineering and Physical Sciences,
Heriot-Watt University, Edinburgh EH4 3QS, Scotland, UK
e-mail: d.m.lane@hw.ac.uk

E. Miguelanez • P. Patron
SeeByte Ltd, Edinburgh, Scotland, UK

M.L. Seto (ed.), *Marine Robot Autonomy*, DOI 10.1007/978-1-4614-5659-9_5,

autonomous embedded recoverability is a key capability to maintain the vehicle's endurance. This can be achieved via adaptation of the vehicle's mission plan.

5.1.2 Mission Planning for Autonomous Systems

Current AUV mission plan solutions are procedural and static [3]. If behaviours are added [4], they are only to cope with possible changes that are known a priori by the operator. In order to achieve higher levels of autonomy, an evolution in plan adaptability from current waypoint-based approaches on trajectory planning to a declarative (i.e. explicitly stated) goal-based solution for adaptive mission planning is required.

Declarative mission planning for unmanned vehicles leverages from the research carried out on autonomous planning generation [5]. However, almost as important as the mission plan generation (if not more) is the capability of the mission to adapt and recover from failures. The aim is to be effective and efficient and a mission plan costs time to prepare. This time has been already invested once (to compute the mission that is now failing), so it might be wise to try to reuse previous effort by repairing the plan (i.e. changing small sections that can address the failure) rather than by regenerating it. Also, commitments might have been made to the current mission plan: trajectory reported to other intelligent agents, assignment of resources, and assignment of parts of a mission plan to executors. Repairing an existing mission ensures that as few commitments as possible are invalidated. Finally, often several planners (usually autonomous and human planners combined) could be working together to achieve the goals. In such cases, it is more likely that a similar mission plan will be accepted by the operator to one that is potentially completely different.

With the growing use of autonomous and semi-autonomous platforms and the increase data flows in modern maritime operations, it is critical that the data is handled efficiently across multiple platforms and domains. At present, the knowledge representations used are embryonic. Targets, for example, are obtained from single platforms only, which limits the potential of multiple coordinated actions between agents. Consequently, the main application for AUVs is information gathering from sensor data. In a standard mission flow, data is collected during a mission and then post-processed offline.

However, as decision-making technologies evolve towards providing higher levels of autonomy, embedded service-oriented agents (SOA) require access to higher levels of data representation. These higher levels of information will be required to provide knowledge representation for contextual awareness, temporal awareness and behavioural awareness.

Two sources can provide this type of information: the *domain* knowledge extracted from an expert a priori or the inferred knowledge derived from processed sensor data. In both cases, it will be necessary for the information to be stored, accessed and shared efficiently by the deliberative agents while performing a mission. These agents, providing different capabilities, might be distributed among the different platforms working in collaboration.

Space operations, and Mars rovers in particular, have driven the motivation behind autonomous planning and adaptability for unmanned vehicles. The remote agent experiment (RAX) [6], which flew on Deep Space 1, demonstrated a three-tier architecture and its ability to autonomously control an active spacecraft. The architecture comprised a planner/scheduler (PS), a smart executive (EXEC) robust plan-execution system, and the mode identification and recovery (MIR) system for model-based fault diagnosis. One problem discovered involved using a batch (i.e. offline) planner on a reactive system. It took hours to re-plan after updates of events through sensors had invalidated the original plan.

Another system, CLEaR [7], used the CASPER system [8] in conjunction with a sequencer/controller in order to provide a planning and execution framework for closed-loop commanding. The deliberative and reactive methods operated in parallel at run time to determine how to best respond to failures and take advantage of opportunities. The system highlighted the difference between "local" conflicts, errors that require changes only to the currently executing part of the plan, and"global" conflicts, errors that occur which require changes to future parts of the plan. Local conflicts were managed by the executive handling events at run time while global ones were passed back to the planner where deliberation was used to change future parts of the plan. Van der Krogt later formalised this two levels of handling events in what was called executive repair and planning repair [9]. His approach combined unrefinement (removing obsolete and obstructing actions) and refinement (adding more detail) stages in order to provide faster performance than planning from scratch. However, it failed to produce the most optimal plan. This could be considered an issue in domains requiring optimality. This is not generally the case in unmanned vehicle mission plans where optimality can be sacrificed for operability. The approach was compared with the GPG [10] and the Sherpa systems. In the GPG, based on the graphplan planner, the unrefinement stage is done only on the initial plan and never on any of the plans produced by a refinement step. The Sherpa [11], based on the LPA_* algorithm used in [2] for adaptive trajectory planning, could only be applied to problems in which actions have been removed from the domain description.

Another system handling a certain level of planning adaptation in a real scenario was the automated schedule and planning environment (ASPEN) [12]. This system classified constraint violations and attempted to repair each by performing a planning operation. However, the set of repair methods was predefined and the type of conflict determined the set of possible repair methods for any given conflict. The system could not guarantee that it would explore all possible combinations of plan modifications or that it would not retry unhelpful modifications.

Recently, Fox proposed an on-board planning assistant to the operator to adjust and repair plans on board [13]. The system was developed to handle idle times due to conservative mission planning and plan failures on the unfortunate Beagle 2 Mars spacecraft. The approach relaxes methodological constraints and fills opportunity gaps only in situations where resources would otherwise go unused.

A wider discussion on other related research that goes from strong executors and formal planning approaches to strong deliberators can be found in Knight's review of the field [14].

In the underwater domain, several challenges have been identified requiring adaptive planning solutions of the mission [15, 16]. The potential benefits of the adaptive planning capabilities have been promoted by Rajan [17]. Together with Fox, they have started using sensor data to adapt the decision making on the vehicle's mission [18]. Throughout, probabilistic approaches have been increasingly evident to represent the uncertainty in plan representation and decision-making. Saigol [19] has demonstrated this for plume detection and tracking and compared it to bio-inspired rheotaxis approaches. To address the computational complexities that can rapidly ensue, Fox [20] has developed a policy learning approach that moves the burden of computation pre-mission, simplifying and speeding the real-time flight code.

In this chapter, an approach for fault tolerant adaptive mission planning for AUVs is presented. It uses the potential of a semantic knowledge representation, or ontology model, that provides a set of logical relationships describing the platform, the local environment and abstract concepts about them. Fault prognosis can be obtained using these in order to observe changes in the platform capabilities and orient the decision-making process [21].

5.1.3 Focus of Chapter

This chapter focuses on the approach of a semantic world model framework for hierarchical distributed representation of knowledge in autonomous underwater systems to support fault tolerance for underwater autonomy. The framework uses a pool of hierarchical ontologies for representation of the knowledge extracted from the expert and the processed sensor data. Its major advantage is that SOA can gain access to the different levels of information and might also contribute to the enrichment of the knowledge. If the required information is unavailable, the framework provides the facility for requesting that information be generated by other agents with the necessary capabilities.

At the decision level, an autonomous planner generates mission plans based on the representation of the mission goals in this semantic framework. As a consequence, we assert that the proposed framework enhances interoperability, independence and situation awareness of the embedded service-oriented agents running on the autonomous platforms. The results obtained specifically impact on mission flexibility, robustness and autonomy.

To the best of our knowledge, this is the first time that an approach to goal-based planning using semantic representation is applied to the adaptation of an underwater mission in order to maintain the operability of the platform. Furthermore, we have demonstrated on a set of sea trials the advantages of this approach in a real scenario.

5.2 Situation Awareness for Autonomy

The human capability of dealing and understanding highly dynamic and complex environments is known as situation awareness (SA_H). SA_H breaks down into three separate levels: perception of the environment, comprehension of the situation and projection of the future status.

According to Boyd, decision-making occurs in a cycle of observe-orient-decide-act, known as OODA loop [22]. The Observation component corresponds to the perception level of SA_H. The Orientation component contains the previously acquired knowledge and understanding of the situation. The Decision component represents the SA_H levels of comprehension and projection. This last stage is the central mechanism enabling adaptation before closing the loop with the final Action stage. Note that it is possible to make decisions by looking only at orientation inputs without making any use of observations.

Based on the autonomy levels and environmental characteristics, SA_H definitions can be directly applied to the notion of unmanned vehicle situation awareness SA_V [23]. The levels of situation awareness for individual unmanned vehicle systems (SA_S) span from full human control to fully autonomous unmanned capabilities (see Fig. 5.1).

In current implementations, the human operator constitutes the decision-making phase. When high-bandwidth communication links exist, the operator remains in the loop during the mission execution. Examples of the implementation of this architecture exist in tethered remotely operated underwater vehicles (ROVs). However, when the communication is poor, unreliable or not allowed, the operator tries, based only on the initial orientation or expertise, to include all possible behaviours to cope with execution alternatives. As a consequence, unexpected and unpredictable situations can cause the mission to abort and might even cause the loss of the vehicle, as happened with Autosub2 under the Fimbul ice sheet in the Antarctic [24]. Current implementations for AUVs are examples of this approach.

In order to achieve an autonomous decision-making loop, two additional components are required: a status monitor and a mission plan adapter. The status monitor reports any changes detected during the execution of a plan. These modifications might change the SA_V perception. When the mission executive is unable to handle the changes detected by the status monitor, the mission planner is called to generate

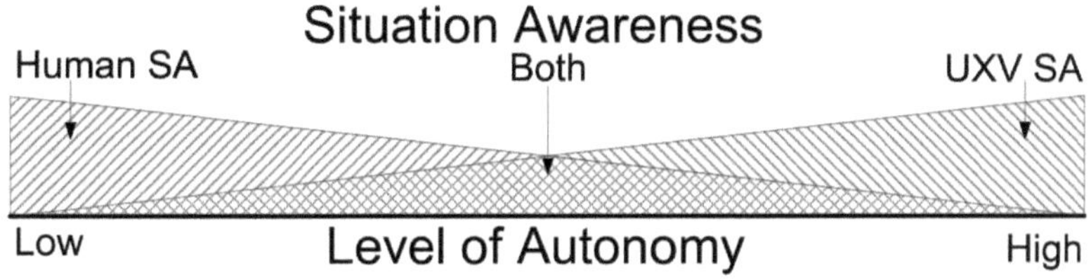

Fig. 5.1 Human and AUV situation awareness levels across the levels of autonomy

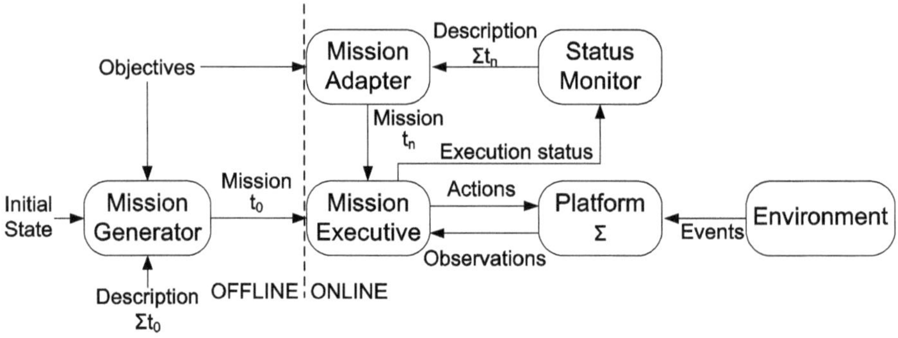

Fig. 5.2 Required OODA loop for $SA_S = SA_V$. Decision stage for adaptation takes place on board based on observations from the status monitor

a new modified mission plan that agrees with the updated knowledge of the world (see Fig. 5.2). The RAX architecture was the first attempt of implementing this type of architecture on a real system [6]. However, tight time deadlines, more restricted communications and different environmental constraints existing in general AUV applications have led us to the research of a new implementation of this approach based on ontological knowledge-based situation awareness.

5.3 Related Work

5.3.1 Knowledge Representation and Middleware for Sharing

Current knowledge representation approaches are only able to provide the perception or observation level of SA_V. State-of-the-art embedded agents make use of different message transfer protocols in order to stay updated with the current status of the platform and the environment. Several approaches can be found in the literature implementing information transfer protocols for robotics.

For example, robotic libraries such as Player [25] and Yet Another Robotic Platform (YARP) [26] support transmission of data across various protocols—TCP, UDP, MCAST (multi-cast) and shared memory. They provide an interface to the robot's sensors and actuators. This allows agents to read data from sensors, to write commands to actuators, and to configure devices on the fly. These two approaches separate the agents from the details of the network technology used, but they do not provide any standardisation about the meaning of the information transferred.

The mission oriented operating suite (MOOS) [27, 28] uses human-readable ASCII messages for communication of data to a centralised database. This centralised topology is vulnerable to "bottle-necking" at the server as the ASCII messages can generate considerable parsing overheads.

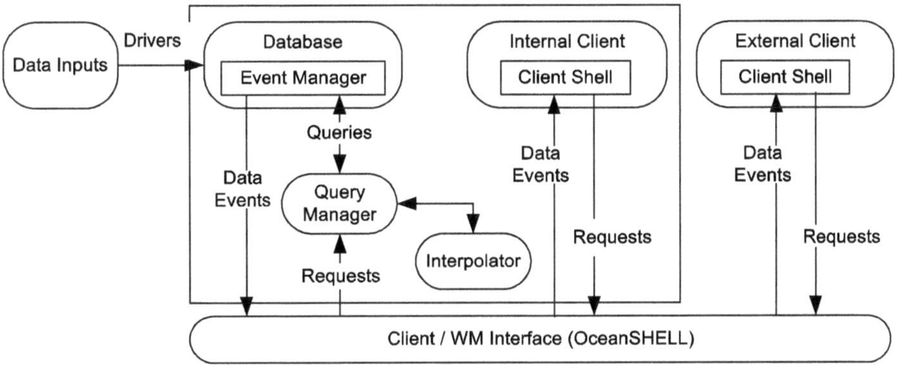

Fig. 5.3 World model architecture for the BAUUV MoD program (Battlespace Access for Unmanned Underwater Systems). Its database handles data inputs and queries from internal and external clients

The OceanSHELL libraries [29] implement UDP (user datagram protocol) broadcast protocol to pass data information between agents. In the latest version, abstraction messages are able to carry semantic information. These libraries are the first attempt to standardise semantic information for unmanned underwater platforms in order to share knowledge between different multidisciplinary agents. However, the approach is still limited to the observation or perception level of SA_V.

More recently, the robot operating system (ROS) [21] is gaining momentum and acceptance. It provides standard operating system services such as hardware abstraction, low-level device control, implementation of commonly used functionality, message passing between processes and package management. Processing takes place in nodes that may receive, post and multiplex sensor, control, state, planning, actuator and other messages. These nodes are referred to as a graph architecture.

A more detailed review of these and other robotic knowledge representation approaches was published by [30].

As an extension of these libraries, the battlespace access for unmanned underwater vehicles (BAUUV) program, sponsored by the UK Ministry of Defence Directorate of Equipment Capability (Underwater Battlespace), showed the benefit of also including the mental or orientation model component of SA_V [31]. A diagram describing the BAUUV dynamic multilayered world model architecture is shown in Fig. 5.3. The approach was capable of integrating the capabilities of different software components and provides them with a common picture of the environment. This picture contained processed sensor data and a priori knowledge. Thus it provided a full SA_V. However, the design was limited to autonomous target recognition (ATR) applications in the underwater domain. In order to provide a truly service-oriented architecture, an evolution from this approach is required providing a generic framework independent of service capability and domain of application.

The joint architecture for unmanned systems (JAUS), originally developed for the unmanned ground vehicles domain only, has recently extended to all domains trying to provide a common set of architecture elements and concepts [32]. In order to handle the orientation and observation phases, it classifies four different sets of knowledge stores: status, world map, library and log.

Our experience has shown that overlap exists between these different sets of knowledge stores. This overlap is provided by the strong interconnection existing between the orientation and observation phases for SA_V. The approach proposed in this project will make use of the concepts proposed by JAUS but enhances the knowledge store set by providing higher flexibility in the way the information can be handled and accessed.

Looking at related work on applying ontologies, recent publications [33–35] show that there is a growing inclination to use semantic knowledge in several areas of robotic systems, playing a key role in order to improve the interoperability of the different robotic agents. From mapping and localisation using semantically meaningful structures to human-robot interaction trying to make the robot understand the human environment of words, gestures and expressions, it has been clear that there are many ways in which the technology behind ontologies may be applied in robotics. However, there is yet a lack of well-defined architecture, which abstracts from low-level real-world information to higher-level information, which is enriched by semantics to output enhanced diagnostic and mission planning information.

5.3.2 Fault Detection and Diagnosis

In the field of diagnostics, the gathering and processing of knowledge in AUVs as in most robotic systems are classified into two categories: (1) model free and (2) model based.

Model-free methods, such as rule-based, use limit checking of sensors for the detection of faults. Rule-based diagnostics are the most intuitive form, where through a set of mathematical rules, observed parameters are assessed for conformance to an anticipated system condition. Knowledge gained is thus explicit as rules are either satisfied or not. Rule-based reasoning is an easy concept to employ and if kept simple requires little development time, provided that expert tacit knowledge (system behaviour awareness) can be straight forwardly transformed to explicit knowledge (rules). However, these rules use knowledge gained from outside observations of the system rather than a representation of any internal mechanisms. In other words, they represent only the relationship between symptoms and failures and cannot provide a coherent explanation of the failure. Furthermore, they exhibit a lack of flexibility as only faults that have been explicitly described can be diagnosed. The main advantage of a rule base system is that execution time is generally much faster than other methods using more sophisticated models.

Model-based diagnosis systems rely on the development of a model constructed from detailed in-depth knowledge (preferably from first principles) of the system. There is a wide range of models available for diagnosis, including mathematical, functional and abstract [36]. The fault detection and isolation (FDI) community has tackled the diagnostic task by comparing simulated results to real results and detects abnormalities accordingly based mainly in analytical redundancy. The main advantage of this approach is that it can be adapted easily to changes in the system environment by changing inputs to the model being simulated. However, the numerical models are based on behaviour of the system, with little knowledge of the structure and functionality of the components. Also, there is no mechanism to detect multiple faults and it requires expensive computation.

Currently, there is an increasing move away from fault detection and diagnosis (FDD) model-based to structure and data-driven methods, because complex dynamic systems are difficult to model, based on analytical redundancy alone. Uppal and Patton argue that the interesting combination of certain aspects of qualitative and quantitative modelling strategies can be made [37]. They further state that qualitative methods alone should be used, if faults cause qualitative changes in system performance and when qualitative information is sufficient for diagnosis. Qualitative methods are essential if measurements are rough or if the system cannot be described by differential equations with known parameters.

5.4 Fault Tolerant Adaptive Mission

Adapting of a mission on the fly in response to events is feasible with embedded planners. However, they are limited to the quality and scope of the available information. For them to be effective, the mission programmer must predict all possible situations, which is clearly impractical. Therefore, to adapt mission plans due to unforeseen and incipient faults, it is required that accurate information is available to recognise that a fault has occurred and the root cause of the failure. For example, if a propeller drive shaft developed too much shaft friction, then the added current load may overburden the drive motor. Identification of such a condition before the shaft bearing fails would allow rectification of the mission before a major breakdown occurs.

AUVs are generally equipped with FDD systems based on damage control that results in the vehicle resurfacing in the event of any fault in the system. But future industrial and military AUVs may require systems that operate even while partially damaged. Hence, it is of importance to develop a system, which not only detects failure in the underwater vehicle but also provides meaningful and reliable information to counterpart modules, such as the mission planner, to adapt and recover from the failures.

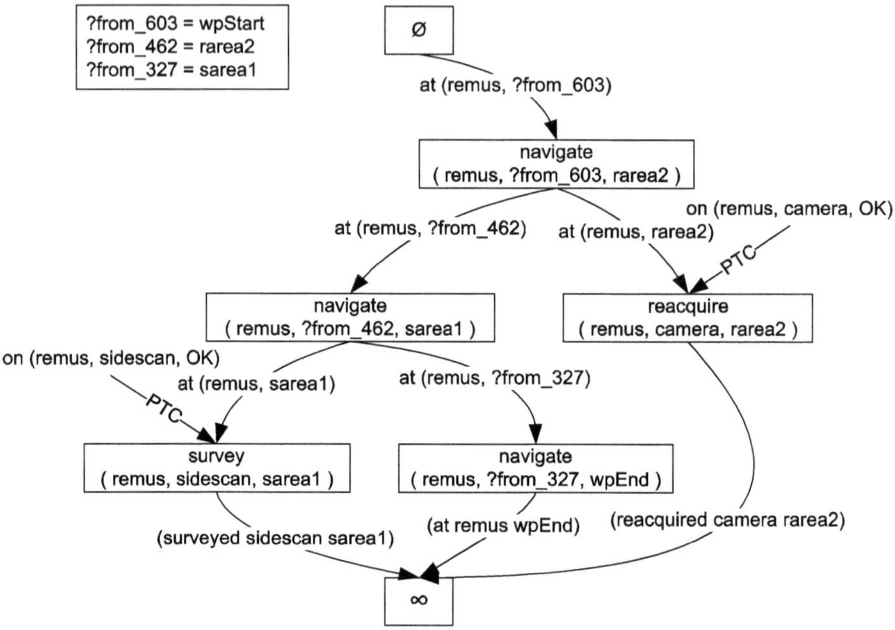

Fig. 5.4 Example of a partial ordered plan representation of an autonomously generated AUV mission for a similar scenario to the one described in Sect. 5.8—MCM scenario demonstrations. Ordering constraints are represented using the graph depth, interval preservation constraints represented with black arrows, point truth constraints represented with PTC-labelled arrows and binding constraints represented in the *top left box*

5.4.1 Mission Plan Adaptation

Following classical planning problem representation, an instance of a vehicle mission problem can be simply defined as $\Pi = \{P,A,I,G\}$, where P is the set of propositions defining the available resources in the vehicle, A is the set of actions, I is the initial platform state and G is the set of possible mission accomplished states. $D = P \cup A$ defines the mission domain and $P = I \cup G$ the mission problem. Given an instance Π, the mission generation problem consists in finding if there exists a mission plan (*mp*), using $ai \in A$, such that satisfies any $g \in G$.

Several approaches exist in the artificial intelligent (AI) literature capable of solving this problem. In a real environment where optimality can be sacrificed by operability, partial ordered planning (*pp*) is seen as a suitable approach because it produces a flexible structure capable of being adapted (see Fig. 5.4). The implemented approach can deal with extensions from the classical representation. It can handle durative actions, fluents, functions and different search metrics in order to minimise resource consumption, such as remaining vehicles battery or total distance travelled.

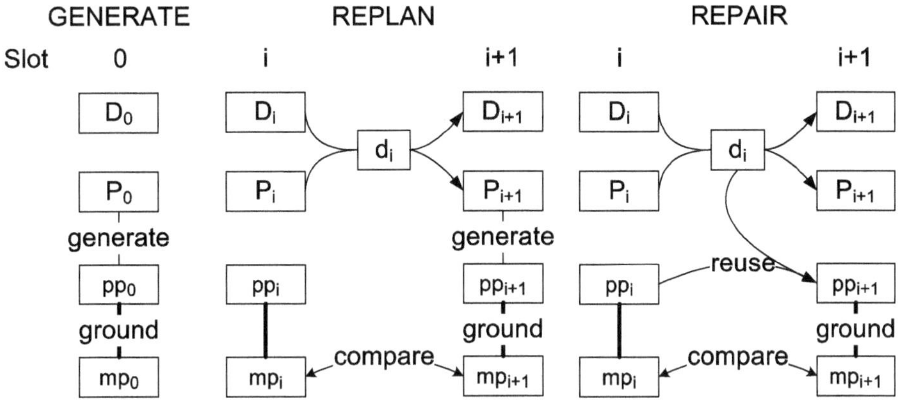

Fig. 5.5 Re-plan and repair processes for mission plan adaptation

Figure 5.5 shows the benefits between re-planning and re-pairing a mission plan. At the initial loop, a partial ordered plan pp_0 (i.e. a plan where only some ordering of actions are specified) is generated satisfying the given mission domain and problem Π_0. The pp_0 is then grounded (i.e. all action orderings are specified) into the minimal mission plan mp_0 including all constraints in pp_0. At an iteration i, the knowledge base is updated by the diagnosis information d_i providing a modified mission domain and problem Π'_{i+1}. The *mission re-plan* process generates a new partial plan pp'_{i+1}, as done at the first stage, based only in the Π'_{i+1} information. On the other hand, the *mission plan repair* process re-validates the original plan by ensuring minimal perturbation of it. Given a mp to Π_i, the mission repair problem produces a solution mission plan mp′ than solves the updated mission problem Π'_{i+1}, by minimally modifying mp.

When a mission failure occurs during the execution, two possible repair levels can be identified: *mission execution repair* and *mission plan repair.* Execution repair changes the instantiation mp such that either an action a_i that was previously instantiated by some execution α is no longer instantiated or an action a_i that was previously instantiated is newly bound by an execution α already part of the mp. Plan repair modifies the partial plan pp itself, so that it uses a different composition, though it still used some of the same constraints between actions. It might also entail the elimination of actions, which have already been instantiated.

Executive repair will be less expensive and it is expected to be executed by the mission executive module (Fig. 5.2). Plan repair, however, will be computationally more expensive and requires action of the mission planner. The objective is to maximise the number of execution repairs over the plan repairs and, at the plan repair level, maximise the number of decisions reused from the old mission.

5.4.2 *Mission Plan Refinement*

A practical approach following the previously described concepts has raised interest recently in the AI community providing a novel solution for all drawbacks identified during the re-planning process. Known as plan recovery methods, they use plan-space searches and are able to adapt the existing plan to the new state of the world. They can be divided into two stages.

The first stage, known as *plan diagnosis*, analyses the effects of the updated platform status on the current mission. According to the new updated constraints received from the status monitor, it identifies the failures and gaps existing in the current mission plan. These plan gaps are the cause of inconsistency between the existing plan and the current status of the platform and the environment. They are, therefore, preventing the correct execution of the mission. The approach developed at this stage is based on unrefinement planning strategies and uses the potential of the knowledge reasoning in order to identify the resources that remain available.

The second stage is known as *plan repair*. The strategy during this stage is to use new partial plans to repair the gaps or failures identified during the plan diagnosis stage. The plan repair stage is based on refinement planning strategies for plan recovery.

In simple terms, when changes in the instances of the planning ontology are sensed (d) that affect the consistency of the current mission plan pp_i, the plan adaptability process is initiated. The plan diagnosis stage starts an unrefinement process that relaxes the constraints in the mission plan that are causing the mission plan to fail. The remaining temporal mission partial plan pp_t is now relaxed to be able to cope with the new sensed constraints. This will be the simplest stage of recovery necessary to continue with the execution of the plan, but it does not guarantee that all the mission goals will be achieved. The plan repair stage then executes a refinement process searching for a new mission plan pp'_{i+1} that is consistent with the new world status D' and P'. By doing this, it can be seen that the new mission plan mp' is not generated again from D' and P' (re-planned) but recycled from pp_i (repaired). This allows reuse of the parts of the plan pp_i that were still consistent with D' and P'.

5.5 Ontologies as Knowledge Representation

Many definitions and descriptions about ontologies can be found in the literature. From a philosophical perspective, the term ontology has been defined as "the study or concern about what kinds of things exist and what entities or things there are in the universe" [38]. From a practical perspective, ontologies are viewed as the working model of entities and interactions either generically (e.g. the SUMO ontology) or in some particular domain of knowledge or practice, such as predictive maintenance or subsea operations.

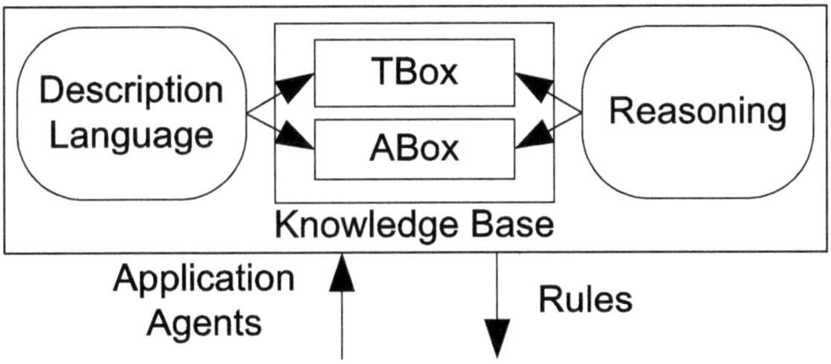

Fig. 5.6 Knowledge representation system

A definition given by Gruber characterises an ontology as the specification of conceptualisations, which are used to help programs and humans share knowledge [39]. The combination of these two terms embraces the knowledge about the world in terms of entities (things, the relationships they hold and the constraints between them), with its representation in a concrete form. One step in this specification is the encoding of the conceptualisation in a knowledge representation language. Ultimately, the main goal behind the concept of ontology is to create an agreed-upon vocabulary and semantic structure for exchanging information about that domain.

The main components of an ontology are concepts, axioms, instances and relationships. A concept represents a set or class of entities or things within a domain. A *fault* is an example of a concept within the domain of diagnostics. A finite set of definitions is called a *TBox*, which includes the set of terminological axioms for every atomic concept. Axioms are used to constrain the range and domain of the concepts, such as the father is a man that has a child or the fault is an error condition that belongs to hardware.

Instances are the specific things represented by a concept; for example, a *FaultySensorXYZ* is an instance of the concept *fault*. The combination of an ontology with associated instances is what is known as a knowledge base (see Fig. 5.6). However, deciding whether something is a concept of an instance is difficult and often depends on the application [40]. This finite set of assertions about individuals is called an *ABox*.

In the *ABox*, besides the concept assertions, one can also specify role assertions or, in other words, the relations describing the interactions between individuals. For example, the property *isComponentOf* might link the individual *SensorX* to the individual *PlatformY*.

Following the definition and characterisation of ontologies, one of the main objectives for an ontology is that it should be reliable and reusable [41]. However, an ontology is only reusable when it is to be used for the same purpose for which it was developed. Not all ontologies have the same intended purpose and may have parts that are reusable and other parts that are not. They will also vary in their coverage and level of detail.

One of the benefits of the ontology approach is the extended querying that it provides, even across heterogeneous data systems. The meta-knowledge within an ontology can assist an intelligent search engine with processing your query. Part of this intelligent processing is due to the capability of reasoning that makes possible the publication of machine understandable metadata, opening opportunities for automated information processing and analysis. For instance a diagnostic system, using an ontology of the system, could automatically suggest the location of a fault in relation to the happening of symptoms and alarms in the system. The system may not even have a specific sensor in that location, and the fault may not even be categorised in a fault tree. The reasoning interactions with the ontology are provided by the reasoner, which is an application that enables the domain's logic to be specified with respect to the context model and executed with the corresponding knowledge, i.e. the instances of the model. A detailed description of how the reasoner works is outside of the scope of this chapter, where the test scenario is described.

5.6 Framework

SA_V consists of making the vehicle autonomously understand the *big picture*. This picture is composed of the experience achieved from previous missions (orientation) and the information obtained from the sensors while on mission (observation). The *TBox* and *Abox* that are the main components in a knowledge base can be assigned to the orientation and observation components of SA_V, respectively. For each knowledge representation, its *TBox–ABox* pair will not only describe the relationships between concepts but also facilitate the decision-making process of the SOA. Reasoning capabilities allow concept consistency to reassure that SA_V remains stable through the evolution of the mission. Also, inference of concepts and relationships allows new knowledge to be extracted or inferred from the observed data.

In addition, a set of ontologies has been developed in order to represent the knowledge information required at SA_V. A key part in the ontology engineering discipline is the construction and organisation of these libraries of ontologies, which should be designed for maximum reusability [39, 42]. A major challenge in building these libraries is to define how these ontologies are constructed and organised and what relations should be among them. Existing approaches propose three discrete levels of vertical segmentation including (1) upper/foundation, (2) core/domain and (3) application (see Fig. 5.7).

The top layer in Fig. 5.7 refers to the foundational ontologies (FOs) (or upper ontologies) which represent the very basic principles and meets the practical needs of a model that has as much generality as possible, to ensure reusability across different domains. There are several standardised upper ontologies available for use, including Dublin Core, GFO, OpenCyc/ResearchCyc, SUMO and DOLCE.

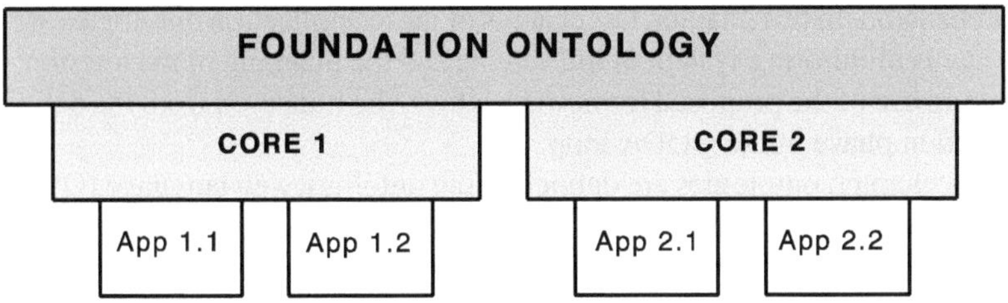

Fig. 5.7 Levels of generality of a domain ontology

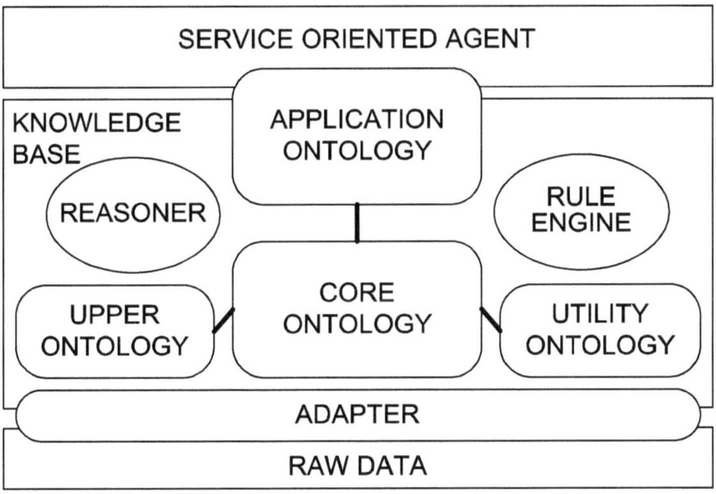

Fig. 5.8 SA$_V$ concept representation (core, application ontologies), instance generation (adapter) and handling (reasoning, inference and decision-making agent)

The second level of the ontology hierarchy represents the core domain ontology, which is arguably another of the building blocks to information integration. The goal of a core ontology is to provide a global and extensible model into which data originating from distinct sources can be mapped and integrated. This canonical form can then provide a single knowledge base for cross-domain tools and services (e.g. vehicle resource/capabilities discovery, vehicle physical breakdown and vehicle status). A single model avoids the inevitable combinatorial explosion and application complexities that result from pairwise mappings between individual metadata formats and/or ontologies (Fig. 5.8).

At the bottom of the stack, an application ontology provides an underlying formal model for tools that integrate source data and perform a variety of extended functions. Here, application concepts are handled at the executive layer and are used to ground the abstract concepts managed by the software agents running in the system. As such, higher levels of complexity are tolerable and the design should be motivated more by completeness and logical correctness than human comprehension. In this chapter, target areas of these application ontologies are found in the status monitoring system of the vehicle and the planning of the mission, which by making use of the proposed framework allows the transition from the deliberative to the action phase of the OODA loop.

In this chapter, ontologies are defined in the ontology web language (OWL) [43]. The OWL language is a specialisation of the resource description framework (RDF) standard [44] and therefore a specialisation of the XML standard. These standards enable the representation of information in a structured format where an abstraction model (i.e. the ontology) is combined with instance data in a repository.

5.6.1 Foundation and Core Ontology

To lay the foundation for the knowledge representation of unmanned vehicles, JAUS concepts have been considered. However, the core ontology developed in this chapter extends these concepts while remaining focused in the domain of unmanned systems. Some of the knowledge concepts identified related with this domain are:

- Platform: static or mobile (ground, air, underwater vehicles)
- Payload: hardware with particular properties, sensors or modules
- Module: software with specific capabilities
- Sensor: a device that receives and responds to a signal or stimulus
- Driver: module for interaction with a specific sensor/actuator
- Waypoint: position in space with coordinate and tolerance
- Coordinate: local frame, global frame, angular
- Velocity: linear, angular
- Attitude: roll, pitch, yaw..

Figure 5.9 represents a snapshot of the core ontology showing the key concepts involved in the test case scenario and the relationships associated with them. In this figure and all the other ontology representations presented in this chapter, concepts and relationships between them are depicted using the expressive nature of OWL, which leads to a much stronger specification for AUV integration. Essentially, the OWL language specifies that all information is represented in the form of a simple triple, named subject, where each component of the triple is represented by a uniform resource identifier (URI). The URI is a unique identifier for the data that is held in a data store. And a triple represents a unit of information through the combination of a subject, a predicate and an object. The representation of the triple with the associated URI reference can be observed in Fig. 5.9 as well as in all ontology representations throughout this chapter, such as

core:Software core:hasPlatform core:Platform core:Software core:hasCapability core:Capability core:Platform core:hasCapability core:Capability

To support generic context-aware concepts, this framework makes use of the SOUPA (standard ontology for ubiquitous and pervasive applications) [45] ontology, which is the core of the context broker architecture (CoBrA) [46], a system for

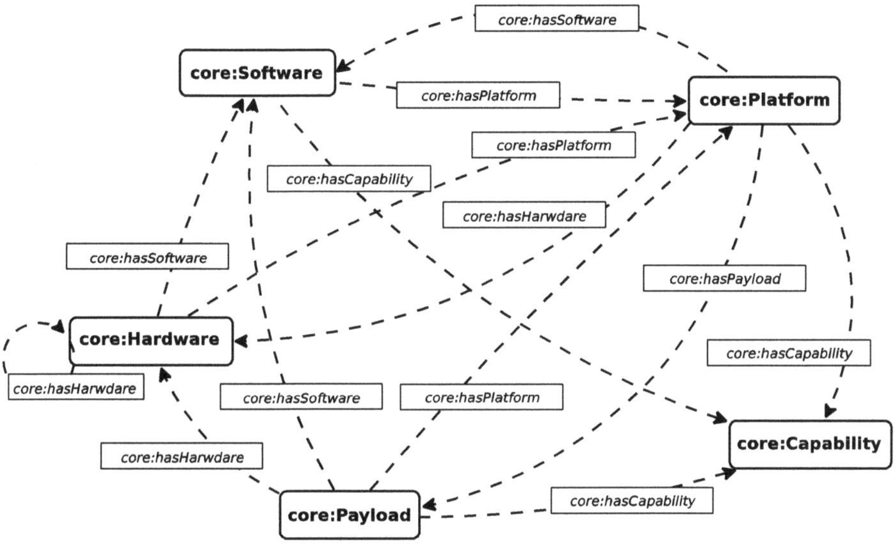

Fig. 5.9 Snapshot of the core ontology representing the environment around the concept platform

supporting context-aware computing. The contribution of the SOUPA ontology is the spatio-temporal representation, which allows the representation of time instants, intervals and temporal relations, as well as space enhanced by high-level concepts such as movable spatial thing, geographical entity or geometry concepts.

5.6.2 *Application Ontology*

Each service-oriented agent has its own application ontology. It represents the agent's awareness of the situation by including concepts that are specific to the expertise or service provided by the SOA.

In the case study presented in this chapter, these agents are the status monitor and the mission planner. Together, they provide the *status monitor* and *mission adapter* components described in Fig. 5.2 required for closing the OODA loop and provide on-board decision-making adaptation.

5.6.3 *Status Monitor Application Ontology*

The status monitor agent considers all symptoms and observations from environmental and internal data in order to identify and classify events according to their priority and their nature (critical or incipient). Based on internal events and context

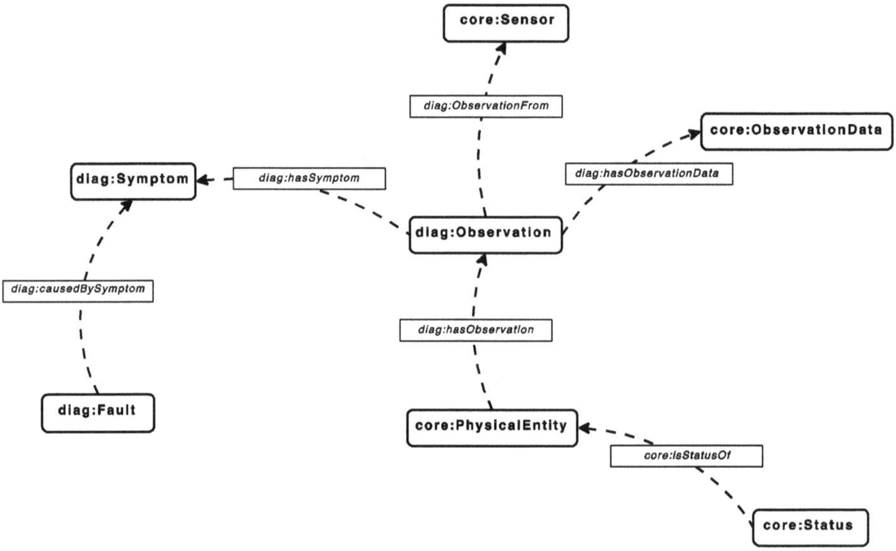

Fig. 5.10 Representation of the system observation pattern

information, this agent is able to infer new knowledge about the current condition of the vehicle with regard to the availability for operation of its components (i.e. status). In a similar way, environmental data is also considered for detecting and classifying external events in order to keep the situation awareness of the vehicle updated.

The status monitoring application ontology is used to express the SA_V of the status monitor agent. Recent developments in defining ontologies as a knowledge representation approach for a domain provide significant potential in model design, able to encapsulate the essence of the diagnostic semantic into concepts and to describe the key relationships between the components of the system being diagnosed. To model the behaviour of all components and subsystems considered from sensor data to possible model outputs, the status monitoring application ontology is designed and built based on ontology design patterns [47]. Ontology patterns facilitate the construction of the ontology and promote reuse and consistency if it is applied to different environments. In this chapter, the representation of the monitoring concepts is based on a system observation design pattern, which is shown in Fig. 5.10. Some of the most important concepts identified for status monitoring are:

- Data: all internal and external variables (gain levels, water current speed)
- Observation: patterns of data (sequences, outliers, residuals)
- Symptom: individuals related to interesting patterns of observations (e.g. low gain levels, high average speed)

- Event: represents a series of correlated symptoms (low power consumption, position drift). Two subclasses of events are defined: *CriticalEvent* for high priority events and *IncipientEvent* for the remaining ones
- Status: links the latest and most updated event information to the systems being monitored (e.g. sidescan transducer)

It is useful to note how some of these concepts are related to concepts of the core ontology (e.g. an *observation* comes from a *sensor).* These core ontology elements are thus the enablers for the knowledge exchange between SOA. This will be shown in the demonstration scenario of Sect. 5.8.

Note that this knowledge representation of diagnostic concepts is linked to concepts already described in the core ontology, such as the status of the system which is the key component in the handshaking of information between the status monitoring and mission planner agents.

5.6.4 Mission Planning Ontology

On the mission planning side, knowledge modelling is implemented by using a language representation. This language is then used to express the input to the planner. Language vocabularies generally include the information concepts and the grammars that are used for describing the relationships and constraints between these concepts.

The STRIPS language from [48] is generally acknowledged as the basis for classical planning. Basic concepts in this language are: an *initial state*, a *goal state* (or set of goal states) and a *set of actions* which the planner can perform. Each action consists of a set of *preconditions*, which must be true for the action to be performed, and a set of *postconditions*, which will be true after the action has been completed. A classical planner then normally attempts to apply actions whose *postconditions* satisfy the goal, recursively applying other actions to meet the preconditions until a complete plan (if available), which can be executed from the initial state and ends at the goal, is formed.

The PDDL language was originally created by [49] and stands for Planning Domain Definition Language. It is the planning language used by the planning community during the biannual International Planning Competition that takes place during the ICAPS Conference [50]. It can be considered as an extension of the original STRIPS language with extra functionality added. PDDL intends to express the physics of a planning domain, which is, what predicates there are, what actions are possible, what the structure of compound actions is and what the effects of actions are. In recent versions, it contains extensions for dealing with extended goals where good quality plans are as valid as optimal quality plans. It is a complex language with complex syntactic features such as specification of safety constraints or hierarchical actions. State-of-the-art plan generators are not able to handle the entire set of PDDL language features. In consequence, several versions of the

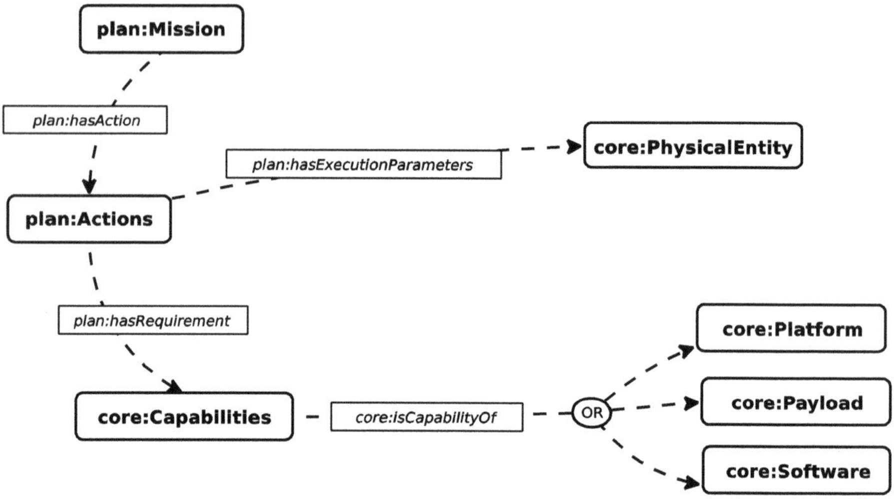

Fig. 5.11 Representation of the mission plan description

language have came out describing a subset of features, called requirements, that are extended every 2 years when the competition takes place.

The adaptive decision-making process requires concepts for generating a mission following a declarative goal-based approach instead of a classic procedural waypoint-based description. This can be achieved by looking at the concepts described by PDDL. However, in order to provide a solution to a mission failure, it also requires concepts capable of representing incidents or problems occurring during the mission. These concepts have been extracted from [24]. Some of the most important concepts identified for mission plan adaptability are:

- Resource: state of an object (physical or abstract) in the environment (vehicle, position, release)
- Action: collection of state of resources. State changes (calibrate, classify, explore)
- Catalyst resource: resources that are not consumed for an action but needed for the proper execution of the action (sensor activation)
- Plan gap: actions that may no longer be applicable. At least two ways of achieving a subgoal but a commitment has not been taken yet
- Gap: a non-executable action
- Execution: when an action is executed successfully
- Mission Failure: an unsuccessful execution

The representation of the planning concepts related to the mission plan and mission actions are shown in Fig. 5.11.

Note that this knowledge representation of planning concepts is linked to concepts already described in the core ontology, such as the list of capabilities required to perform each of the mission acts.

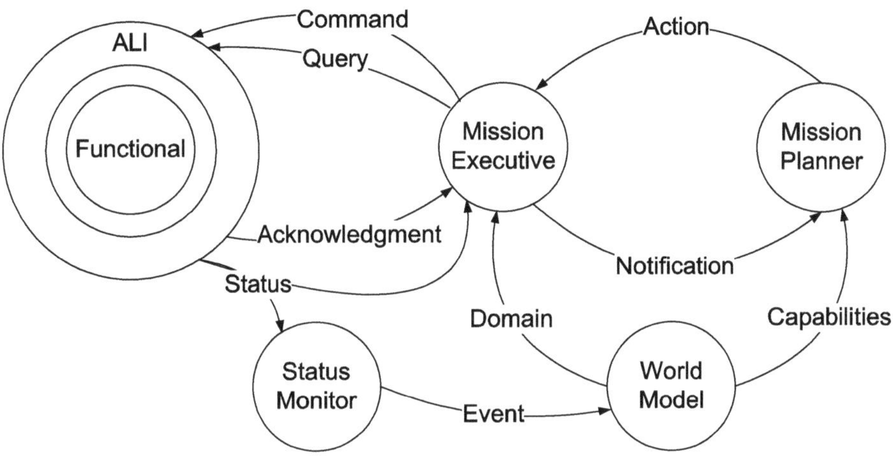

Fig. 5.12 Modules and interconnection messages implementing the three-tier architecture of the system. The mission planner provides the mission adapter function of the OODA loop in Fig. 5.2

5.7 System Architecture

As described earlier, the system implements the four stages of the *OODA* loop (see Fig. 5.12). The status monitor reports changes in the environment and the internal status of the platform to the world model. The world model stores the ontology-based knowledge provided by the a priori expert orientation and the observation of events received from the status monitor. A mission planner module generates and adapts mission plans based on the notifications of updates by the world model and capabilities reported by the mission executive. The mission executive module is in charge of executing the mission commands in the functional layer based on the actions received from the mission planner. In order to provide independency of the architecture with the vehicle's functional layer, an abstract layer interface (ALI) has been developed.

The translation from the mission planner to the mission execution is done via a sequence of instances of action executions. An action execution is described by the domain ontology *TBox* and gets instantiated by the action grounded on *mp*. The instance contains the script of commands required to perform the action. The action execution contains a timer, an execution counter, a timeout register and a register of the maximum number of executions. The success, failure or timeout outputs control the robust execution of the mission and the executive repair process.

Once *mp* is obtained and the list of α_i is generated for *mp*, the mission plan gets transformed into a state machine of action execution instances. An action execution graph is generated that contains all the possible options for the new plan. This directed graph shows not only the forward execution path through the plan actions to reach the goal but also the recovery actions to employ when predefined exceptions (or failures) in action execution take place. In this way, the number of calls to the mission planner can be reduced, particularly for dealing with commonly occurring and deterministic failures. This in turn reduces the system response time reacting to a failure.

5.8 Test Case Scenario

This section demonstrates the benefits of the approach through application to a real mine countermeasures (MCM) scenario. The demonstration shows the benefits of using an ontological representation to describe the SA_V and how the status monitoring and adaptive planner agents are capable of interchanging knowledge in order to dynamically adapt the mission to the changes occurring in the platform and the environment.

In this scenario, AUVs support and provide solutions for mine detection and neutralisation. The operation involves high levels of uncertainty due to the noisy nature of the sensor data and risk of damage in the vehicle. Navigating in such a hazardous environment is likely to compromise the vulnerability of the platform. Additionally, if a vehicle is damaged or some of its components fail, mission adaptation will be required to cope with the new restricted capabilities.

The performance of the system has been evaluated on a REMUS 100 AUV platform in a set of integrated in-water field trial demonstration days at Loch Earn, Scotland (56°23.1 N,4°12.0 W). A PC/104 1.4 GHz-payload-running Linux has been installed in the vehicle. The payload system is capable of communicating with the vehicles control module and taking control of it via the manufacturer's *remote control* protocol.

Figure 5.14 shows one of the original mission plan waypoints as described by the operator following the vehicles procedural waypoint-based mission description specification. The mission plan consists of a starting waypoint and two waypoints describing a simple north to south pattern at an approximate constant longitude (4°16.2 W). The mission leg was approximately 250 m long and it was followed by a loiter recovery waypoint. This mission plan was loaded to the vehicle control module. The track obtained after executing this mission using the vehicle control module is shown in Fig. 5.14 with a dark line. A small adjustment of the vehicles location can be observed on the top trajectory after the aided navigation module corrects its solution to the signals received from the long baseline (LBL) transponders previously deployed in the area of operations.

On the other hand, the operator using a priori information about the area of operation and a declarative description of the mission plan oriented the adaptive system in the payload computer. All this knowledge is represented using concepts from the core ontology and the planning ontology, respectively. The original scenario is displayed in Fig. 5.13. Additionally, the a priori information of the area is provided based on automatic computer-aided classification knowledge generated from previous existing data [51]. These areas are shown in Fig. 5.14. For the goal-oriented mission plan, the requirements of the mission are specified using the

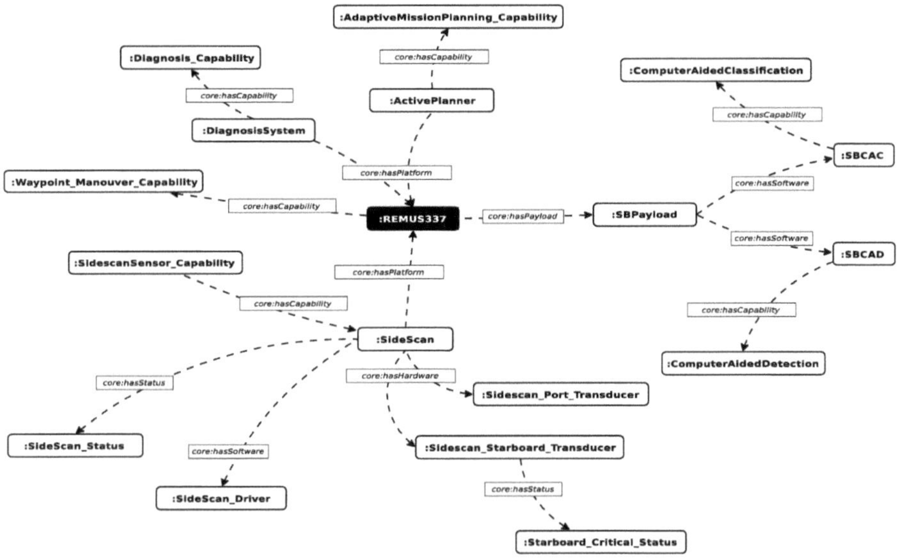

Fig. 5.13 Instances in the knowledge base representing the demonstration scenario. The diagram represents the main platform, its components and their capabilities

planning ontology in a way that can be referred to as "survey all known areas". The benefits of the approach and its performance in this scenario are described in the following sections.

5.8.1 In-Mission Adaptation

The payload system is given a location in which to take control of the host vehicle. At this point the mission planner agent generates the mission based on the knowledge available, including the mission requirements. The instantiation of the mission with the list of actions to be executed is described in Fig. 5.15 using the planning ontology elements. The mission is then passed to the executive agent that takes control of the vehicle for its execution.

While the mission is being executed the status monitor maintains an update of the knowledge stored in the world model. In this example, the status monitor reports status of hardware components, such as batteries and sensors and external parameters such as water currents.

When the executive is about to execute the next action in the mission plan, it looks at the knowledge stored in the world model in order to generate the adequate list of commands to the vehicle. In this case, the list of waypoints generated for the lawnmower pattern survey of the areas is related to the measured water current at the moment of initiating the survey. Under this framework, information

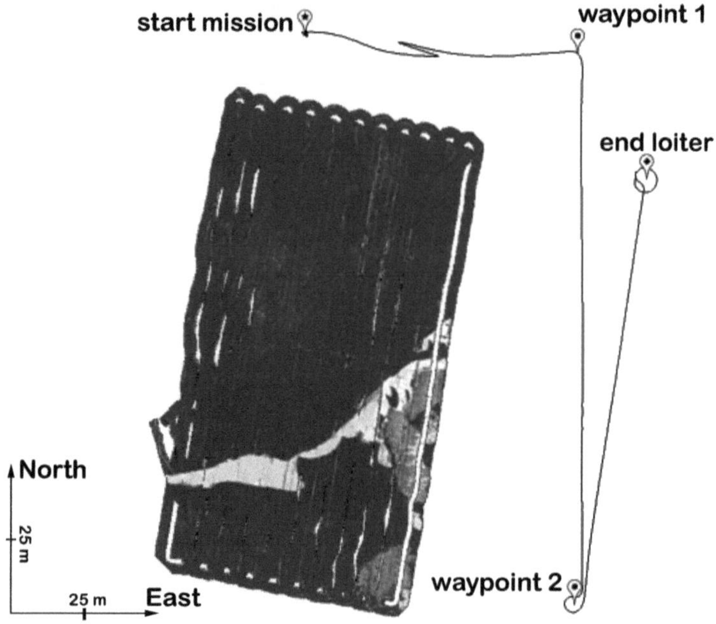

Fig. 5.14 Procedural mission uploaded to the vehicle control module and a priori seabed classification information stored in the knowledge base

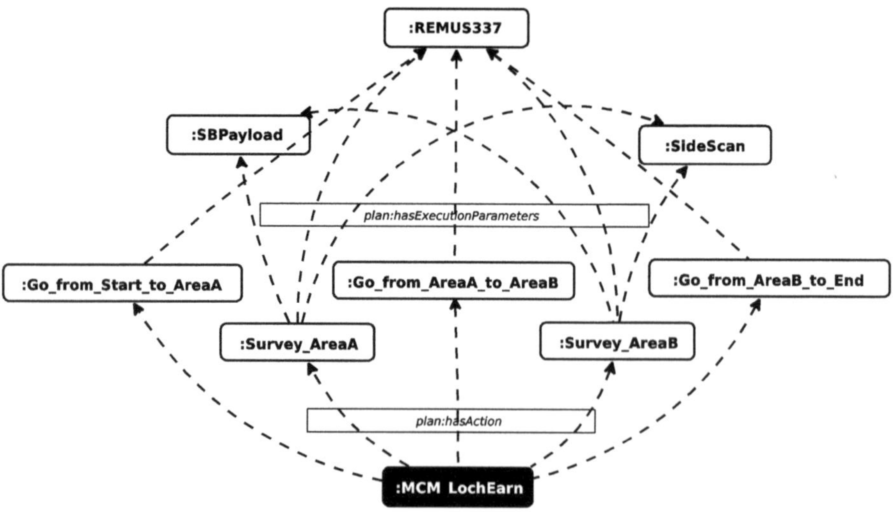

Fig. 5.15 Representation of the mission planning action execution parameters

and knowledge are able to transfer between the status monitoring system and the adaptive mission planner while on mission. When the observations coming from the environment being monitored by the status monitor agent indicate that the mission under execution is affected, the mission planner is activated and the mission plan gets adapted. The transfer of knowledge between agents is possible by providing a solution to answer the following question: *Are the observations coming from the environment affecting the mission currently under execution?*

In order to explain the reasoning process involved, a component fault has been simulated in the vehicle while performing the mission. For this case, the gains of the starboard transducer of the sidescan sonar of the vehicle were modelled to drop to their minimum levels halfway through the execution of the seabed survey action.

The first step is to deal with the information coming from the sensor, which is signalling a low gain in the starboard transducer. This signal triggers a symptom instance, which has an associated event level. This plays a key role in the classification of the instances in the *fault* concept between being critical or incipient. This classification is represented axiomatically in Eqs. 5.1 and 5.2

$$\begin{aligned} &diag : CriticalFault \sqsubseteq ... \\ &\text{diag} : Fault \sqcap \ni \text{diag} : causedBySymptom... \\ &(diag : Symptom \sqcap \ni diag : hasEventLevek... \\ &(diag : Level \sqcap diag : High)\ ) \end{aligned} \tag{5.1}$$

$$\begin{aligned} &diag : IncipientFault \sqsubseteq ... \\ &\text{diag} : Fault \sqcap \ni \text{diag} : causedBySymptom... \\ &(diag : Symptom \sqcap \ni diag : hasEventLevek... \\ &(diag : Level \sqcap \ni diag : Med)\ ) \end{aligned} \tag{5.2}$$

Once the fault individuals are reclassified, the status of the related system is instantiated using the most updated fault information. Therefore, a critical status of the sidescan starboard transducer is caused by a critical fault. However, because the sidescan sonar is composed of two transducers, one malfunctioned transducer only causes an incipient status of the sidescan sonar.

The characteristics of the status concepts and reusability presented here supports the transfer of knowledge between the two involved agents. Now, the mission planner agent is responsible to adapt the mission according to this new piece of information.

Equation 5.3 reports to the mission planner that the two survey actions in the mission are affected by the incipient status of the sidescan sonar. In this case, the sensor required by the action is reporting an incipient fault. The action can therefore only be modified by adapting the way it is being executed, an execution repair. If both transducers were down and the status monitor agent would have reported a critical status of the sidescan sensor, a plan repair adaptation of the mission plan would have been required instead. In this case, it would have been necessary to look for redundant components or platforms with similar capabilities in order to be able

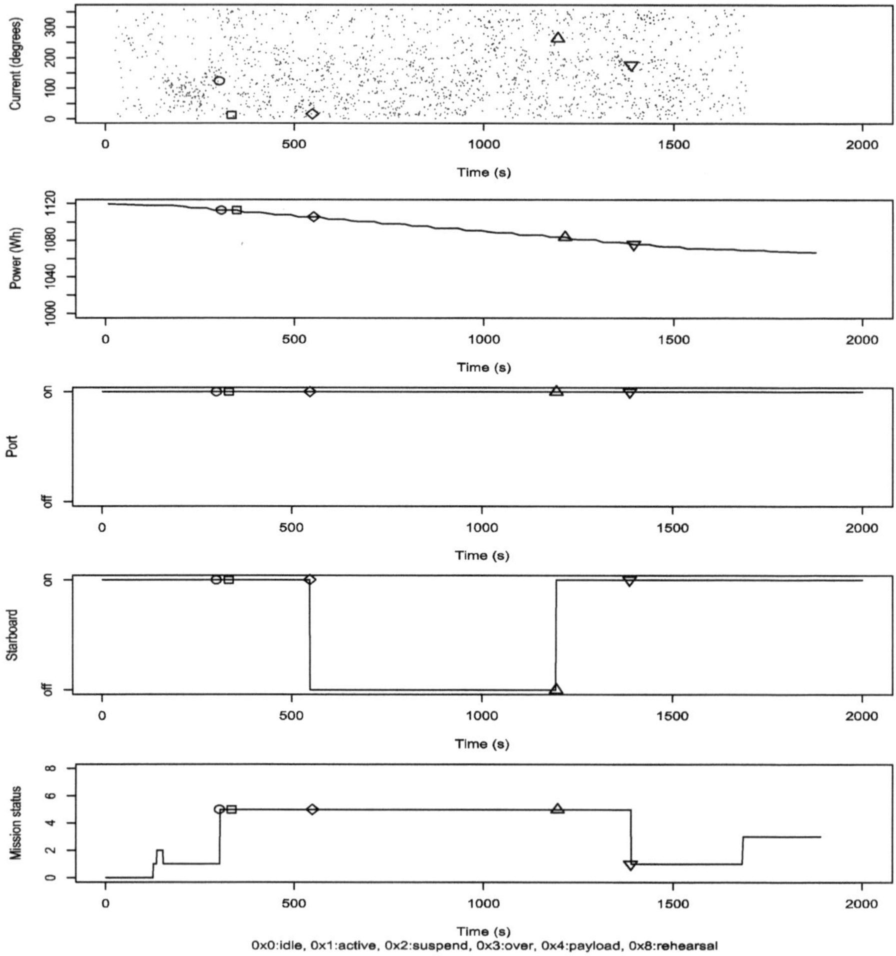

Fig. 5.16 Status monitoring (*top to bottom*): direction of water current (degrees), battery power (Wh), sidescan sensor port, starboard transducers availability (on/off) and mission status binary flag, all plotted against mission time (s)

to perform the action. The same procedure is used once the transducer is reported as recovered and the mission plan commands change back to the original pattern.

$$
\begin{aligned}
&SLRECTWHERE\{\,?Mission?Action?Param?Status\\
&plan: hasAction(?Mission, ?Action)\wedge\\
&plan: hasExecParam(?Action, ?Param)\wedge\\
&plan: hasStatus(?Param, ?Status)\,\}.
\end{aligned}
\tag{5.3}
$$

The mission performance timeline is described in Fig. 5.16, which shows the most representative subset of the variables being monitored by the status monitor

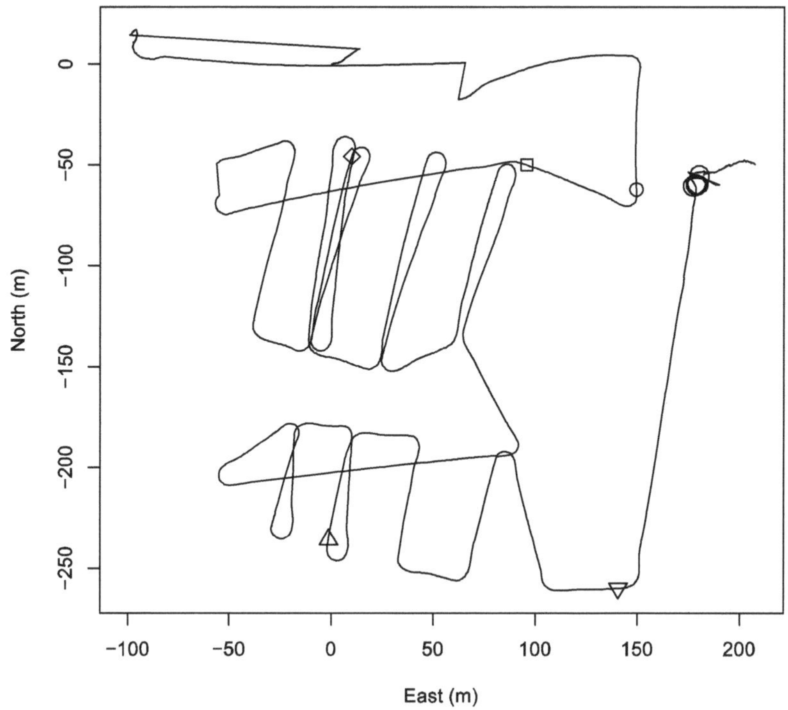

Fig. 5.17 Vehicle's track during mission in north-east local coordinate frame

agent. Each of the symbols on the aforementioned figures represents a point during the mission where an event has occurred.

- The circle symbol represents the point where the payload system takes control of the vehicle. Note a change on the mission status flag that becomes 0*x*05, i.e. mission active and payload in control.
- The square symbol represents the point where the vehicle arrives to perform the survey of the area. At this point, the action survey gets instantiated based on the properties of the internal elements and external factors.
- The diamond symbol represents the point when the diagnosis system reports an incipient status in the starboard transducer of the sidescan sonar. It can be seen how the lawnmower pattern was adapted to cope with the change and use the port transducer to cover the odd and even spacing of the survey and avoiding gaps on the sidescan data (see Fig. 5.17). This adaptation allows maximising sensor coverage for the survey while the transducer is down.
- The symbol Δ indicates the point where the diagnosis system reports back the correct functionality of the starboard transducer. It can be observed how the commands executing the action are modified in order to optimise the survey pattern and minimise distance travelled. Although also being monitored, the

power status does not report any critical status during the mission that requires modification of the actions.
- The symbol ∇ shows the location where all the mission goals are considered achieved and the control is given back to the vehicle.

5.9 Conclusions

Nowadays, autonomous vehicles in the underwater domain employ a knowledge representation that is embryonic and targets are simple mono-platform and mono-domain applications, therefore limiting the potential of multiple coordinated actions between agents.

In order to enhance interoperability, independence of operation and situation awareness, we have presented a semantic-based framework that provides the core architecture for knowledge representation for embedded service-oriented agents in autonomous vehicles. The framework combines the initial expert orientation and the observations acquired during the mission to improve the situation awareness of the vehicle. This approach has direct impact on the knowledge distribution between embedded agents at all levels of representation, providing a more capable and holistic system and involving semantic interoperability among all involved information sources.

The fully integrated framework has achieved the following:

- Formal platform view: Three different ontology models have been produced representing the concepts related to an underwater platform, and to the diagnostic and mission planning domain. And, as a consequence, the ontological commitments made by these interrelated domains are made explicit.
- Interoperability: The proposed framework supports the communication and cooperation between systems that are normally working in isolation.
- Model-based knowledge acquisition: before and during the mission, these ontology models are able to acquire knowledge of a real world situation. Based on the data obtained during the mission, the SOA provide to each other the requested information by the execution of predefined queries, in order to manage unforeseen events.

We have tested and evaluated this proposed framework to the problem of fault tolerant adaptive mission planning in a MCM scenario, where the ontological representation of knowledge, model-based diagnosis and adaptive mission plan repair techniques are combined. To demonstrate the benefit of this approach, the mission adaptation capability and its cooperation with the diagnosis unit has been shown during an in-water field trial demonstration. In this scenario, the approach has shown how the status monitoring and the mission planner agents can collaborate together in order to provide a solution to mission action failures when component faults are detected in the platform.

Acknowledgements Our thanks to members of the Ocean Systems Laboratory for their inputs and helpful critique. The work reported in this chapter is partly funded by the Project SEAS-DTC-AA-012 from the Systems Engineering for Autonomous Systems Defence Technology Centre and by the Project RT/COM/5/059 from the Competition of Ideas, both established by the UK Ministry of Defence.

References

1. Benjamin M, Curcio J, Leonard J, Newman P (2006) Navigation of unmanned marine vehicles in accordance with the rules of the road. In: International conference on robotics and automation (ICRA), May 2006
2. Petres C, Pailhas Y, Patron P, Petillot Y, Evans J, Lane DM (2007) Path planning for autonomous underwater vehicles. IEEE Trans Robot 23(2):331–341
3. Hagen P (2001) Auv/uuv mission planning and real time control with the hugin operator system. In: IEEE OCEANS 2001, vol 1, Nov 2001
4. Pang S, Farrell J, Arrieta R, Li W (2003) AUV reactive planning: deepest point. In: IEEE OCEANS 2003, vol 4, Sept 2003
5. Ghallab M, Nau D, Traverso P (2004) Automated planning. Morgan Kaufmann, San Francisco
6. Pell B, Gat E, Keesing R, Muscettola N, Smith B (1997) Robust periodic planning and execution for autonomous spacecraft. In: International conference artificial intelligence
7. Fisher F, Knight R, Engelhardt B, Chien S, Alejandre N (2000) A planning approach to monitor and control for deep space communications. In: Proceedings of the IEEE aerospace conference
8. Chien S, Knight R, Stechert A, Sherwood R, Rabideau G (1999) Integrated planning and execution for autonomous spacecraft. In: IEEE aerospace conference, Aspen, CO, Mar 1999
9. van der Krogt R (2005) Plan repair in single-agent and multi-agent systems. Ph.D. dissertation, Netherlands TRAIL Research School
10. Gerevini A, Serina I (2000) Fast plan adaptation through planning graphs: local and systematic search techniques. In: AAAI conference on AI planning systems, pp 112–121
11. Koenig S, Likhachev M, Furcy D (2004) Lifelong planning a*. Artif Intell 155(1–2):93–146
12. Rabideau G, Knight R, Chien S, Fukunaga A, Govindjee A (1999) Iterative repair planning for spacecraft operations in the ASPEN systems. In: Proceedings of the 5th international symposium on artificial intelligence robotics and automation in space, pp 99–106. [Online]. Available: citeseer.ist.psu.edu/rabideau99iterative.html
13. Fox M, Long D, Baldwin L, Wilson G, Wood M, Jameux D, Aylett R (2006) On-board timeline validation and repair: a feasibility study. In: Proceedings of 5th international workshop on planning and scheduling in space, pp 22–25. Baltimore, USA, Oct 2006
14. Knight R, Fisher F, Estlin T, Engelhardt B, Chien S (2001) Balancing deliberation and reaction, planning and execution for space robotic applications. In: International conference on intelligent robots and systems (IROS 2001). Maui, HI, Nov 2001
15. Turner R (2005) Intelligent mission planning and control of autonomous underwater vehicles. In: ICAPS, workshop on planning under uncertainty for autonomous systems
16. Bellingham J, Kirkwood B, Rajan K (2006) Tutorial on issues in underwater robotic applications. In: ICAPS
17. Rajan K, McGann C, Py F, Thomas H (2007) Robust mission planning using deliberative autonomy for autonomous underwater vehicles. In: ICRA workshop on robotics in challenging and hazardous environments
18. Fox M, Long D, Py F, Rajan K, Ryan J (2007) In situ analysis for intelligent control. In: IEEE OCEANS conference, Vancouver
19. Saigol Z (2011) Automated planning for hydrothermal vent prospecting using AUVs. PhD Thesis, University of Birmingham, UK

20. Fox M, Long D, Magazzeni D (2012) Plan-based policy-learning for autonomous feature tracking. In: ICAPS 2012, the 22nd international conference on automated planning and scheduling Atibaia. Sao Paulo Brazil, June 25–29
21. Patron P (2010) Semantic-based adaptive mission planning for unmanned underwater vehicles. PhD Thesis, Ocean Systems Lab, Heriot-Watt University
22. Boyd, J (1995) The Essence of Winning and Losing. Available: http://www.danford.net/boyd/essence.htm
23. Adams JA (2007) Unmanned vehicle situation awareness: a path forward. In: Human systems integration symposium
24. Strutt JE (2006) Report of the inquiry into the loss of autosub2 under the fimbulisen. In: National Oceanography Centre Southampton. Southampton, UK. Technical Report
25. Collett TH, MacDonald BA, Gerkey BP (2005) Player 2.0: toward a practical robot programming framework. In: Proceedings of the Australasian conference on robotics and automation (ACRA 2005). Sydney, Australia, Dec 2005
26. Fitzpatrick P, Metta G, Natale L (2008) YARP user manual. Yet Another Robotic Platform (YARP), Technical Report
27. Newman P (2009) Introduction to programming with MOOS. Oxford Robotics Research Group, Technical Report
28. Newman P (2009) Under the hood of the MOOS communications API. Oxford Robotics Research Group, Technical Report
29. OceanSHELL: An embedded library for distributed applications and communications. Ocean Systems Laboratory, Heriot-Watt University, Technical Report, Aug 2005
30. Somby M (2007) A review of robotics software plat- forms. http://www.windowsfordevices.com/c/a/Windows-For-Devices- Articles/A-review-of-robotics-software-platforms/, August 2007. [On- line]. Available: http://www.windowsfordevices.com/c/a/Windows-For- Devices-Articles/A-review-of-robotics-software-platforms/
31. Arredondo M, Reed S, Petillot YR (2006) Battlespace access for unmanned underwater vehicles - dynamic multi-dymensional world modelling - final report. SeeByte Ltd. - Ministry of Defence, Technical Report, 2006
32. Society of automotive engineers AS-4 AIR5665 JAUS architecture framework for unmanned systems. SAE International Group, Technical Report, 2008. [Online]. Available: http://www.jauswg.org/
33. Bouguerra A, Karlsson L, Saffiotti A (2008) Monitoring the execution of robot plans using semantic knowledge. Robot Auton Syst 56(11):942–954. Semantic Knowledge in Robotics. [Online]. Available: http://www.sciencedirect.com/science/article/B6V16–4T7XGSM- 1/2/04f9b80a1f7cdebb975141bc910cd594
34. Galindo C, Fernndez-Madrigal J-A, Gonzlez J, Saffiotti A (2008) Robot task planning using semantic maps. Robot Auton Syst 56(11):955–966. Semantic Knowledge in Robotics. [Online]. Available: http://www.sciencedirect.com/science/article/B6V16–4T9CCW6- 2/2/5bf373f40885a5995dbcf60b0a48ae80
35. Hertzberg J, Saffiotti A (2008) Using semantic knowledge in robotics. Robot Auton Syst 56(11):875–877. Semantic Knowledge in Robotics. [Online]. Available: http://www.sciencedirect.com/science/article/B6V164T72WWW-1/2/9edd0eb7357cb93ab0a7f0285979c469
36. Chantler M, Cogill G, Shen Q, Leitch R (1998) Selecting tools and techniques for model based diagnosis. Artif Intell Eng 12:81–98
37. Uppal FJ, Patton RJ (2002) Fault diagnosis of an electro-pneumatic valve actuator using neural networks with fuzzy capabilities. In: Proceedings of European symposium on artificial neural networks, pp 501–506
38. Blackburn S (1996) The Oxford dictionary of philosophy. Oxford University Press, Oxford
39. Gruber TR (1995) Towards principles for the design of ontologies used for knowledge sharing. Int J Hum-Comput Stud 43:907–928
40. Brachman RJ, McGuinness DL, Patel-Schneider PF, Resnick LA, Borgida A (1991) Living with classic: when and how to use a KL-ONE- like language. Principles of Semantic Networks: Explorations in the representation of knowledge, pp 401–456

41. Uschold M, King M, Moralee S, Zorgios Y (1998) The enterprise ontology. Knowl Eng Rev 13(1):31–89
42. van Heijst G, Schreiber A, Wielinga B (1996) Using explicit ontologies in kbs development. Int J Hum-Comput Stud 46(2–3):183–292
43. Smith MK, Welty C, McGuinness DL. Owl web ontology language guide, w3c recommendation, February. [Online]. Available: http://www.w3.org/TR/owl-guide/
44. Becket D, McBride B (2004) Rdf syntax specification (revised), resource description framework, February 2004. [Online]. Available: http://www.w3.org/TR/rdf-syntax-grammar/
45. Chen H, Perich F, Finin T, Joshi A (2004) Soupa: standard ontology for ubiquitous and pervasive applications. In: International conference on mobile and ubiquitous systems: networking and services. Boston, MA
46. Chen H, Finin T, Joshi A (2003) Using owl in a pervasive computing broker. In: Workshop on ontologies in open agent systems, pp 9–16. Melbourne, Australia
47. Blomqvist E, Sandkuhl K (2005) Patterns in ontology engineering classification of ontology patterns. In: 7th international conference on enterprise information systems. Miami, USA
48. Fikes R, Nilsson N (1971) STRIPS: a new approach to the application of theorem proving to problem solving. Artif Intell 2:189–208
49. Ghallab M, Howe A, Knoblock C, McDermott D, Ram A, Veloso M, Weld D, Wilkins D (1998) PDDL: the planning domain definition language. Yale center for computational vision and control, Technical Report
50. ICAPS (2008) [Online]. Available: http://www.icaps-conference.org/
51. Reed S, Ruiz I, Capus C, Petillot Y (2006) The fusion of large scale classified side-scan sonar image mosaics. IEEE Trans Image Process 15(7):2049–2060

Chapter 6
Cooperation Between Underwater Vehicles

Signe Redfield

Abstract The underwater environment intensifies difficulties inherent in cooperation between autonomous vehicles by dramatically reducing their ability to communicate with each other. Lack of communication drives the need for decentralized control which in turn requires a shared view of the tasks and their status as the mission progresses. Decision-making, especially task allocation, requires low-bandwidth mechanisms for negotiation and achieving consensus. Strategies to obtain this low-bandwidth decision-making and control rely on both the availability of significant a priori information about the mission and tasks and careful design of the system.

6.1 Introduction

Cooperation among underwater vehicles can be even more complex than cooperation among other system types; the underwater environment imposes different constraints than the environments in which ground, aerial, and surface vehicles operate, and lacks cues available in other environments. Nobody hangs the underwater equivalent of windsocks on convenient structures to enable them to detect the strength and direction of current, there are no stars or GPS to navigate by, and in deeper water and even some shallow environments, landmarks are few and far between. Unless they are operating in extremely close proximity or defining their trajectories specifically for good communications, the vehicles also have severe limitations on how much information they can communicate to each other and how often. Sensing each other and each other's position is just as difficult for underwater

---

S. Redfield (✉)
Office of Naval Research Global, London UK
Naval Surface Warfare Center, Panama City Division, FL, USA
e-mail: signe@ieee.org

M.L. Seto (ed.), *Marine Robot Autonomy*, DOI 10.1007/978-1-4614-5659-9_6,

Fig. 6.1 Navigation and environmental estimation underwater. Natural aids (such as rocks or vegetation features) can allow a vehicle to estimate its position underwater. Artificial aids (such as windsocks) are rare

vehicles as for any other vehicles, but as long as commensurate sensing technology is available, similar techniques can be used across domains (Fig. 6.1).

This section will cover background information about the tasks and environment relevant to the development of cooperating underwater systems. Cooperation is a very broad term, covering everything from reactive systems which appear to cooperate simply because they interact with the same objects in the same environment to complex systems capable of allocating tasks among group members to optimize group performance. Sections 6.2 and 6.3 cover the basic components and design trade-offs necessary to implement cooperation among underwater vehicles, while Sect. 6.4 covers software frameworks and approaches that can be used to integrate these components. Section 6.5 describes examples of tasks and behaviors for cooperating vehicles that have been developed and demonstrated.

6.1.1 Definitions

Cooperation is a generic term indicating any action where the robots work together or in concert to accomplish their task or mission. Any combination of multiple robots working together to accomplish a goal will be referred to as a *team*. On one end of the spectrum of possible varieties of cooperation, *collaboration* will indicate a situation where the vehicles are physically linked or tight cross-vehicle timing constraints exist. On the other end, *coordination* will refer to situations where the vehicles have no or loose cross-vehicle timing constraints and where their physical actions are only loosely related.

A few additional terms will be used throughout this chapter to describe various aspects of a cooperative system:

- *Task.* Any intermediate desired result of the team's actions, any intermediate desired result of an individual's actions, an action or state available to a single vehicle.
- *Mission.* Overall desired result of the team's actions. Three main missions will be used as examples:

  - A scientific mission, where a team of vehicles gathers information about and possibly tracks an oceanographic feature like an algae plume or volcanic vent (a primarily collaborative mission) [40]
  - A military mission, where a team of vehicles is deployed to search an area, getting a second close-range look at anything that shows up as interesting on a low-resolution survey (a primarily coordinated mission) [65]
  - An industrial mission, where a team of vehicles is inspecting the substructure and possibly maintaining equipment below an at-sea platform (coordinated in some tasks, collaborative in others) [48]

- *System.* The complete set of components required for the team of vehicles to complete their mission; this includes not only the vehicles but also the operator and any operator control software or hardware.
- *Plan.* Refers to the representation of how various tasks or actions are allocated to team members. It includes the entire team, not just a single vehicle, and whatever information is relevant for the successful completion of the assigned tasks such as timing or sequential nature of assignments, vehicle IDs, and tasks (represented via whatever means is most appropriate for the task). The plan may cover all the tasks required until the mission is complete, or it may cover only near-term or immediate contingencies.
- *Designer.* The individual or team responsible for determining how the vehicles attempt to complete the mission, the way in which the mission is separated into tasks, and the best way to balance the limitations of the platforms and their environment against the goals of the mission.

6.1.2 Impact of Underwater Environment

Making multiple underwater vehicles cooperate with each other combines cooperative robotics problems with the problems associated with operating underwater vehicles. Environmental constraints, primarily communications

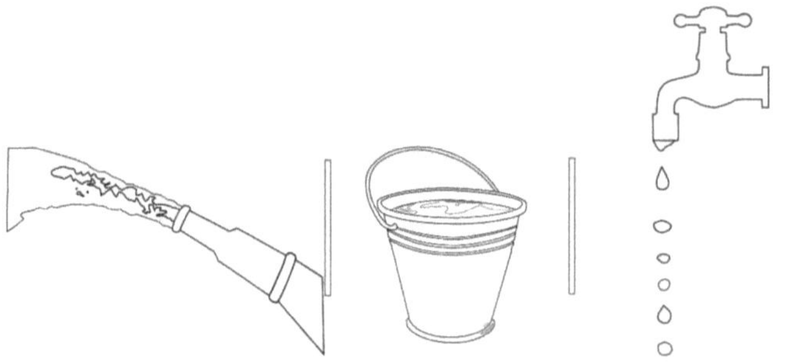

Fig. 6.2 If a single UUV message of 32 bytes is equivalent to a single droplet, then a second's worth of space messages is roughly equivalent to two five-gallon bucketfuls of water, and in-air wireless networking is equivalent to 40 gallons of water per second (about 15 s of fire hose output)

limitations, significantly affect the dynamic of cooperation. Lack of certainty in one's position in the environment makes map registration and task synchronization complicated. Without communication, any apparent coordination or cooperation between vehicles must be carefully designed in advance.

6.1.2.1 Communications

Ground vehicles and aerial vehicles can use standard wireless networking to communicate with a minimum bandwidth of 1 Mbit/s (often much higher), several orders of magnitude higher than that available to underwater vehicles (similar to a hose pouring water out constantly) [33]. Vehicles in space have limited windows in which information can be transferred and a moderate data rate: NASA's maximum rate to the Mars rovers during a window is 256 kbit/s (like getting water in intermittent bucketfuls) [7]. Vehicles operating underwater typically have intermittent but frequent access to a very low-bandwidth acoustic channel (like the constant dripping of a broken tap) on the order of 5 kbit/s at the high end (transmitted by a vehicle to a buoy with a vertical array) [49] and 256 bit/message on the lower end (several seconds per message, between vehicles) [74] for autonomous vehicles in shallow water. Furthermore, because communications uses active acoustics, it is very difficult to balance the requirements of multiple vehicles and still get sensing and communications capability on each platform equivalent to that of a single vehicle operating alone. Typically, multiple vehicles share the available communications bandwidth on a given frequency, usually by transmitting only during their assigned temporal block, further reducing the speed with which information can be transmitted (32 bytes every N = number of vehicles temporal blocks with a minimum duration of 15 s each [65]) (Fig. 6.2).

6.1.2.2 Navigation

Underwater systems often do not have access to the variety of sensory information that enables other types of platforms to know where they are in the world. Their environment can be relatively featureless (in the water column away from the sea floor [39]) or full of features that are not sufficiently distinguishable to identify a specific place with certainty compared to other places (the loop closure problem) or that change over time (potentially due to tides or major storms). Both relative and global localization are very important when multiple vehicles attempt to cooperate on a single task.

Underwater navigation uncertainty in the global frame is unbounded without some form of surface expression or environmental constraints. Surface expression is the most common method used to bound navigation uncertainty. An underwater vehicle's uncertainty can be bounded to within a few meters [22] using a surface craft with GPS to provide frequent updates and acoustic ranging as it follows the underwater vehicle. This level of navigation error is negligible (but impractical) in the scientific mission because the information gathered and the models used tend to work on the order of kilometers and because the vehicles are able to surface when navigation uncertainty increases. In the military mission, the vehicles operate over moderate distances with a time limit and therefore need to minimize surfacing. A trailing surface craft providing navigation information is a practical solution for this mission. Uncertainty on the order of a few meters becomes very important when the pose estimate must be precise enough to be able to find a target the size of a bicycle marked by another vehicle as potentially important. In the industrial mission, when vehicles work together to manipulate an object or coordinate their coverage of a small inspection region, reducing navigation uncertainty within their local frame becomes most important and error on the order of a few meters may be catastrophic. The vehicles in this situation are more likely to rely on locally sensed information about the position of their teammates than on external information from surface craft. In any case where the vehicles must rendezvous or base their actions on the position of other vehicles, the accuracy of that shared frame becomes critical.

6.1.2.3 Platform Limitations

Underwater vehicles operate in a three-dimensional environment and come in a variety of shapes with dramatically different mobility, from hovering platforms [37] that can stop and rotate in place, to large cylindrical submersibles with huge turning radii [24], to gliders that must change depth in order to move forward [40]. These differences in mobility need to be taken into account during the decision-making and planning stages or handled at the vehicle level to ensure that the chance of a collision and time lost dealing with mobility issues are minimized.

The limitations on decision-making and planning imposed by maneuverability constraints can be severe, but they are not dissimilar to those in effect in many other environments. In any system of cooperating autonomous vehicles, planning horizons based on mobility constraints can dramatically impact both the rate at which plans must be revised and how far ahead plans must be solidified. For example, a torpedo-shaped underwater vehicle has a larger turn radius than most laboratory robots, but a smaller turn radius than an automobile or a fixed-wing aerial platform. Its sensor range is between that of the car and the aerial vehicle. At an individual level, the main difference is that the underwater vehicle's slow speed increases the time required to respond.[1] At the group level, the main difference is that while ground and aerial vehicles have ample time (multiple communications cycles) in which to negotiate and replan before a decision needs to be made, the underwater vehicle's team members would not find out about the change in trajectory until the vehicle has already chosen a course of action and begun to act upon it.[2] Available time for planning and negotiating can be increased by increasing the range of the sensors used to detect objects requiring changes in the plan, but planning time of more than a single underwater communications cycle would require close to double the sensor range.

Other sensing limitations on group operation include the ability to sense other platforms and their actions. Sensing is generally a primary goal of underwater vehicles. While the mission-specific sensors can often detect other vehicles, they generally lack perceptual algorithms for identifying other platforms and are unlikely to be able to tell which vehicle is which within a group. Exceptions to this include the following:

- Situations where each vehicle in the group is physically unique in a way that can be detected by the other vehicles' sensors (if the vehicles do not have color vision, painting them different colors will not make them unique)
- Situations where they are working in a sufficiently confined space or in sufficiently clear water that vision processing can be used
- Situations where there are only two vehicles (so the vehicle you see must be your teammate)
- Situations where they use communications to initiate behavioral mechanisms (e.g., if vehicle 1 sends out a message asking vehicle 2 to turn left and then looks for the vehicle turning left in order to determine their relative positions)

Active sensors can interfere with other platforms, requiring complex timing mechanisms (for sonar) or geometric mechanisms (for cameras) to reduce interference. Passive sensors may be blinded or confused by active sensors. However, active sensors can also allow systems to collaborate in more complex ways, as researchers enable teams of vehicles to use a single active pulse from one system to gather data at other platforms [57, 60].

---

[1]Assumes a ground vehicle with a turn radius of 30 m and sensor range of 60 m traveling at 8.94 m/s (20 mph) and an aerial vehicle with a turn radius of 50 m and sensor range of 400 m traveling at 13.41 m/s (30 mph), which both require that a decision be made at around 3.5 s from impact, and an underwater vehicle with a turn radius of 15 m and sensor range of 60 m traveling at 2.06 m/s (4 kts), which requires that a decision be made at 7.3 s from impact.

[2]Total planning time for all three platform types is less than 30 s.

6.1.3 *Elements of Cooperation*

There are three main components to a system of cooperating agents:

- A shared task view
- A decision-making process
- The behaviors that enable the agents to perform the tasks

All three of these are affected by the approach the designer takes in solving his or her problem. If the designer opts for a purely local solution, the shared view becomes implicit in the responses of each vehicle and can only be seen by an operator monitoring their states. Decision-making is abstracted into the responses of the vehicles to sensory data about the environment and merged with the behaviors the vehicles perform. For a centralized solution, each vehicle contributes part of the shared view, but only the centralized controller has the entire picture and is able to determine the way in which tasks are broken down and assigned to the vehicles. Most underwater systems operate somewhere between these two extremes. Each vehicle typically performs some onboard processing of its sensory data to develop a local shared view of the status of the tasks and some onboard decision-making about the group's performance with respect to their mission, while the operator or centralized controller manages higher-level shared views and decision-making involving both external requirements and information not available to the vehicles.

6.2 Shared Task View

The shared task view consists of:

- The definition of the mission or problem the vehicles must address
- Whatever tasks and subtasks may be necessary in order to complete it
- All the information available relevant to the performance of the tasks and subtasks

The structure and content of the shared task view determines what information needs to be available to each vehicle, what information needs to be transferred between the vehicles in the team, what information needs to be made available to the operator, and how that information is encoded to enable it to be transmitted between vehicles. The shared task view can be broken up into *goal representation* and *data management*.

6.2.1 Goal Representation

There are many different ways goals can be represented, but the most common include the following:

- Representation as some form of modified field [72]
- The use of objective functions to define the relative effect of specific actions [58] (for reactive systems and problems where multiple objectives are active simultaneously)
- Representation as a set of behavioral states or tasks utilizing waypoints for specific task descriptions (for missions whose tasks are readily broken down into specific behaviors that may or may not require low-level control, e.g., tasks where an appropriate solution is the use of a TSP (traveling salesman problem) solver to sort a series of waypoints) [23, 59, 63]

Internally, underwater vehicles can use any representations for their goals. Between teammates, however, communication limitations drive goals and tasks to a compact form. The communications overhead required for efficient team operation can be significantly reduced in some scenarios by adapting the mission to enable transmission of more compact information.

For example, in the military mission, the team is judged on how long it takes them to complete the survey (or how much of the area they cover in the time allowed), on how accurately they identify the objects, and on how certain they are that they have found all the interesting ones. The team performs a fast preliminary survey until the first potentially interesting object is found. As each object is found, the team must determine who should perform the identification step and on which of the available objects, and how many vehicles (if any) should continue the fast survey to ensure the coverage goal is met. Even something as simple as a coverage goal can be represented in many different ways.

In the simplest case, the operator wants to survey some fraction of the area sufficient to predict the likelihood of interesting objects being found in the area that was not surveyed. This could be represented as:

- A single percentage (e.g., "[search $p\%$ of the region's area]" or $[p]$)
- A percentage combined with a maximum size of any unsearched area (e.g., "[search $p\%$ of the region's area], [but no unsearched region greater than $a\%$ of the total area]" or $[p,a]$)
- A time limit (e.g., "[maximize coverage of the region in h hours]" or $[h]$)
- A probability distribution with a time limit (e.g., "[search the region *uniformly*], [to whatever coverage level can be managed in h hours]" or $[U,h]$)

The first option is conceptually simple and straightforward to design. Each vehicle can simply be given $p/N\%$ of the total area to survey with a scaling factor to manage differences in mobility and sensor range to evenly balance the workload. In the trivial case, the vehicles will search the first $p\%$ of the region fully and leave the rest unsearched. If the designer wishes to ensure that the entire region is searched, the second option limits the size of the unsearched region to enforce a more even spatial distribution of effort. However, this takes longer as the vehicles require more transit time to cover the full extent of the region.

Searching to a time limit (the third option) is another conceptually simple approach and ensures the team finishes on time. It provides no guarantees about whether or not the system will meet the coverage requirement. Adding a probability distribution to the parameters enables the operator or designer to use additional *a priori* information about the environment to ensure higher coverage of regions of particular difficulty. The operator or designer can tell the vehicles that certain areas are more important than others by using a non-uniform probability distribution.

However, *what the operator actually wants* is information about how many interesting objects are present in a region. In addition to the four approaches listed above, there is one more way to represent the team's goal:

- The actual information desired from the survey, taking into account regions where the environmental conditions change the certainty associated with whether or not an object is present (e.g., "[the certainty c the team has about its estimate of how many interesting objects n_i are present in the region]" or $[c]$)

With this option, the actual information the operator wants is used as the parameter that drives the team's behavior, theoretically resulting in improved performance and higher user satisfaction. In practice, however, this option requires significant on-board processing for each vehicle and increases complexity in the design and implementation of the system. The previous definitions of the team's goals are based on specifications provided by users or on information that is available to the vehicles as they perform the task and therefore can be implemented with a simpler system.

None of these options requires a specific methodology for the individual search behavior used by the vehicles; they work equally well with vehicles using a waypoint-based planning tool or reactive entropy-based trajectory optimization. The important characteristic of the individual behaviors used on the vehicles to accomplish the tasks is that there is a mechanism to translate their parameters to and from the mission parameters stated as part of the goal representation. The decision-making mechanism must be able to allocate individual tasks on the basis of the shared goals.

6.2.2 Data Management

Vehicles operating underwater typically use an acoustic modem for communications which severely restricts the ways in which they can share information. Acoustic modems have extremely limited bandwidth with slow physical transmission speed,

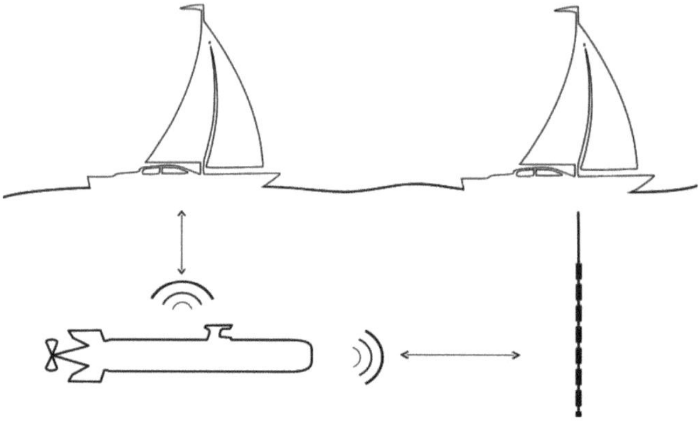

Fig. 6.3 Good communication channels maximize the difference between direct signals and reflected signals. The long array of transducers dropped from the right-hand surface vessel enables improved signal-to-noise ratio and higher bandwidth for long-distance communications

and even that limited bandwidth can be severely degraded by poor relative placement of the transducers within the water column. The acoustic bandwidth can increase if the vehicle is communicating with a surface or stationary object [42] transmitting and receiving vertically through the water column or using a lengthy array on one end of the horizontal acoustic channel (Fig. 6.3). Between vehicles moving through the environment independently, the options are to surface and communicate over the air or to remain submerged and use the limited acoustic channel. High-bandwidth communication is possible underwater if the vehicles are in very close proximity. The tight physical constraints of the industrial mission may both require and allow higher-bandwidth communication.

6.2.2.1 Surfacing

Even with systems that periodically surface to enable higher-bandwidth communications, there is no guarantee that the team members will be on the surface at the same time and therefore able to communicate. These systems can use a buoy as a centralized clearing house for large messages to be relayed from one vehicle to another as they surface, but these systems are capable of only loose coordination. The information from any given teammate will be out of date by an unspecified amount of time, and therefore, the estimate of the actual state of the mission will be outdated and fragmentary. If closer cooperation is required, these systems generally devolve to a centralized controller receiving all the messages, using its single shared view to generate tasks for the vehicles, and relaying the current task allocation to each vehicle as it surfaces and makes contact [17, 25, 40].

6.2.2.2 Underwater

When they cannot schedule their surfacing times to ensure sufficiently tight task coordination, the vehicles must use the acoustic channel to coordinate their actions. Each vehicle must develop its own shared task view in order to determine available tasks and allocate them efficiently. Over an acoustic channel, this information must be compressed and encoded to minimize the amount of data transferred. Because of the limitations of the acoustic channel, the vehicles typically take turns communicating. With message sizes between 32 and 256 bytes and single-message send/receive duration of 15 s [74], a team of only four vehicles in a coastal environment can take a full minute to each send 32 bytes of data. The communication limitations associated with underwater operations mean that data management is critically important.

6.3 Decision-Making

Cooperative systems requires decision-making on several levels. On the internal level, each individual vehicle determines the best way to prosecute the tasks assigned to it. Group decision-making determines which tasks are assigned to which vehicles and occasionally how to break up the larger problem into tasks or behaviors that can be performed by a single vehicle. On the mission level, the operator or user determines which vehicles and capabilities are available to make up the group.

This section focuses on the decision-making that a group or centralized controller must make about how the team should best accomplish its task. This topic encompasses how communications constraints affect task decomposition and task allocation, the sources of error that result in incompatible plan generation across platforms, and negotiation techniques to enable a team of vehicles to agree on a course of action.

There are several ways to approach the problem of ensuring all the tasks the team needs to perform are completed. Assume tasks are arriving at random intervals and communications are unreliable (lossy and slow). One solution is to require that in order to return to the retrieval point, the vehicle must ensure that all tasks it knows about are complete. This approach makes the system operate less efficiently but more surely (since one task may be duplicated by several vehicles but no tasks will be skipped). Alternately, the system can operate opportunistically so that the discovery of a new task results in a decision-making session to generate a new plan for the team. The generation of this new plan relies on a shared view of the current status of other tasks assigned to all the vehicles (to ensure that completed or partially completed tasks are not reassigned) and of the current list of tasks to be done (including the newly discovered task(s)). In some cases, the most efficient approach is to simply append the new task to a vehicle's list without altering the allocation of current tasks. In other cases, it is much more efficient to reallocate all the tasks to ensure that the distribution of tasks and the matching of tasks to platforms is as even as possible. However, the more complex the planning and task reallocation problem becomes, the more important it is that the shared view and negotiation process be manageable from a communications standpoint.

6.3.1 Task Allocation

Many missions in the underwater environment have task decomposition performed by the designer or operator, either *a priori* during the mission definition stage or during the mission via specific algorithms to balance the load on the vehicles. Since there are very few ways to break up any of the sample missions into tasks (search, cover, or track; identify target or manipulate object), the primary problem faced by the team is one of task allocation rather than task decomposition.

Most approaches to task allocation are predicated either on easy and frequent availability of information about team status (either through communication [68] or observation [41]) or on substantial *a priori* information about the costs and benefits (to the individual or the team) associated with each possible action the vehicle may take [66]. According to Matarić, Sukhatme, and Östergaard [44], the best task allocation strategy is a function of the task at hand; there is no one task allocation strategy that will consistently provide the best result across a range of tasks.

Biological systems without long-range communication techniques tend to rely on opinion polling and quorum approaches [26]. In these systems, the group decides on a single choice in a group, taking advantage of the immediate interaction possible because of close physical proximity to other group members. Since it eliminates communications lag, this approach is likely to result in better decision-making. Unfortunately, the time delays introduced by the rendezvous stage will often overwhelm the gains from improved decision-making. This approach is unlikely to prove beneficial for underwater robots.

Utility-based approaches are common in the multi-robot cooperation literature. Gerkey and Matarić's [29] analysis and taxonomy of multi-robot task allocation defined a nonnegative utility measure for evaluating the expected impact of a given action by a given vehicle. This method of weighting the relative benefit of different assignments enables a robot to determine the best task allocation for a given scenario, using whatever strategy is appropriate. It requires that the vehicles know the relative cost of that action and the improvement in the quality of the solution given that action. The advantage of this approach is that the tools of operations research become available for solving the problem.

In practice, defining these values and linking them back to the mission requirements can be extremely difficult. As it says in the paper, "it is vital that all relevant aspects of the state of the robots and their environment be included in the utility calculation." Externalities like increases in certainty associated with sensing from different viewpoints during transits over already-surveyed areas or minor variations in skill or capabilities on the part of individual robots are critical. If they are not included, the system can end up using a tie breaker to choose between solutions that could be easily ranked by a human operator (based on the excluded externalities) from most helpful to least helpful. For example, assume the utility computation includes only the characteristics of the vehicles and fails to include the time and energy spent launching and retrieving each additional vehicle. The algorithm is likely to allocate small tasks to many vehicles, even if the entire mission can be done with just two. Later uses of this approach merged the quality and cost components into a single cost number and allocated tasks based on that alone [67].

6.3.2 Sources of Dissent

Even when designers ensure that their systems are all running the same algorithm to perform the decision-making and planning, individual vehicles still occasionally generate plans that differ from those of their neighbors. This dissent can be the result of discrepancies in the shared task view due to loss or latency in the communications channel or from discrepancies in the projected planning mechanism due to learning or adaptation.

Some researchers compensate for lossy and inadequate communications channels by interpreting the behavior of the other vehicle(s) to infer their state or intent [61, 69]. This is then used to inform the models on each individual vehicle and produce similar plans while reducing the need for communications. This is not quite a stigmergic[3] approach, as the vehicles are not marking the environment in any way, but still uses implicit communication rather than explicit communication. Solutions that explicitly address lossy communications rely on redundancy (of action or information) to compensate for the lack of explicit information.

Typically, information from more distant team members lags slightly behind information from closer vehicles. Information about one's self is of course more current than information about anyone else. Without careful attention, even vehicles that are theoretically reasoning about the same situation can produce mission plans that vary significantly from one robot to another. A *consensus* problem arises when the vehicles are attempting to jointly solve a problem incrementally (reach consensus on the solution). A *reconciliation* problem arises when each vehicle independently develops a plan to solve the problem and then must reconcile that plan with the other team members' plans.

What mechanism should be used to ensure the team reaches consensus about the appropriate solution to a problem? Alternatively, how can the team reconcile their disparate solutions into a single solution? There are several ways to solve this problem, including careful design of the system, brute force, auction mechanisms, and game theory.

[3] *Stigmergic*. Communication via alteration in the environment, as ant pheromone trails provide information about ant behavior to other ants.

Occasionally the system can be designed so that the problem can be solved without explicit negotiation. Instead of proposing solutions directly to one another, the vehicles rely on an external system monitoring the performance of the group [64]. In this scenario, platforms plan individual trajectories on the basis of messages from the external system, to create a platoon of vehicles that operate as a group. This approach scales to an arbitrary number of vehicles because the external system is broadcasting one message to the group and the vehicles do not communicate with each other. However, it does require careful design and a mission that can be framed and implemented with something to play the role of an external observer.

When each vehicle is independently preparing its own solution to the current task allocation problem (a reconciliation problem), the hierarchical approach provides a simple mechanism for determining which plan to follow. In general, vehicles on a team are assigned identifying symbols. If these symbols are numeric, an adaptive leader/follower approach [21] can be implemented where the plan generated by the vehicle with the lowest number (or highest, or closest to some arbitrary value) is the plan adopted by the group. When communications are intermittent, the *group* is defined as the group capable of communicating. This leads to inefficiencies (as two subgroups of the team may settle on different plans with some duplication of effort), but is very robust.

An alternate method involves fixing each vehicle's starting conditions for the planning process at whatever value was sent to or from the other vehicles. Even though each vehicle plans with slightly delayed information, they are working from the same initial conditions and should develop the same plan. The hierarchical approach can then be applied to handle any inconsistencies between these plans.

If data is lost during an update to the shared view, the vehicle with the additional information will come up with a different plan than the vehicles operating with incomplete information. During the task planning stage, if that information is significant to the planning process, a discrepancy in the plans can result. Discrepancies should be identified during planning and drive a reconciliation stage either to ensure that all vehicles are operating from the same initial conditions in their planning or to achieve consensus on the correct allocation of tasks. In the DEMiR-CF architecture, the types of errors generated by lossy communications between vehicles are explicitly handled [55]. For example, if lossy data results in individual tasks allocated to multiple vehicles, the robot with the minimum cost is assigned to that task and the other(s) cancel their operation. In this case, the reconciliation process would include sharing of cost information from each robot working on the task.

6.3.3 Negotiation and Consensus

Negotiation is the means by which a group agrees upon the best action for the team to take among the set of actions generated by the individuals within the group. Group decisions rely on the interests of the individuals, externalities such as time limits, and information [11]. The communications limitations of the underwater environment prevent the timely acquisition of information in the form of an accurate and up-to-date shared view. If the vehicles are simply responding to tasks assigned by a central controller or if they are operating purely independently and any coordination is an emergent property of their behavior, negotiation is not an issue. But if the vehicles are using a decentralized or partially decentralized approach, each vehicle will contribute to the shared view and to the decision-making. If there is any discrepancy in the shared view or if there is disagreement about the best way to perform the tasks due to variation in the onboard algorithms through design or adaptation, negotiation between the vehicles will be required in order to achieve consensus about the appropriate actions to take. In the artificial intelligence community, solutions can be found under "negotiation techniques," encompassing methods for tasks to be allocated and for ensuring that the team agrees on which tasks go to which vehicles (known as the task allocation problem in the robotics community). In the multi-agent systems community, the "consensus problem" indicates that individuals disagree about something (usually related to a plan or task allocation solution) and must reach agreement in order to continue with their mission.

In the aerial vehicle community, game theory has been used to develop negotiation strategies, but the easy availability of relatively high-speed communications channels has resulted in negotiation techniques that are only tested for large numbers of vehicles and are evaluated in terms of the CPU time required to perform the calculations, not the number of communications cycles required to achieve consensus. In one scenario with 200 vehicles where iterations (communications cycles) were reported, this approach required 100 iterations to achieve locally optimal solutions and up to 1,000 iterations to fully converge [2]. This work also ignores potential delays in communications. A game theoretic approach to potential games demonstrated even more communication-intensive negotiations [43], requiring up to 600 iterations to learn a consensus solution to a single problem with four vehicles and nine possible alterations of the solution for each vehicle at each step. Even with only 100 iterations, a four vehicle underwater team would require 100 min to reach agreement. As this is not yet a practical scenario for underwater systems, the designer must develop alternate ways to reach agreement, ideally in only one or two communication cycles. These include hierarchical approaches, the use of performance metrics, and market-based approaches.

6.3.3.1 Hierarchy

Most cooperative systems use hierarchy at some point in their negotiation structure. The vehicles could be randomly assigned positions in the hierarchy, or have inherent characteristics that fit them to make certain decisions, or simply be in a position to take control. Hierarchy can be used:

- As a tie breaker
- To ensure a fast solution when other methods of negotiation fail
- To ensure consistency in the way information is processed by having vehicles follow a specific sequence when interacting
- To provide more robust centralized control, where if the leader fails, the next vehicle in the hierarchy takes over [69]

Several groups have developed hierarchy-based techniques for multi-robot control, including dynamic hierarchy trees [32], to create and manage communications relays.

6.3.3.2 Auctions

Auctions and market-based approaches are commonly used for task allocation and design in decentralized autonomous teams [4, 28]. One major problem in market-based approaches to task allocation is ensuring that the tasks are appropriately priced. Ideally, the market determines the prices based on each agent's valuation of that task, either through estimates of the time and resources needed to complete it or through some reward or utility function based on the designer's understanding of the importance of the task [4]. In practice, however, the designer has significant input into how the vehicles in the market determine the cost of a given task. For time-constrained underwater vehicles, this is often some function of distance traveled [55]. In one of the few papers addressing the communications overhead associated with auction techniques [28], seven robots auctioned nineteen individual tasks over 15 min with an average message size of 36 bytes. Actual messaging requirements, however, alternated between times with no messaging required (no auction underway) and data transfers of 400–500 bytes per second (single vehicle communications while auction underway), with a maximum burst transmission of 4.74 kb/s. Even with an auction consisting of only a handful of robots and only two simple tasks to balance, the communications overhead for this approach was still far beyond the capability of most underwater systems.

6.3.3.3 Performance Metrics

Performance metrics can be used to quantify how well a team has performed a task or mission. They can be used to compare the effectiveness of different approaches to a given problem and are often independent of mission-specific parameters that may affect implementation, focusing instead on general properties of the desired solution. The US National Institute of Standards and Technology (NIST) [1] has defined test procedures used for an autonomous ground vehicle (with potential human intervention in the form of additional waypoints and teleoperation where necessary) that consist of comparison to a human-equivalent benchmark. Within the human-robot interaction community [13], sets of metrics addressing different aspects of human interaction are used to quantify overall system effectiveness. Performance metrics are typically task specific and fail to generalize across different situations or uses.

Pricing mechanisms and cost and utility functions are often used within an autonomous vehicle to drive task allocation, while performance is typically judged after the mission is completed and all relevant information is available. Where the performance metric is known, the designer can use it to develop cost or utility functions or to tune the parameters of the pricing mechanism to encourage the desired behavior.

Occasionally neither cost functions nor pricing mechanisms are appropriate to drive task allocation. In these circumstances a performance metric can be combined with extensive run-time simulations of approximate future states to evaluate potential solutions. When the internal workings of the system are opaque to users, performance metrics can provide a more intuitive and transparent option for controlling team behavior than utility functions or pricing mechanisms, especially for systems requiring complex combinations of potential tasks. Linking a performance metric to desirable vehicle behavior is, however, a computationally expensive solution.

6.3.4 Specific Coordination Mechanisms

There are several different approaches to coordination of vehicles, depending on the problem domain and the communications available. Underwater glider fleets engaged in oceanographic missions have used potential fields for formation control [25] (with an external waypoint generator) and for ocean sampling (to maximize information gained with human input for changes in the assigned plan [40]). Theoretical approaches include game theoretic solutions [43] and physics-based particle simulation [62].

6.3.4.1 Stoplight

In this approach, each vehicle is assigned a series of tasks. As each vehicle finishes its first task, it stops and waits for the rest to complete their tasks [61]. When everyone has completed their first task, the entire group starts their second task. No new task is started until every previous task is complete. This approach is inefficient and time-consuming, but ensures synchronization between systems and enables the use of platforms that can only communicate with each other while surfaced [25]. It is susceptible to failures (if a vehicle malfunctions, the rest of the team may wait indefinitely to start their next tasks).

6.3.4.2 Implicit Synchronization

In implicit synchronization, task decomposition is performed in such a way that the information required to enable cooperation is inherent in the act of performing the tasks. If a vehicle is assigned to do task X when there is a task X to be performed (and it can sense whether task X is available or not) and another vehicle is assigned to do a task that results in the availability of task X, explicit communication and coordination is unnecessary. Since it can be very efficient and does not rely on any explicit communication between vehicles, this approach is particularly appropriate for underwater vehicles. However, each vehicle must be able to determine the availability of its tasks, which often implies a requirement for communications of some sort.

6.3.4.3 Explicit Synchronization

Explicit synchronization relies on communication between the vehicles and enables more complex interactions. Simple approaches include broadcasting of tasks available, while in complex approaches the vehicles negotiate among themselves to determine both how to break up tasks and which vehicles should perform which tasks. This approach is useful when tasks are very complex, when the appropriate allocation of tasks will depend heavily on the circumstances when new tasks are discovered, and when relatively high-bandwidth communications are available (e.g., when systems are in close proximity). Most techniques for complex missions fall into this category, including game-theoretic and physics-based particle approaches.

6.4 Architectures

For our purposes, architectures are software frameworks that enable a system designer to ensure that multiple vehicles attempting to cooperate are working within a consistent framework that has at least some guarantees on the mechanisms used to arrive at a consistent sets of actions. Within the ground robot community, researchers have developed architectures based on reactive behaviors [8], based on sensorimotor schemata [67] and on hybrid low-level reactive and high-level deliberative systems [51], and based on languages [69], grammars [9], and ontologies [35]. Architectures can also utilize either bottom-up (derived from work into emergent behavior and behavior-based systems) or top-down (derived from deliberative and algorithmic approaches to multi-robot systems) approaches.

There is a trade-off between communications (which drives designers to a bottom-up approach) and mission requirements (which drives designers to a top-down approach) [15]. The top-down approach is characterized by mission specifications (or system requirements) which are used to develop a centralized solution to the decision-making problem. This centralized solution is distributed, first by assuming perfect communications and therefore perfect knowledge and then by incorporating realistic communications constraints to develop approximate local models on each vehicle. Finally, the solution is calibrated, typically via learning or adaptation.

The bottom-up approach, on the other hand, starts with the specifications of the vehicles in the team and the desired task. The task drives the development of individual behaviors, which include the vehicle's actions, its interactions with other vehicles, and its interactions with the environment. These behaviors are combined with a controller (often either a finite state automaton (FSA) or grammar) to ensure they are activated in the correct sequence for the task(s) to be completed. A probabilistic model of the system is generated to ensure the collective behavior achieves the desired goal, and the system is optimized by adjusting the parameters that control either the performance of the behavior or transitions between behaviors.

The bottom-up approach works well for missions where the team is expected to either operate over long periods of time [25] (surveillance, survey, or patrol missions) or perform a tightly coupled collaborative task (bistatic sensing or simultaneous manipulation of an object). In these cases, the task itself is easily translated into vehicle states. Adjustments to any given vehicle's behavior are easily defined locally in terms of the sensor data of the robot itself.

However, many practical missions for underwater vehicles have requirements such as externally imposed time limits, within which the team needs to accomplish the maximum work possible. The bottom-up approach, with its analysis and optimization in terms of aggregate and probabilistic behavior and inability to enumerate a worst-case scenario, is not suited for missions with hard, definite constraints on performance. The trade-off between communications (driving a bottom-up approach) and performance constraints (driving a top-down approach) causes problems for the underwater vehicle team developer.

Researchers have attempted to combine the top-down and bottom-up approaches to meet these conflicting goals, resulting in composite systems. In these systems, a top-down approach is used to ensure that the vehicles address all the tasks within the time limit, while a bottom-up approach is used where feasible. To take the military mission as an example, the vehicles must survey a certain region to a given percent coverage and must get a second look at targets found during the survey, all within a given time limit.

In a top-down approach, the developer would ensure that every time a new target was detected, the vehicles would follow the same sequence: communicate to ensure they are all computing their behavior from the same initial conditions, run the centralized algorithm to determine the best solution (including both which vehicle will get the second look and at what point), communicate again to ensure all vehicles have arrived at the same task allocation and that the solution met all the mission completion requirements, and then enact that solution.

In a bottom-up approach, as a vehicle detected a target, it would choose, on the basis of internal parameters, whether to immediately stop surveying and get a second look, or wait and get a second look later in its survey, or communicate the presence of the target to the other vehicles in the team so that one of them could get a second look. These internal parameters would also define the relative priority of specific tasks (given an approaching time limit, determine the relative importance of continuing to survey against the importance of getting a second look at one or more targets) and attempt to ensure that the tasks were completed before time ran out.

In a composite approach, a centralized algorithm would be used to allocate tasks to vehicles. Those tasks would be structured as either their own top-down missions (survey this region as a group, ensuring minimal overlap but specific area coverage) or as bottom-up missions (jointly sense this target). In underwater operations, the top-down algorithms must be carefully defined in order to minimize the communications burden.

6.4.1 Planning

In many cases, a well-defined planning process can decrease the communications burden. The vehicles first initialize their planning algorithms, using whatever mechanism the designer has chosen. Once the group has synchronized their initial conditions for the current round of planning, they each devise a group plan by choosing the best from the subset of the problem or search space that they work on. These group plans should be characterized by an output compact enough to send to the other vehicles. This output should indicate both the value associated with the plan and the role of each vehicle in performing the plan. As each vehicle determines its best plan, it communicates that plan to the group. The plan with the best value overall is the plan each vehicle chooses. If extra certainty is required, an additional communications cycle can ensure all the vehicles have chosen the same plan.

In the top-down approach, the designer could use a version of this process where the vehicle that finds the target performs all the planning calculations on the basis of current state data from each vehicle. This reduces communications (only the computing vehicle sends out a new plan) but fails when more than one target is found within a single communications cycle by two different vehicles.

The choice of how to manage this planning process is where the designer's skill and forethought can minimize complexity and communications requirements. In the military problem:

- It could be framed as minimizing the total travel time or distance travelled by the vehicle with the maximum path length, with a heuristic used to determine how the targets are allocated.
- It could be framed as choosing the best solution among a set of heuristic solutions to the problem of allocation or to the problem of maximum path length or to the problem of maximum duration or all three.
- It could be framed as how to allocate the targets to minimize mission completion time, using a heuristic to determine at what point in their search the vehicle breaks off to look at all the targets in sequence.

Each of these approaches results in very different planning techniques and in different plans, but effectively they all optimize the same variable: how long it takes the team to complete the mission. This approach minimizes the communications burden as long as it is possible to break the problem up so each vehicle is searching part of the space and as long as the vehicles have a way of conveying the relative benefit and the details of the plan generated such that the other vehicles can replicate the planning and generate their own actions. It also produces a solution as effective as the algorithms in question can deliver. However, in this case one of these approaches is easier for the designer to apply than the others.

If the designer attempts the first approach, there is no clear way to break up the search into subspaces such that the vehicles will easily know what elements are their responsibility and such that there are guarantees that the entire space of possible solutions is searched. Unless the designer wants to use additional communication cycles during the planning process or have a single vehicle compute the best plan and propagate it to the other vehicles, defining components of a solution framed in this way is extremely complex.

If the designer takes the second approach and breaks up the problem by heuristics, a heuristic to determine which vehicle computes the subspace for which heuristic planner will be required. This additional planner must be able to compensate for a failed vehicle or one that is withdrawn from the group or even just for the case where the number of heuristics to be searched fails to match the number of vehicles available.

If the designer attempts the third approach, the permutations of target allocation can be easily shared among the vehicles, with each vehicle searching the subspace of all the permutations that result from that vehicle taking the most recently added target. This evenly distributes the work without any additional communication cycles. It also both produces a very simple way for each vehicle to know which part of the subspace is its responsibility and ensures that the entire space is searched. If a vehicle fails, clearly that vehicle will not be taking any new tasks, and the lack of information from that vehicle can be used to start a new truncated planning process without that vehicle.

6.5 Sample Tasks

There are several specific tasks and missions that have been used to demonstrate teamwork, both among underwater vehicles and between underwater vehicles and other systems (including humans [19] and surface craft [18]).

In many of these specific tasks, the limitations posed by communication bandwidth and navigation accuracy are thrown into high relief. In the simultaneous localization and mapping (SLAM) or feature-based navigation (FBN) communities, the number of features in each image that must be available in order for it to be used to close a loop and enable the vehicle to recognize a previously visited location must be larger than one [46]. In many situations, underwater vehicles have no or only one or two features that could be used for loop closure. However, in feature-rich environments SLAM is used effectively to improve individual vehicles' localization accuracy. Some researchers are using underwater vehicles' teammates as features to improve both the group's overall localization accuracy and their estimates of individuals' relative positions within the group.

Mobility differences between team members can also affect the quality of the solution. In the military mission, if a single, fast vehicle is assigned to perform the initial survey mission with a long-range sensor, while smaller, slower vehicles are used to get a second look at any targets found, the speed discrepancy can create an inefficient solution where larger areas result in the fast vehicle spending a larger proportion of the mission time idling compared to the time the smaller vehicles spend identifying objects.

6.5.1 Group Navigation

There are several approaches that achieve good relative position information in the water column based on time of flight for the message signal between acoustic modems. This enables the vehicles to maintain position within formations [16] (using analysis of one-way travel time of the signal from a long baseline array of bottom-mounted transponders) and to localize vehicles in arbitrary positions within a team [10] (using a surface GPS node for a global reference point and including reflections at the boundaries of the water column). The bottom-mounted transponders used an *a priori* scheduled broadcast cycle and relied heavily on accurate timing, while the surface-relative group used ray tracing and an *a priori* depth-dependent sound velocity profile. This technique has also been used to bound the navigation uncertainty by having a surface vehicle follow the underwater vehicle, providing frequent updates and acoustic ranging. This bounds the uncertainty to within a few meters [22], but requires a surface vehicle whose trajectory is dependent on improving the position estimate of the underwater platform.

SLAM, while useful in many ways, has difficulty in regions with very few features. (The loop closure mechanism that gives SLAM its power is ineffective if there are too few features for recognition of regions that have been visited before or if the features are of low saliency.) It can be used with a downward-looking camera in regions with a complex bottom (coral reefs in Florida [3] and Australia [73]). As part of a moving baseline network, SLAM can be used to improve the navigation error of an individual by enabling it to integrate relative pose information from teammates [71] (this was tested in ground platforms with sensing and communications capability analogous to underwater systems, not in actual underwater vehicles).

If the vehicles can recognize characteristic patterns in teammates' motion, they can use that to follow one another without requiring specific position information. This has been demonstrated on underwater robots, using vision to follow divers based solely on the divers' motions [56].

6.5.2 *Rendezvous Planning*

Some form of rendezvous planning will be necessary in all underwater vehicle operations, if only to ensure the vehicles surface at an appropriate retrieval point. If the vehicles are going to be operating at a significant distance from each other, or if the task requires tight interactions between vehicles, they will have to rendezvous with each other as well as with their support vessel. The current state of underwater vehicle rendezvous mirrors that of human rendezvous planning before the advent of the mobile phone.

With a mobile phone, individuals can easily arrange a rendezvous as a response to a particular event (e.g., "I'll call for you to pick me up when I've finished" or "Let's meet at the store"). Before mobile phones, rendezvous were arranged in advance, in more detail ("I'll be waiting for you to pick me up in 2 h, right here" or "Let's meet at the store at 2pm. I'll be wearing a blue dress and standing under the clock"). The extra initial detail eliminates the need for the follow-on conversation; the "Let's meet at the store" rendezvous is often followed by "I'm under the clock. Where are you?" "Is that you in the blue dress? I see you! I'm waving, do you see me?" Humans use detail about time, detail about place, detail about physical characteristics of the participants, and the use of recognizable landmarks to improve their chances of meeting their partner.

Applying this to underwater vehicles, researchers at McGill University [45] have demonstrated that landmark quality affects rendezvous time and have added the ability to identify appropriately salient landmarks in the environment to their robot's repertoire.

Instead of relying on landmarks that are part of the existing environment, homing beacons can be used as man-made landmarks. They are used when a specific position cannot be defined in advance or when the local environment in which the rendezvous is desired lacks appropriate landmarks. Homing beacons are designed to have extremely high saliency in any underwater environment and have been effective at ranges in excess of 50 km [14].

6.5.3 *Formations*

Formation development is an active research area, with approaches varying from potential fields (which relies on the availability of information about the relative or global position and type of each other object to be taken into account) to purely reactive pair-wise sensing (which relies on the availability of information about the relative position of a single other vehicle) to systems that include surface craft or buoys to enable sufficiently accurate position information for the team to maintain their formation even under uncertainty about their own and others' positions.

Potential fields have been used in simulation to demonstrate trajectory control in the MOOS[4] [5] architecture and with glider fleets using virtual bodies to configure the structure of the fleet. In these fleets, potential fields were used to control the relationships between each vehicle in the fleet and its relationship to the controlling virtual bodies [25].

In ground vehicles constrained to low-bandwidth communications and local sensing, researchers were able to use a single sensor with a 90-degree field of view and the ability to identify a single other robot in the group with that sensor to create a predetermined geometric shape [27]. The desired shape determines the angle at which the sensor is oriented relative to the vehicle, while the vehicle's only job is to keep the other robot centered in the field of view of the sensor and listen for any change in the desired formation. This approach only works without communication for cases where the designer has specified the potential formations in advance so that they are known to all the participants before starting the mission.

A control approach based on redundant robotic manipulator techniques used broadcast communication from an external source to enable the underwater vehicles participating in the formation to generate trajectories consistent with maintaining their formation [64]. In this case the vehicles did not respond to the broadcast message with their own messages, so this approach is independent of the number of vehicles; the same bandwidth is required for many vehicles as for two. The technique was implemented in gliders, but due to communications limitations, the control algorithm was implemented off-line at a shore-based station which computed the next trajectory for each surfaced vehicle on the basis of its most recent trajectory data.

A decentralized leader/follower mechanism [21] was demonstrated using simulated torpedo-shaped platforms traveling in formation as they performed a lawn-mower survey pattern. If the leader failed, another vehicle took over as the new leader [47].

6.5.4 Coverage

One use for team formations is to achieve the ability to cover large areas efficiently. Underwater vehicles are commonly asked to perform survey missions [30] on a variety of scales for a variety of reasons. Goals include finding objects (black boxes, mines [55]), measuring fields (power cables, ship hulls [50]), locating cracks (dams [54], post-disaster safety) and plumes (hydrothermal [75], chemical [52]), tracking organisms (krill, algae [25], coral reef monitoring [3]), and oceanographic

---

[4]MOOS, the "Mission Oriented Operating Suite," is a software architecture for autonomous mobile robots, used on ground, surface, and underwater systems. More detail is available at the Oxford Mobile Robotics Group: http://www.robots.ox.ac.uk/~mobile/MOOS/wiki/pmwiki.php/Main/Introduction.

data collection [25]. Only some of these tasks have been attempted with teams of cooperating vehicles; techniques used include swarms, potential fields, entropy calculations (most suitable for situations where time constraints are less important than overall coverage, and where multiple views of the same region are beneficial rather than inefficient), waypoint-based planners, and state machines. Some work [36] specifically compensates for "holidays" within the searched area (spaces that are not ensonified by an acoustic sensor because of variation from the planned trajectory).

6.5.5 Networks

A taxonomy of underwater networking regimens [49] plots node population against the geographic area covered and breaks them down into single-hop networks, multi-hop networks, disruption-tolerant networks, and "not a single network" situations with fragmented isolated nodes which are distant enough that they may never come into contact with one another. The majority of current implementations operate either underwater in the single-hop regime [16, 64] for survey tasks or as a surface network of disruption-tolerant nodes [17, 25, 40] for tracking and sensor network coverage tasks. Implementations of underwater sensor networks may also employ stationary bottom nodes and/or towed arrays [6]. A multi-hop network has been proposed for drifting nodes (rather than vehicles) [34], while an ad hoc self-organizing process was used to register fixed nodes and integrate mobile nodes [53]. This self-organizing network was tested with anchored sensors rather than autonomous vehicles, and while the network was able to handle collisions and manage packets using a 64-bit header, the bandwidth associated with this network was still only 50 bit/s.

One method to create an underwater network uses high-data-rate optical communications between nodes in close physical proximity, with a communications relay vehicle traveling from one node to the next [20, 70]. This approach was tested in a limited environment with clear visibility and used a simple color-coding scheme to enable the relay vehicle to detect individual (for this test, stationary) nodes in the network.

6.5.6 Bistatic Sensing

Instead of using multiple vehicles to sense a larger area than a single vehicle, multiple vehicles can also be used to obtain more information about a single object. In bistatic sensing, a single acoustic source insonifies a target and listens for a direct return while other receivers position themselves to listen for reflections at other angles. The resulting data is merged to form a more detailed model of the target than that available by listening for returns only at one location. Researchers have been able to perform tests in a test tank [12] and in the ocean using a fixed source, a fixed target, and a single vehicle as receiver [31,38]. In these tests, the single vehicle adapted its path to the sensor data (from the direct source rather than the echo) and demonstrated the sensor processing against the echo from the target, rather than combining the autonomy and the sensor processing in a single test. The accuracy of the source position and timing are critical to ensuring the accuracy of the bistatic sonar processing.

6.6 Conclusions

Communications limitations drive much of the demonstrated work using underwater vehicle teams. While the lack of communications indicates that bottom-up, decentralized solutions are preferable, many of the missions these vehicles are asked to do require top-down, centralized solutions to ensure the team meets the mission requirements. In cases where decentralized solutions are appropriate, underwater vehicle teams have demonstrated capability. Where communications are necessary in order to synchronize effort, or to rendezvous, or to ensure task completion, careful design can enable even top-down solutions to operate. Even though some of the more complex scenarios have only been tested in simulation, there is ongoing work to translate these efforts into physical demonstrations of capability.

References

1. Albus JS (2002) Metrics and performance measures for intelligent unmanned ground vehicles. In: Proceedings of the performance metrics for intelligent systems (PerMIS) workshop
2. Arslan G, Marden JR, Shamma JS (2007) Autonomous vehicle-target assignment: a game-theoretical formulation. J Dyn Syst Meas Control 129(5):584–596
3. Aulinas J, Carreras M, Lladó X, Salvi J, García R, Prados R, Petillot YR (2011) Feature extraction for underwater visual SLAM. In: Proceedings of OCEANS 2011-Europe
4. Balcan MF, Blum A, Hartline JD, Mansour Y (2008) Reducing mechanism design to algorithm design via machine learning. J Comput Syst Sci 74(8):1245–1270. doi:10.1016/j.jcss.2007.08.002
5. Barisic M, Vukic Z, Omerdic E (2009) Introduction of rotors to a virtual potentials uuv trajectory planning framework. In: Proceedings of IFAC (International Federation of Automatic Control) conference
6. Been R, Hughes DT, Vermeij A (2008) Heterogeneous underwater networks for ASW: technology and techniques. In: Undersea Defense Technology-Europe
7. Bridges A (2011) Mars rovers get bandwidth boost. Associated Press, New York. http://www.msnbc.msn.com/id/4269545/ns/technology_and_science-space/
8. Brooks R (1986) A robust layered control system for a mobile robot. IEEE J Rob Autom 2(1):14–23. doi:10.1109/JRA.1986.1087032
9. Brumitt B, Stentz A, Hebert M, Group TCU (2001) Autonomous driving with concurrent goals and multiple vehicles: Mission planning and architecture. Auton Robots 11(2):103–115. doi:10.1023/A:1011266907321

10. Casalino G, Turetta A, Simetti E, Caiti A (2010) RT^2: a real-time ray-tracing method of acoustic distance evaluations among cooperating AUVs. In: Proceedings of OCEANS 2010 IEEE-Sydney, pp 1–8. Sydney, NSW, Australia. doi: 10.1109/OCEANSSYD.2010.5603869
11. Conradt L, List C (2009) Group decisions in humans and animals: a survey. Philos Trans R Soc B 364:719–742. doi: 10.1098/rtsb.2008.0276
12. Cosci M, Caiti A, Blondel P, Jayasundere N (2005) A potential algorithm for target classification in bistatics sonar geometries. In: Pace N, Blondel P (eds) Boundary influences in high frequency, shallow water acoustics, pp 367–374. University of Bath Press, UK
13. Crandall JW, Cummings ML (2007) Developing performance metrics for the supervisory control of multiple robots. In: HRI '07 Proceedings of the ACM/IEEE international conference on human-robot interaction. doi: 10.1145/1228716.1228722. http://doi.acm.org/10.1145/1228716.1228722
14. Crees T, Kaminski C, Ferguson J, Laframboise JM, Forrest A, Williams J, MacNeil E, Hopkin D, Peterson R (2010) UNCLOS under ice survey-an historic AUV deployment in the canadian high arctic. In: Proceedings of OCEANS 2010, pp 1–8. Seattle, WA. doi: 10.1109/OCEANS2010.5664438
15. Crespi V, Galstyan A, Lerman K (2008) Top-down vs bottom-up methodologies in multi-agent system design. Auton Robots 24:303–313. doi: 10.1007/s10514-007-9080-5
16. Crosbie BP, Anderson M, Wolbrecht E, Canning J, Edwards DB (2010) Synchronous navigation of AUVs using WHOI micro-modem 13-bit communications. In: Proceedings of OCEANS 2010, pp 1–5. doi: 10.1109/OCEANS.2010.5664459
17. Das J, Py F, Maugham T, Messié M, Ryan J, Rajan K, Sukhatme GS (2010) Simultaneous tracking and sampling of dynamic oceanographic features with autonomous underwater vehicles and lagrangian drifters. In: 12th international symposium on experimental robotics
18. Djapic V, Nad D (2010) Using collaborative autonomous vehicles in mine countermeasures. In: Proceedings of OCEANS 2010 IEEE-Sydney, pp 1–7. doi: 10.1109/OCEANSSYD.2010.5603969
19. Dudek G, Roy N (2007) Multi-robot rendezvous in unknown environments or what to do when you're lost at the zoo. In: Proceedings of AAAI 2007
20. Dunbabin M, Corke P, Vascilescu I, Rus D (2006) Data muling over underwater wireless sensor networks using autonomous underwater vehicles. In: IEEE international conference on robotics and automation (ICRA), pp 2091–2098
21. Edwards DB, Bean TA, Odell DL, Anderson MJ (2004) A leader-follower algorithm for multiple AUV formations. In: 2004 IEEE/OES autonomous underwater vehicles, pp 40–46. doi: 10.1109/AUV.2004.1431191
22. Fallon MF, Kaess M, Johannsson H, Leonard JJ (2011) Efficient AUV navigation fusing acoustic ranging and side-scan sonar. In: IEEE international conference on robotics and automation (ICRA), pp 2398–2405. doi: 10.1109/ICRA.2011.5980302
23. Ferguson D, Stentz AT (2006) Using interpolation to improve path planning: the field d* algorithm. J Field Robot 23(2):79–101
24. Ferguson J (2009) Under-ice seabed mapping with AUVs. In: Proceedings of OCEANS 2009-Europe, pp 1–6. Bremen. doi: 10.1109/OCEANSE.2009.5278204
25. Fiorelli E, Leonard NE, Bhatta P, Paley D (2004) Multi-AUV control and adaptive sampling in Monterey Bay. In: Proceedings of IEEE autonomous underwater vehicles 2004: workshop on multiple AUV operations (AUV04)
26. Franks NR, Pratt SC, Mallon EB, Britton NF, Sumpter DJT (2002) Information flow, opinion polling and collective intelligence in house-hunting social insects. Philos Trans: Biol Sci 357:1567–1583 (1427, Information and Adaptive Behavior). http://www.jstor.org/stable/3558092
27. Fredslund J, Matarić MJ (2002) A general algorithm for robot formations using local sensing and minimal communication. IEEE Trans Robot Autom 18(5):837–846. doi: 10.1109/TRA.2002.803458
28. Gerkey BP, Matarić MJ (2002) Sold!: auction methods for multirobot coordination. IEEE Trans Robot Autom 18(5):758–768

29. Gerkey BP, Matarić MJ (2004) A formal analysis and taxonomy of task allocation in multi-robot systems. Int J Rob Res 23(9):939–954. doi: 10.1177/0278364904045564
30. Hagen PE, Fossum TG (2005) Applications of AUVs with SAS. In: Proceedings of OCEANS 2005, pp 1–4. Quebec City, QC, Canada. doi: 10.1109/OCEANS.2008.5152013
31. Hamilton MJ, Kemna S, Hughes D (2010) Antisubmarine warfare applications for autonomous underwater vehicles: the GLINT09 sea trial results. J Field Robot 27(6):890–902. doi: 10.1002/rob.20362
32. de Hoog J, Cameron S, Visser A (2010) Dynamic team hierarchies in communication-limited multi-robot exploration. In: 2010 IEEE workshop on safety security and rescue robotics, pp 1–7. doi: 10.1109/SSRR.2010.5981573
33. IEEE: 802.11-2007 IEEE standard for information technology-telecommunications and information exchange between systems-local and metropolitan area networks-specific requirements-part 11: wireless LAN medium access control (MAC) and physical layer (PHY) specifications. IEEE Standard (2007). doi: 10.1109/IEEESTD.2007.373646. http://ieeexplore.ieee.org/servlet/opac?punumber=4248376
34. Jaffe J, Schergers C (2006) Sensor networks of freely drifting autonomous underwater explorers. In: WUWNet '06 proceedings of the 1st ACM international workshop on underwater networks. ACM Press. doi: 10.1145/1161039.1161058
35. Johnson M, Bradshaw J, Feltovich P, Jeffers R, Jung H, Uszok A (2006) A semantically rich policy based approach to robot control. In: Proceedings of the international conference on informatics in control, automation and robotics
36. Jung YS, Lee KW, Lee BH (2008) Advances in sea coverage methods using autonomous underwater vehicles (AUVs). In: Recent advances in multi-robot systems, pp 69–100. I-Tech Education and Publishing, Vienna, Austria
37. Kaess M, Johannsson H, Englot B, Hover F, Leonard J (2010) Towards autonomous ship hull inspection using the bluefin HAUV. In: Ninth international symposium on technology and the mine problem. http://www.cc.gatech.edu/~kaess/pub/Kaess10istmp.pdf
38. Kemna S, Hamilton MJ, Hughes DT, LePage KD (2011) Adaptive autonomous underwater vehicles for littoral surveillance. Intell Serv Robot 4:245–258. doi: 10.1007/s11370-011-0097-4
39. Kinsey JC, Eustice RM, Whitcomb LL (2006) A survey of underwater vehicle navigation: recent advances and new challenges. In: IFAC conference of manoeuvering and control of marine craft. Lisbon, Portugal
40. Leonard NE, Paley DA, Davis RE, Fratantoni DM, Lekien F, Zhang F (2010) Coordinated control of an underwater glider fleet in an adaptive ocean sampling field experiment in monterey bay. J Field Robot 27(6):718–740. doi: 10.1002/rob.20366
41. Lerman K, Jones C, Galstyan A, Matarić MJ (2006) Analysis of dynamic task allocation in multi-robot systems. Int J Robot Res 25(3):225–241. doi: 10.1177/0278364906063426
42. Li B, Huang J, Zhou S, Ball K, Stojanovic M, Freitag L, Willett P (2009) MIMO-OFDM for high rate underwater acoustic communications. IEEE J Oceanic Eng 34(4):634–644. doi: 10.1109/JOE.2009.2032005
43. Marden JR, Arslan G, Shamma JS (2009) Cooperative control and potential games. J Syst Man Cybern 39(6):1393–1407
44. Matarić MJ, Sukhatme GS, Østergaard EH (2003) Multi-robot task allocation in uncertain environments. Auton Robots 14:255–263
45. Meghjani M, Dudek G (2011) Combining multi-robot exploration and rendezvous. In: Proceedings of the 8th Canadian conference on computer and robot vision (CRV'11), pp 80–85. St. John's, Newfoundland, Canada
46. Newman P, Ho K (2005) Slam-loop closing with visually salient features. In: Proceedings of the 2005 IEEE international conference on robotics and automation (ICRA), pp 635–642. doi: 10.1109/ROBOT.2005.1570189
47. Okamoto A, Feeley JJ, Edwards DB, Wall RW (2004) Robust control of a platoon of underwater autonomous vehicles. In: Proceedings of OCEANS 2004-North America

48. Palomeras M, Garcia JC, Prats M, Fernández JJ, Sanz PJ, Ridao P (2010) A distributed architecture for enabling autonomous underwater intervention missions. In: 4th annual IEEE international system conference. San Diego, CA, USA
49. Partan J, Kurose J, Levine BN (2006) A survey of practical issues in underwater networks. In: Proceedings of the ACM international workshop on underwater networks, pp 17–24. ACM Press. doi: http://doi.acm.org/10.1145/1161039.1161045
50. Pentzer J, Crosbie BP, Bean T, Canning J, Frenzel J, Anderson M, Edwards D (2010) Measurement of magnetic field using collaborative AUVs. In: Proceedings of OCEANS 2010 IEEE-Sydney, pp 1–7. doi: 10.1109/OCEANSSYD.2010.5603655
51. Pirjanian P, Huntsberger TL, Trebi-Ollennu A, Aghazarian H, Das H, Joshi SS, Schenker PS (2000) CAMPOUT: a control architecture for multi-robot planetary outposts. In: Proceedings of the SPIE symposium on sensor fusion and decentralized control in robotic systems III. Boston, MA
52. Porfiri M, Roberson DG, Stilwell DJ (2006) Environmental tracking and formation control of a platoon of autonomous vehicles subject to limited communication. In: IEEE international conference on robotics and automation (ICRA), pp 1050–4729. doi: 10.1109/ROBOT.2006.1641775
53. Rice J, Creber B, Fletcher C, Baxley P, Rogers K, McDonald K, Rees D, Wolf M, Merriam S, Mehio R, Proakis J, Scussel K, Porta D, Baker J, Hardiman J, Green D (2000) Evolution of seaweb underwater acoustic networking. In: Proceedings of OCEANS 2000, vol 3, pp 2007–2017. Providence, RI. doi: 10.1109/OCEANS.2000.882235
54. Ridao P, Carreras M, Ribas D, García R (2010) Visual inspection of hydroelectric dams using an autonomous underwater vehicle. J Field Robot 27(6):759–778. doi: 10.1002/rob.20351
55. Sariel S, Balch T, Stack J (2006) Distributed multi-AUV coordination in naval mine countermeasure missions. Technical report, Georgia Institute of Technology. http://hdl.handle.net/1853/8722
56. Sattar J, Dudek G (2009) Underwater human-robot interaction via biological motion identification. In: Proceedings of robotics: science and systems V, Seattle
57. Schmidt H, Leonard J, Edwards JR, Liu TC (2004) Sub-seabed mapping using AUV-based multi-statics acoustic sensing and adaptive control. In: International congress on the applications of recent advances in underwater detection and survey techniques to underwater archaeology
58. Shafer AJ, Benjamin MR, Leonard JJ, Curcio J (2008) Autonomous cooperation of heterogeneous platforms for sea-based search tasks. In: Proceedings of MTS/IEEE OCEANS 2008. Quebec City, QC, Canada
59. Smith RN, Chao Y, Li PP, Caron DA, Jones BH, Sukhatme GS (2010) Planning and implementing trajectories for autonomous underwater vehicles to track evolving ocean processes based on predictions from a regional ocean model. Int J Robot Res 29(12):1475–1497. doi: 10.1177/0278364910377243. http://cres.usc.edu/cgi-bin/print_pub_details.pl?pubid=646
60. Song HC, Sabra K, Kuperman WA, Hodgkiss WS (2009) Multi-static detection and localization of buried targets using synthetic aperture iterative time-reversal processing. Technical report, Scripps Institution of Oceanography / DTIC. http://handle.dtic.mil/100.2/ADA494990
61. Sotzing CC, Lane DM (2010) Improving the coordination efficiency of limited communication multi-AUV operations using a multi-agent architecture. J Field Robot 27(4):412–429. doi: 10.1002/rob.20340
62. Spears WM, Spears DF, Kerr W, Hettiarachchi S, Zarzhitsky D (2005) Optimizing interaction potentials for multi-agent surveillance: final report. Technicalreport, UW-CS-AP-WMS-ARO-1, University of Wyoming Computer Science Department. http://www.stormingmedia.us/92/9294/A929434.html
63. Stack JR, Smith CM, Hyland JC (2004) Efficient reacquisition path planning for multiple autonomous underwater vehicles. In: Proceedings of OCEANS 2004-North America, vol 3, pp 1564–1569. Kobe. doi: 10.1109/OCEANS.2004.1406355
64. Stilwell DJ, Bishop BE, Sylvester CA (2005) Redundant manipulator techniques for partially decentralized path planning and control of a platoon of autonomous vehicles. IEEE Trans Syst Man Cybern-Part B: Cybern 35(4):842–848. doi: 10.1109/TSMCB.2005.846643

65. Stokey RP, Freitag LE, Grund MD (2005) A compact control language for auv acoustic communication. In: Proceedings of OCEANS 2005-Europe, pp 1133–1137. doi: 10.1109/OCEANSE.2005.1513217
66. Tang F, Parker LE (2005) Coalescent multi-robot teaming through asymtre: a formal analysis: a formal analysis. In: Proceedings of the 12th international conference on advanced robotics, 2005, ICAR '05, pp 817–824. doi: 10.1109/ICAR.2005.1507502
67. Tang F, Parker LE (2007) A complete methodology for generating multi-robot task solutions using ASyMTRe-D and market-based task allocation. In: Proceedings of the 2007 IEEE international conference on robotics and automation (ICRA). doi: 1-4244-0602-1/07
68. Tardioli D, Mosteo AR, Riazuelo L, J.L.V (2010) Enforcing network connectivity in robot team missions. Int J Robot Res 29(4):460–480. doi: 10.1177/0278364909358274
69. Tye C, Kinney M, Frenzel J, O'Rourke M, Edwards D (2009) A MOOS module for autonomous underwater vehicle fleet control. In: Proceedings of OCEANS 2009, MTS/IEEE Biloxi, pp 1–6. http://www.mrc.uidaho.edu/cisr/pubs/jff_biloxi_09.pdf
70. Vasilescu I, Kotay K, Rus D, Dunbabin M, Corke P (2005) Data collection, storage, and retrieval with an underwater sensor network. In: SenSys '05 proceedings of the 3rd international conference on embedded networked sensor systems. ACM Press, New York, NY. doi: 10.1145/1098918.1098936
71. Walter M, Leonard J (2004) An experimental investigation of cooperative SLAM. In: Proceedings of 5th IFAC (International Federation of Automatic Control) symposium on intelligent autonomous vehicles. Lisbon, Portugal
72. Warren CW (1990) A technique for autonomous underwater vehicle route planning. IEEE J Oceanic Eng 15(3):199–204. doi: 10.1109/48.107148
73. Williams S, Mahon I (2004) Simultaneous localisation and mapping on the great barrier reef. In: IEEE international conference on robotics and automation (ICRA), pp 1771–1776. doi: 10.1009/ROBOT.2004.1308080
74. Woods Hole Oceanographic Institution: WHOI acoustic communications: Micro-modem (2006). http://acomms.whoi.edu/umodem/. (Hardware specifications)
75. Yoerger DR, Jakuba M, Bradley AM, Bingham B (2007) Techniques for deep sea near bottom survey using and autonomous underwater vehicle. Int J Robot Res 26(1):41–54. doi: 10.1177/0278364907073773

Chapter 7
Behavior Adaptation by Means of Reinforcement Learning

Marc Carreras, Andrés El-fakdi, and Pere Ridao

Abstract Machine learning techniques can be used for learning the action-decision problem that most autonomous robots have when working in unknown and changing environments. Reinforcement learning (RL) offers the possibility of learning a state-action policy that solves a particular task without any previous experience. A reinforcement function, designed by a human operator, is the only required information to determine, after some experimentation, the way of solving the task. This chapter proposes the use of RL algorithms to learn reactive AUV behaviors and therefore not having to define the state-action mapping to solve the task. The algorithms will find the policy that optimizes the task and will adapt to any environment dynamics encountered. The advantage of the approach is that the same algorithms can be applied to a range of tasks, assuming that the problem is correctly sensed and defined. The two main methodologies that have been applied in RL-based robot learning for the past 2 decades, value-function methods and policy gradient methods, are presented in this chapter and evaluated in two AUV tasks. In both cases, a well-known theoretical algorithm has been modified to fulfill the requirements of the AUV task and has been applied with a real AUV. Results show the effectiveness of both approaches, each of them with some advantages and disadvantages, and point out the further investigation of these methods for making AUVs perform more robustly and adaptively in future applications.

M. Carreras (✉) • P. Ridao
Computer Vision and Robotics Group, University of Girona, Edifici PIV, Campus Montilivi, 17071 Girona, Spain
e-mail: marc.carreras@udg.edu; pere.ridao@udg.edu

A. El-fakdi
Control Engineering and Intelligent Systems Group, University of Girona, Edifici PIV, Campus Montilivi, 17071 Girona, Spain
e-mail: andres.elfakdi@udg.edu

M.L. Seto (ed.), *Marine Robot Autonomy*, DOI 10.1007/978-1-4614-5659-9_7,

7.1 Introduction

Over the past few years, considerable research has been conducted to extend the range of AUV applications. Industries are interested in several underwater scenarios, such as environmental monitoring, oceanographic research, or maintenance of underwater structures, where AUVs could play an important role. AUVs are attractive for utilization in these areas because of their reduced operation cost and their non-reliance on human operators. However, in comparison with humans in terms of efficiency and adaptability, they are still very rudimentary. The development of autonomous control systems able to deal with such issues becomes a priority. The use of AUVs for unknown dynamic underwater environments is a very complex problem, mainly when the AUV is required to react in real time to unpredictable changes in the environment. Machine learning techniques can be applied to face this uncertainty and to adapt in real time to changes. This chapter proposes the use of robot learning techniques for improving the autonomy and robustness of AUVs. The purpose is to demonstrate the feasibility of learning algorithms in some basic AUV tasks.

A commonly used methodology in robot learning is *reinforcement learning* (RL) [41]. RL is a very suitable technique to learn in unknown environments. Unlike supervised learning methods, RL does not need any database of examples. Instead, it learns from interaction with the environment and according to a scalar value. This scalar value, or *reward*, evaluates the environment state and the last taken action with reference to a given task. The final goal of RL is to find an *optimal state/action mapping* which maximizes the sum of future rewards whatever the initial state is. The learning of this optimal mapping or *policy* is also known as the *reinforcement learning problem* (RLP).

The specific features in which RL algorithms work make this learning theory useful for robotics. There are parts of a robot control system which cannot be implemented without experiments. For example, when implementing a reactive robot behavior, the main strategies can be designed without any real test. However, for the final tuning of the behavior, there will always be parameters which have to be set with real experiments. A dynamics model of the robot and environment could avoid this phase, but it is usually difficult to achieve this model with reliability. RL offers the possibility of learning the behavior in real time and avoid the tuning of the behaviors with experiments. RL automatically interacts with the environment and finds the best mapping for the proposed task, which in this example would be the robot behavior. The only necessary information which has to be set is the *reinforcement function* which gives the rewards according to the current state and the past action. It can be said that by using RL, the robot designer reduces the effort required to implement the whole behavior to the effort of designing the reinforcement function. This is a great improvement since the reinforcement function is much simpler and does not contain any dynamics. There is another advantage in that an RL algorithm can be continuously learning, and, therefore, the state/action mapping will always correspond to the current environment. This is an important feature in changing environments.

The dominant approach 10 years ago was to apply RL using the value-function (VF) approach. As a result, many RL-based control systems have been applied to robotics. In [37], an instance-based learning algorithm was applied to a real robot in a corridor-following task. For the same task, in [21], a hierarchical memory-based RL was proposed, obtaining good results as well. In [13], an underwater robot learns different behaviors using a modified Q-learning algorithm [47]. VF methodologies have worked well in many applications, achieving great success with discrete lookup table parameterization but giving few convergence guarantees when dealing with high-dimensional domains due to the lack of *generalization* among continuous variables [41]. RL is usually formulated using finite Markov decision processes (FMDP). This formulation implies a discrete representation of the state and action spaces. However, in some tasks, the states and/or the actions are continuous variables. A first solution can be to maintain the same RL algorithms and discretize the continuous variables. If a coarse discretization is applied, the number of states and actions will not be too high and the algorithms will be able to learn. However, in many applications, the discretization must be fine in order to assure a good performance. In these cases, the number of states will grow exponentially, making the use of RL impractical. The reason is the high number of iterations necessary to update all the states or state/action pairs until an optimal policy is obtained. This problem is known as the *curse of dimensionality*. In order to solve this problem, most RL applications require the use of generalizing function approximators such as artificial neural networks (ANN), instance-based methods, or decision trees. Another feature of VF methods is that these approaches are oriented to finding deterministic policies. However, stochastic policies can yield considerably higher expected rewards than deterministic ones as in the case of partially observable FMDP (POFMDP), selecting among different actions with specific probabilities [35]. Furthermore, some problems may appear when the state space is not completely observable; small changes in the estimated value of an action may or may not cause it to be selected, resulting in convergence problems [12].

Rather than approximating a VF, policy gradient (PG) search techniques approximate a policy using an independent function approximator with its own parameters, trying to maximize the future reward expected [29]. The advantages of PG methods over VF-based methods are various. The main advantage is that using a function approximator to represent the policy directly solves the generalization problem. Working this way should represent a decrease in the computational complexity, and, for learning systems which operate in the physical world, the reduction in time consumption would be enormous. Furthermore, learning systems should be designed to explicitly account for the resulting violations of the Markov property. Studies have shown that stochastic policy-only methods can obtain better results when working in POFMDP than those obtained with deterministic value-function methods [35]. In [3] a comparison between a policy-only algorithm [8] and a value method [47] is presented where the value method oscillates between the optimal and a suboptimal policy, while the policy-only method converges to the optimal policy. Attempts to apply PG algorithms to real robotic tasks have shown slow convergence, in part caused by the small gradients around plateau landscapes of the expected return [28]. In order to avoid these situations, studies presented in [2] pointed that the natural gradient may be a good alternative to the *typical* PG. Being easier to estimate than regular PG, it was expected to be more efficient and therefore accelerates the convergence process. Natural gradient algorithms have found a variety of applications over the last few years, as in [34] with traffic-light system optimization and in [46] with gait optimization for robot locomotion. On the other hand, PG applications share a common drawback, gradient estimators used in these algorithms may have a large variance [22, 25], learning much more slowly than RL algorithms using a VF (see [42]), and they can converge to local optima of the expected reward [26], making them less suitable for online learning in real applications. In order to decrease convergence time and avoid local optima, the newest applications combine PG search with VF techniques, adding the best features of both methodologies [7, 22]. These techniques are commonly known as actor critic (AC) methods. In AC algorithms, the critic component maintains a VF, and the actor component maintains a separate parameterized stochastic policy from which the actions are drawn. The actor's PG gives convergence guarantees, while the critic's VF reduces variance of the policy update, improving the convergence rate of the algorithm [22].

This chapter presents the results of applying a VF method and a PG method for learning the state-action mapping of a reactive AUV behavior for solving a simple task. For each method, the algorithm, the application, and the results are detailed. The goal of this chapter is to present the application of the two main approaches for robot learning with RL (VF and PG) on a real AUV task. Although results show the effectiveness of the system, further work will consist on applying the methods in a more realistic environment and in a more challenging task, such as intervention with AUVs. To the authors best knowledge, this research constitutes the state of the art of robot learning for AUVs.

The structure of the chapter will be as follows. In Sect. 7.2, the basic theory about RL is presented, explaining basic concepts, the use of FMDP, and the basic idea of value-function methods based on temporal differences and policy gradient methods. Details on these two approaches are given in the case studies presented in Sects. 7.3 and 7.4. Section 7.3 is concerned with VF methods. It details the Q-learning algorithm [47] and how it is modified to solve the generalization problem for its application in a target-following task with URIS AUV. Details of the implementation and results are given with real experiments in a water tank. Section 7.4 is concerned with PG methods. It details the natural actor critic algorithm [29] and how it is implemented to speed up the learning process. The algorithm is applied to solve a cable-tracking task with the Ictineu AUV. Details of the implementation and results are also given with real experiments. Finally, the chapter is concluded in Sect. 7.5 pointing out the further work to be accomplished in order to make AUVs exhibit learning capabilities in their future applications.

7.2 The Reinforcement Learning Problem

Reinforcement learning (RL) is an approach to learning by trial and error in order to achieve a goal. Basic to the learning process is the interaction between the RL algorithm and an environment. An RL algorithm does not use a set of examples which show the desired input/output response, as do *supervised learning* techniques. Instead, a reward given by the environment is required. This reward evaluates the current state of the environment. The *reinforcement learning problem* (RLP) consists of maximizing the sum of future rewards. The goal to be accomplished by RL is encoded in the received reward. To solve the problem, an RL algorithm acts over the environment in order to yield maximum rewards. Any algorithm able to solve the RLP is considered an RL algorithm. This section is an overview of RL theory; for a deeper understanding of this field, refer to the book "Reinforcement Learning: An Introduction," written by Sutton and Barto [41].

7.2.1 *Basic Concepts*

Several elements appear in all RLPs. The first is the RL algorithm or the *learner*. The learner interacts with an *environment*. The environment comprises everything outside the learner. The learner can usually observe the *state* of the environment and interacts with it by generating an *action*. The environment reacts to the action with a new state and a *reward*. The reward is a scalar value generated by the *reinforcement function* which evaluates the current state and/or the last executed action according to the RLP. This learner/environment interaction can be represented in a discrete time base as in the diagram shown in Fig. 7.1. In each time iteration t, the learner observes the state s_t and receives the reward r_t. Following the rules of the RL algorithm, the learner generates an action a_t. The environment reacts to the action changing to state s_{t+1} and generates a new reward r_{t+1}. Both will be taken during the next iteration. The state space $\{S\}$ can contain different variables (n-dimensional) and the action space $\{A\}$ different actions (m-dimensional). The time notation used for the states, rewards, and actions is interpreted as: being at state s_t and applying action a_t, the environment reacts with a reward r_{t+1} and a new state s_{t+1}.

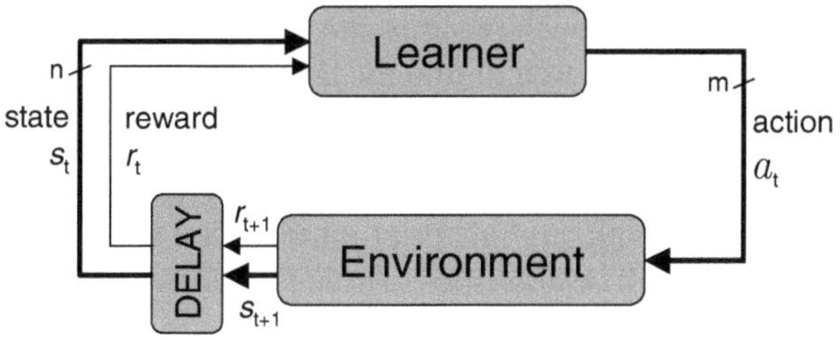

Fig. 7.1 Diagram of the learner/environment interaction

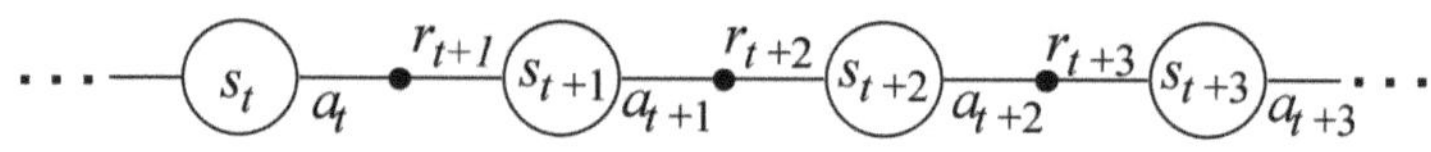

Fig. 7.2 Sequence of states, actions, and rewards

A sequence of states, actions, and rewards is shown in Fig. 7.2, in which the states are represented as circles, the actions as lines, and the rewards as dots. The most important features of the learner and environment are enumerated hereafter:

Learner

- Performs the learning and decides the actions
- Input: the state s_t and reward r_t (numerical value)
- Output: an action a_t
- Goal: to maximize the amount of rewards $\sum_{i=t+1}^{\infty} r_i$

Environment

- Everything outside the learner.
- It reacts to actions with a new state.
- Contains the reinforcement function which generates the rewards.

The learner interacts with the environment in order to find the action which maximizes the sum of future rewards. The action to be applied will always depend on the current state. Therefore, the learner needs a *mapping function* which relates each state with an action. Two processes are commonly used to find the best actions: *exploration* and *exploitation*. Exploration means that the learner will have to select the actions which have not yet been used for a particular state. On the other hand, in order to solve the RLP, the best known actions must be exploited. The compromise between exploration and exploitation is a common feature in RL. It is important to note that the final goal of RL is to maximize the sum of future rewards. The best state/action mapping can propose actions which do not maximize immediate rewards, but actions which maximize the long-term sum of rewards. The effect of an action can also be delayed in terms of the received reward. An action can be the best one to solve the RLP, but high rewards may not come until after some iterations.

The main keys of reinforcement learning which have been related are the following:

- Learning through rewards, not through examples.
- Trial-error search to find the best state/action mapping.
- Exploration/exploitation dilemma.
- Delayed rewards, the effect of the actions is not immediately rewarded.

Before finalizing this brief introduction to reinforcement learning, the basic RL functions will be described. These functions constitute the definition of the RLP and the tools needed to find the solution.

- Policy Function. A policy function indicates the action to be taken at each moment. A policy is a mapping between the states and the actions and is contained in the learner. When this mapping is stochastic, the policy is usually represented with the $\pi(s,a)$ function, which contains the probability of choosing action a from state s. When a policy solves the RLP, that is, the action related to each state maximizes the sum of future rewards, it is said that the policy is *optimal* π^*. A policy can be *random*, in which case the action will be chosen randomly. A *greedy* policy contains the best mapping known by the learner, that is, the actions supposed to best solve the RLP. A *greedy action* is an action taken from a greedy policy. Finally, a ε-greedy policy selects random actions with ε probability and greedy actions with $(1-\varepsilon)$ probability. The policy, in the examples that will be presented in the following sections, can also be considered as the controller that is learned and that guides the robot.
- Reinforcement Function. The reinforcement function is located in the environment and indicates the goal to be accomplished by the learner. A reinforcement function $r(s,a)$ is a deterministic function which relates a state and/or an action to a single numerical value or reward. The learner uses the reward to learn the optimal policy. The reinforcement function gives an immediate evaluation of the learner, which is not the criterion to be improved in reinforcement learning.
- Value Function. The value function contains the expected sum of future rewards following a particular policy. In general the policy will be stochastic, although it would also apply for deterministic policies. This function is used and modified by the learner to learn the optimal policy. There are two kinds of value functions. The *state-value* function $V^\pi(s)$ contains the expected sum of future rewards from state s followed by the policy π. The *action-value* function $Q^\pi(s,a)$ contains the expected sum of future rewards from state s, applying the action a, and following afterwards the policy π. The action-value function $Q^\pi(s,a)$ has a second dimension as input, the action. When the action chosen is the same than the one proposed by a deterministic policy π, $Q^\pi(s,\pi(s))$ will be equal to the state-value function $V^\pi(s)$. Unlike the reinforcement function, the value functions give the long-term evaluation of the learner following a particular policy. These functions directly show the performance of the learner in solving the RLP. When following the optimal policy π^*, the value functions $V^*(s)$ and $Q^*(s,a)$ show the maximum sum of expected rewards, and therefore, the functions are also optimal. In the same way, if the learner achieves the optimal value functions, the optimal policy can be extracted from it.
- Dynamics Function. This function is located in the environment and is usually not known. A dynamics function is usually a stochastic function which generates the future state of the environment from its previous state and the applied action. Although this function is not known, the state transitions caused by the dynamics are contained in the value functions.

A RL algorithm uses these functions to solve the RLP. The learning process is performed over one of the value functions, the state, or the action-value functions.

The algorithm proposes a learning update rule to modify the value function and also proposes a policy to be followed by the learner-environment interaction. If the RL algorithm converges after some iterations, the value function has changed to the optimal value function from which the optimal policy can be extracted. The solution of the RLP is accomplished by following the state-action mapping contained in this optimal policy π^*.

7.2.2 RL with Finite Markov Decision Processes

Reinforcement learning theory is usually based on *finite Markov decision processes* (FMDP). The use of FMDP allows a mathematical formulation of the RLP; therefore, the suitability of RL algorithms can be mathematically demonstrated. In order to express an RLP with FMDP, the environment has to accomplish the *Markov property*. An environment which accomplishes the Markov property shows in its state all the relevant information to predict the next state. This means that in order to predict the next state of the environment, the knowledge of the past states and actions is not necessary if the present state and action are known. At the same time, the future reward does not depend on past states and actions. Therefore, in a Markovian state, the prediction of the next state and reward is accomplished by knowing only the present state and action. Equation 7.1 formally defines the Markov property equaling the probability of predicting the next state and reward with and without knowing the past states, actions, and rewards:

$$\begin{aligned} Pr\{s_{t+1} = s', r_{t+1} = r' | s_t, a_t, r_t, s_{t-1}, a_{t-1}, \ldots, s_0, a_0\} \\ = Pr\{s_{t+1} = s', r_{t+1} = r' | s_t, a_t\}. \end{aligned} \tag{7.1}$$

If the environment accomplishes the Markov property, it is considered a Markov decision process (MDP). Moreover, if the state and action spaces are finite, the environment is considered a finite MDP (FMDP). Having discrete state and action spaces, the stochastic dynamics of the environment can be expressed by a transition probability function $P^a_{ss'}$. This function contains the probability of reaching the state s' if the current state is s and action a is executed. Equation 7.2 formally defines this function. Also, the reinforcement function can be expressed as a deterministic function $R^a_{ss'}$. In this case, the function contains the expected reward received when being at s, a is applied and s' is reached; see Eq. 7.3:

$$P^a_{ss'} = Pr\{s_{t+1} = s' | s_t = s, a_t = a\}. \tag{7.2}$$

$$R^a_{ss'} = E\{r_{t+1} | s_t = s, a_t = a, s_{t+1} = s'\}. \tag{7.3}$$

$P^a_{ss'}$ and $R^a_{ss'}$ specify the dynamics of the environment. One of the learner's functions, the value function, is highly related with this dynamic. For a particular learner policy π, the value function (V^π or Q^π) can be expressed in terms of $P^a_{ss'}$ and $R^a_{ss'}$ and also in terms of the expected future reward R_t. In the case of the *state-value* function, the value of a state s under a policy π, denoted $V^\pi(s)$, is the expected discounted sum of rewards when starting in s and following π thereafter. For the *action-value* function, the value of taking action a in state s under policy π, denoted $Q^\pi(s,a)$, is the expected discounted sum of rewards when starting in s, the action a is applied, and following π thereafter. Equations 7.4 and 7.5 formally define these two functions:

$$V^\pi(s) = E_\pi\{R_t|s_t = s\} = E_\pi\left\{\sum_{k=0}^{\infty}\gamma^k r_{t+k+1}|s_t = s\right\}. \tag{7.4}$$

$$Q^\pi(s,a) = E_\pi\{R_t|s_t = s, a_t = a\} = E_\pi\left\{\sum_{k=0}^{\infty}\gamma^k r_{t+k+1}|s_t = s, a_t = a\right\}. \tag{7.5}$$

Optimal value functions can be obtained using the Bellman equation [9]. This equation relates the value of a particular state or state/action pair with the value of the next state or state/action pair. To relate the two environment states, the dynamics of the FMDP ($P^a_{ss'}$ and $R^a_{ss'}$) is used. The *Bellman optimality equations* for the state and action-value functions are found in Eqs. 7.6 and 7.7:

$$V^*(s) = \max_a \sum_{s'} P^a_{ss'}[R^a_{ss'} + \gamma V^*(s')]. \tag{7.6}$$

$$Q^*(s,a) = \sum_{s'} P^a_{ss'}[R^a_{ss'} + \gamma \max_{a'} Q^*(s',a')]. \tag{7.7}$$

The Bellman optimality equations offer a solution to the RLP by finding the optimal value functions V^* and Q^*. If the dynamics of the environment is known, a system of equations with N equations and N unknowns can be written using Eq. 7.6, being N the number of states. This nonlinear system can be solved getting the V^* function. Similarly, the Q^* function can be found.

Once V^* is known, the optimal policy can be easily extracted. For each state s, any action a which causes the environment to achieve a state s' with maximum state value with respect to the other achievable states can be considered as an *optimal action*. The set of all the states with its corresponding optimal actions constitutes an optimal policy π^*. It is important to note that to find each optimal action, it is only necessary to compare the state value of the next achievable states. This is due to the fact that the state-value function V^* already contains the expected discounted sum of rewards for these states. In the case where the Q^* is known, the extraction of the optimal policy π^* is even easier. For each state s, the optimal action will be the action a which has a maximum $Q^*(s,a)$ value; see Eq. 7.8:

$$\pi^*(s) = \arg\max_{a \in A(s)} Q^*(s,a). \tag{7.8}$$

7.2.3 Temporal Difference Learning

There are several methodologies to solve the reinforcement learning problem formulated as an FMDP: dynamic programming, Monte-Carlo methods and temporal difference learning [41].

Temporal difference learning (TD) is the most suitable methodology for solving the RLP when the dynamics is not known and when the learning has to be performed online, as in most robotics tasks. TD learning is able to learn online; it uses bootstrapping to estimate the unknown value functions. As a consequence, it can be applied in continuing tasks. Moreover, TD learning does not need the dynamics of the environment. Instead, it uses the experience with the environment. The general update rule for the state-value function can be seen in Eq. 7.9. The updating of the state value $V^{\pi}_{k+1}(s_t)$ is accomplished by comparing its current value with the discounted sum of future rewards. This sum is estimated with the immediate reward r_{t+1} plus the discounted value of the next state. This bootstrapping allows the learning to be online. Also, the update rule does not require the dynamics transition probabilities, which allow TD algorithms to learn in an unknown environment.

$$V^{\pi}_{k+1}(s_t) \leftarrow V^{\pi}_{k}(s_t) + \alpha[r_{t+1} + \gamma V^{\pi}_{k}(s_{t+1}) - V^{\pi}_{k}(s_t)]. \tag{7.9}$$

A very important feature of RL algorithms is the policy they must follow in order to guarantee the convergence. There are algorithms that can only follow a particular policy. That is, at each time step, the action to be executed by the learner is determined by a fixed rule. If the learner does not follow this action, the algorithm may not converge to the RLP solution. These algorithms are called *on-policy* algorithms. Exploration in these algorithms is also done, but not changing randomly the action in the middle of an episode. On the other hand, there are algorithms which do not impose a particular policy to ensure the convergence. Instead, these algorithms usually need to visit all the states and actions regularly. The second kind of algorithm is called an *off-policy* algorithm. It is usually desirable to use an off policy algorithm in a real system, since the actions can be externally modified by a supervisor without interrupting the learning process. The algorithm described in Sect. 7.3, the Q-learning, is an off-policy algorithm that has been widely used in a lot of robotic applications.

7.2.4 Policy Gradient Algorithms

Most of the methods proposed in the reinforcement learning community are not applicable to high-dimensional systems as these methods do not scale beyond systems with more than three or four degrees of freedom and/or cannot deal with parameterized policies [28]. Policy gradient (PG) methods are a notable exception to this statement. Starting with the work in the early 1990s [11,19], these methods have been applied to a variety of robot learning problems ranging from simple control tasks [10] to complex learning tasks involving many degrees of freedom [30]. The advantages of Policy Gradient (PG) methods for robotics are numerous. Among the most important are that they have good generalization capabilities which allow them to deal with big state spaces, that their policy representations can be chosen so that it is meaningful to the task and can incorporate previous domain knowledge, and that often fewer parameters are needed in the learning process than in value-function (VF)-based approaches. Also, there is a variety of different algorithms for PG estimation in the literature which have a rather strong theoretical underpinning. In addition, PG methods can be used model-free [41] and therefore also be applied to robot problems without an in-depth understanding of the problem or mechanics of the robot [29]. Studies have shown that approximating a policy directly can be easier than working with Value Functions (VFs) [3, 42], and better results can be obtained. Informally, it is intuitively simpler to determine *how to act* instead of *value of acting* [1]. So, rather than approximating a VF, new methodologies approximate a stochastic policy using an independent *function approximator* with its own parameters, trying to maximize the future reward expected. In Eq. 7.10 we can see that in the PG approach, the eligibility $e_t(\theta)$ is represented by the gradient of a given stochastic policy $\pi(a_t|s_t,\theta_t)$ with respect to the current parameter vector θ_t as

$$e_t(\theta) = \frac{\mathrm{d}\log\pi(a_t|s_t,\theta_t)}{\mathrm{d}\theta_t}. \tag{7.10}$$

In VF, the eligibility e_t represents a record of the number of visits of the different states while following current policy π. In the PG approach, $e_t(\theta)$ is bound to the policy parameters. It specifies a correlation between the associated policy parameter θ_i and the executed action a_t while following current policy π. Therefore, while the VF's eligibility indicates *how much* a particular state is *eligible* for receiving learning changes, PG's eligibility indicates *how much* a particular parameter of our policy is *eligible* to be improved by learning. The final expression for the policy parameter update is

$$\theta_{t+1} = \theta_t + \alpha r_{t+1} e_t(\theta). \tag{7.11}$$

Here, α is the learning rate of the algorithm, and r_t is the immediate reward perceived. At every iteration step, the algorithm proposes a learning update rule to modify a parameterized policy function $\pi(a|s,\theta)$. This rule is based on the policy derivative with respect to the policy parameters and the immediate reward r_{t+1}. Once the PG algorithm converges, the current policy π becomes the optimal policy π^*. The solution of the reinforcement learning problem (RLP) is accomplished by following optimal policy π^*.

7.3 Value-Function Algorithms: Target-Following Behavior

7.3.1 Q-Learning

The Q-learning algorithm [47] is a temporal difference algorithm that uses the action-value function Q^{π} to find an optimal policy π^*. The $Q^{\pi}(s,a)$ function has an advantage with respect to the state-value function V^{π}. Once the optimal function Q^* has been learned, the extraction of an optimal policy π^* can be directly performed without the requirement of the environment dynamics or the use of an actor. The optimal policy will be composed of a mapping relating each state s with any action a which maximizes the $Q^*(s,a)$ function.

Another important feature of Q-learning is the off-policy learning capability. That is, in order to learn the optimal function Q^*, any policy can be followed. The only condition is that all the state/action pairs must be regularly visited and updated. This feature, together with the simplicity of the algorithm, makes Q-learning very attractive for a lot of applications. In real systems, as in robotics, there are many situations in which not all the actions can be executed. For example, to maintain the safety of a robot, an action cannot be applied if there is any risk of colliding with an obstacle. Therefore, if a supervisor module modifies the actions proposed by the Q-learning algorithm, the algorithm will still converge to the optimal policy.

The Q-learning algorithm can be shown in Algorithm 1. As can be seen, the policy applied is the ε-greedy policy, although any other policy could have been used as well. Recovering the definition of the action-value function, the $Q^*(s,a)$ value is the discounted sum of future rewards when action a is executed from state s, and the optimal policy π^* is followed afterward. To estimate this sum, Q-learning uses the received reward r_{t+1} plus the discounted maximum value of the future state s_{t+1}. The first step of the algorithm is to initialize the values of the Q-function for all the states s and actions a randomly . After that, the algorithm starts interacting with the environment in order to learn the Q^* function. In each iteration, the update function needs an initial state s_t, the executed action a_t, the new state s_{t+1} which has been achieved, and the received reward r_{t+1}. The algorithm updates the value of $Q(s_t,a_t)$, comparing its current value with the sum of r_{t+1} and the discounted

---

Algorithm 1: Q-learning algorithm

---

1. Initialize $Q(s,a)$ arbitrarily.
2. Repeat until Q is optimal:
   (a) $s_t \leftarrow$ the current state.
   (b) Choose action a_{max} that maximizes $Q(s_t,a)$ over all a.
   (c) $a_t \leftarrow (\varepsilon - greedy)$ action, carry out action a_{max} in the world with probability $(1-\varepsilon)$, otherwise apply a random action (exploration).
   (d) Observe the reward r_{t+1} and the new state s_{t+1}.
   (e) $Q(s_t,a_t) \leftarrow Q(s_t,a_t) + \alpha[r_{t+1} + \gamma \cdot max_{a_{max}} Q(s_{t+1},a_{max}) - Q(s_t,a_t)]$.

---

maximum Q value in s_{t+1}. The error is reduced with a learning rate and added to $Q(s_t, a_t)$. When the algorithm has converged to the optimal Q^* function, the learning process can be stopped. The parameters of the algorithm are the discount factor γ, the learning rate α, and the ε parameter for the random actions.

7.3.2 Generalization Problem

The FMDP formulation implies a discrete representation of the state and action spaces. However, in some tasks, the states and/or the actions are continuous variables. A first solution can be to maintain the same RL algorithms and to discretize the continuous variables. If a coarse discretization is applied, the number of states and actions will not be too high and the algorithms will be able to learn. However, in many applications, the discretization must be fine in order to assure a good performance. In these cases, the number of states will grow exponentially, making the use of RL impractical. The reason is the high number of iterations necessary to update all the states or state/action pairs until an optimal policy is obtained. This problem is known as the *curse of dimensionality*. To make RL feasible, *generalization* must be applied among the state and actions. To solve the generalization problem, RL algorithms modify the classical algorithms replacing the discrete spaces with a *function approximator*.

The generalization problem is a very common problem in robotics and in control tasks in general. However, there are plenty of successful approaches in which function approximation techniques have been applied. The reason why there are several techniques to solve the generalization problem is because any one of them offers a perfect solution. Some techniques have a higher generalization capability, while others are computationally faster. However, the most important feature is their capability to converge into an optimal policy. Convergence proofs have only been obtained for algorithms which use a linear function approximator [17, 36, 38, 45], that is, a function which is linear with respect to a set of parameters. In addition, in order to maintain the stability, the learning must be done on policy. This means that the Q-learning algorithm, for example, cannot profit from this convergence proof [31]. TD algorithms estimate the value function based on the immediate rewards and on the same value function (bootstrapping feature). When using a function approximator with TD algorithms, the value function will always be an approximation of the discrete function. It has been verified that off-policy learning algorithms with linear function approximators can cause divergence. Some classical examples of this divergence can be found in [5,44]. Several methods were developed to deal with this divergence problem, for example, residual algorithms [5], averages [18], and interpolative representations [44]. These methods use special update rules and/or function approximators to ensure the convergence. However, their use is very limited in practice since their convergence is much slower, and they have a lower generalization capability.

Despite convergence proofs, there are many successful examples including linear and nonlinear approximators with on-policy and off-policy algorithms. Some references are [16, 24, 40, 49]. The most important breakthrough was the use of a neural network to learn the game of backgammon [43]. Using this nonlinear function approximator, the algorithm was able to play at the same level as the best human players in the world. These successful applications have motivated the use of a diverse set of function approximation techniques for real systems.

7.3.3 *SONQL Algorithm*

In order to deal with the generalization problem, the semi-online neural-Q-learning algorithm (SONQL) has been proposed as a modification of the Q-learning able to work with continuous states and actions [15]. The SONQL algorithm has two main components: (1) a neural network that approximates the Q-function, which is called NQL, and (2) a database of learning samples that allows the execution of the NQL in semi-online mode, thus getting the SONQL algorithm.

7.3.3.1 Neural Q-Learning

In order to approximate the Q-function, a feed-forward multilayer neural network has been used. This architecture allows the approximation of any nonlinear function assuming that the number of layers and neuron and the activation functions are appropriated. The input variables of the NN are the environment states and the actions, which have n and m dimensions, respectively. The output of the network has only one dimension and corresponds to the Q value for the current states and actions. The number of hidden layers will depend on the complexity of the Q-function. Figure 7.3 shows a schema of the network.

The use of neural networks in the Q-learning algorithm is known as neural Q-learning (NQL). There are several approaches in which NN can be applied. In particular, the approximation of the Q-function using a feed-forward NN is known as *direct* Q-learning [5]. This is the most straightforward approach since the whole function is approximated in only one NN. This implementation is affected by the interference problem [48], which will be solved in Sect. 7.3.3.2. The technique used to learn the Q-function is the *backpropagation* algorithm [20]. This algorithm uses the error between the output neuron and the desired response to adapt the weights of the network.

To compute the desired response of the NN, the update equation of Q-learning algorithm is used. The Q value for a given state and action is equal to Eq. 7.12. This means that the desired response of the NN has to be equal to Eq. 7.12, and the committed error will be used to update the weights of the network:

$$Q(s_t, a_t) = r_{t+1} + \gamma \cdot \max_{a_{\max}} Q(s_{t+1}, a_{\max}). \tag{7.12}$$

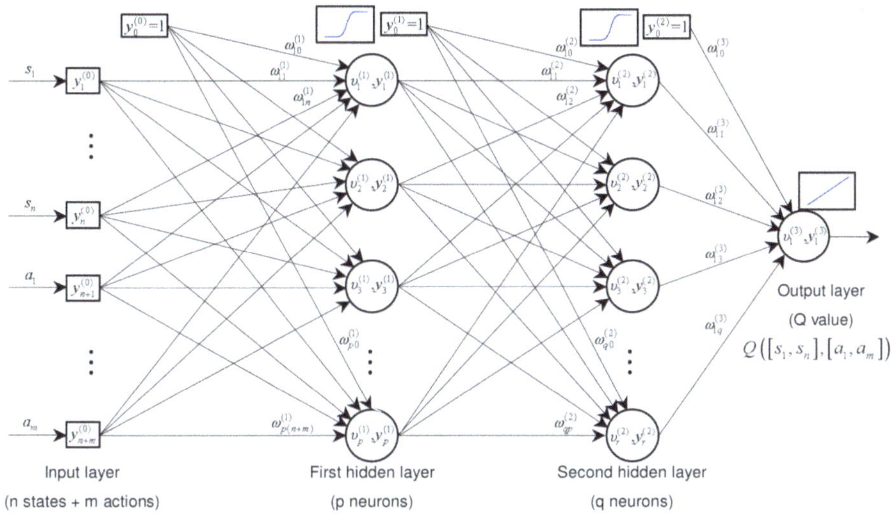

Fig. 7.3 Graph of the multilayer NN which approximates the Q-function. In this case, two hidden layers are used. The input layer is composed of the states and the actions. The output layer has only one neuron which contains the Q value for the current input values

As can be seen, the computation of Eq. 7.12 requires the use of the same NN to calculate $Q(s_{t+1}, a_{\max})$. In order to find the action $a_{\max}$, which maximizes the Q value for state s_{t+1}, a simple strategy is used. The action space is discretized in a set of actions according to the smallest resolution distinguished in the environment. For each action, the Q value is computed, and the maximum Q value is used. If the SONQL algorithm is applied in a task in which there is only one continuous action, the computational cost associated with the searching of $a_{\max}$ does not represent a problem. However, if more continuous actions are present, the required time can increase appreciably and the feasibility of the algorithm decreases.

It is very important to note that two different learning processes are simultaneously in execution and with direct interaction. First of all, the Q-learning algorithm updates the Q values in order to converge to the optimal Q-function. On the other hand, the NN algorithm updates its weights to approximate the Q values. While the Q-function is not optimal, both processes are updating the weights of the network to fulfill its learning purposes. It is clear that the stability and convergence of the NQL algorithm can be seriously affected by this dual learning. The next section will focus on this issue.

Finally, Eq. 7.12 shows that the necessary variables to update the neural Q-function are the initial state s_t, the taken action a_t, the received reward r_{t+1}, and the new state s_{t+1}. These four variables (the states and actions can be multidimensional) constitute a *learning sample*. This term will be used in the following subsections.

7.3.3.2 Semi-online Learning

The problem neural networks have when used to generalize with an RL algorithm is known as the *interference problem*. Interference in NN occurs when learning in one zone of the input space causes loss of learning in other zones. It is specially prevalent in online applications where the learning process is done according to the states and actions visited rather than with some optimal representation of all the training data [48]. The cause of this problem is that two learning processes are actuating at the same time, and each process is based on the other. Q-learning uses the NN to update the Q values, and the NN computes the error of the network according to Q-learning algorithm. This dual learning makes the NQL algorithm very unstable, which is known as the interference problem [48]. An important problem is that each time the NN updates the weights, the whole function approximated by the network is slightly modified. If the NQL algorithm updates the network using learning samples, which are all located in the same state/action zone, the non-updated state/action space will be also be affected. The result is the state/action zones which have been visited and learned are no longer remembered. If the NQL algorithm is updating different state/action zones but with no homogeneity, the interaction between Q-learning and the NN can cause instability.

The solution to the interference problem is the use of a network which acts locally and assures that learning in one zone does not affect other zones. Approaches with radial basis functions have been proposed to this end [48]; however, this implies abandoning the high-generalization capability of a multilayer NN with backpropagation. The solution used in this approach proposes the use of some memory or *database* of *learning samples*. This solution was suggested in [32]; although to the author's best knowledge, there are no proposals which use it. The main goal of the database is to include a representative set of visited learning samples, which is repeatedly used to update the NQL algorithm. The immediate advantage of the database, is the stability of the learning process and its convergence even in difficult problems. Due to the representative set of learning samples, the Q-function is regularly updated with samples of the whole visited state/action space, which is one of the conditions of the original Q-learning algorithm. A consequence of the database is the acceleration of the learning. This second advantage is most important when using the algorithm in a real system. The updating of the NQL is done with all the samples of the database, and, therefore, the convergence is achieved with less iterations.

It is important to note that the learning samples contained in the database are samples which have already been visited. Also the current sample is always included in the database. The use of the database changes the concept of online learning which Q-learning has. In this case, the algorithm can be considered as *semi-online*, since the learning process is based on current as well as past samples.

Each learning sample, as defined before, is composed of the initial state s_t, the action a_t, the new state s_{t+1}, and the reward r_{t+1}. During the learning evolution, the learning samples are added to the database. Each new sample replaces older samples previously introduced. The replacement is based on the geometrical distance

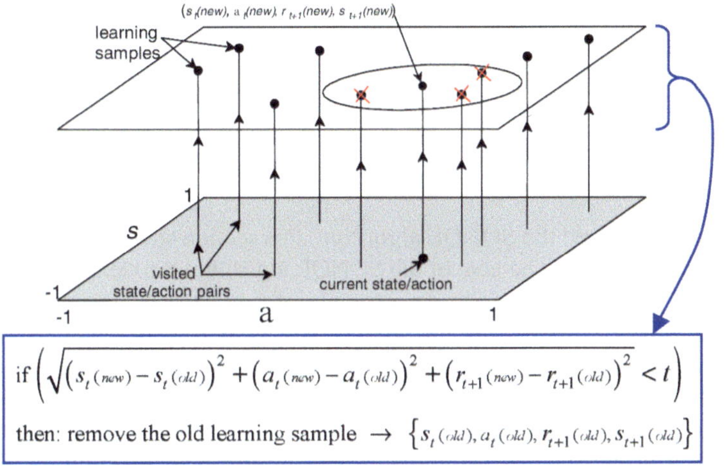

Fig. 7.4 Representation of the learning sample database. The state and action have only one dimension. The replacement rule for all the old samples is also shown

between vectors (s_t, a_t, r_{t+1}) of the new and old samples. If this distance is less than a *density parameter t* for any old sample, the sample is removed from the database. The size of the database is, therefore, controlled by this parameter which has to be set by the designer. Once the algorithm has explored the reachable state/action space, a homogeneous, and therefore, representative set of learning sample is contained in the database. Figure 7.4 shows a representation of the learning samples database in a simple case in which the state and action are only one dimensional.

7.3.3.3 Action Selection

After updating the Q-function with the learning samples, the SONQL algorithm must propose an action. As the algorithm is based on Q-learning, which is an off-policy algorithm, any policy can be followed. In practice, the policy followed is the ε-greedy policy. With probability $(1-\varepsilon)$, the action will be the one which maximizes the Q_function in the current state s_{t+1}. Otherwise, an aleatory action is generated. Due to the continuous action space in the neural Q-function, the maximization is accomplished by evaluating a finite set of actions. The action space is discretized with the smallest resolution that the environment is able to detect. In the case of a robot, the actions would be discretized in a finite set of velocity values considered to have enough resolution for the desired robot performance. If more than one action is present, the search of the optimal action can require a lot of computation. In that case, the search of the greedy action is necessary to obtain the $Q(s, a_{\max})$ value. The search of the greedy action is one of the drawbacks of continuous functions, as pointed out in [6]. However, the SONQL algorithm was designed to learn only one DOF of the robot, and, therefore, this problem is avoided in the results presented in this section.

7.3.4 *Target-Following Behavior*

After having analyzed the SONQL algorithm, this section shows the application for which it was designed. The goal of the SONQL algorithm is to learn the state/action mapping of a reactive behavior. An example of a reactive behavior is the *target-following* behavior. In this case, the goal of the behavior is to generate the control actions which, according to the position of the target, make the robot follow the target. In this kind of application, the states and actions are usually continuous, which causes the generalization problem. The use of the SONQL algorithm permits the learning of the state/action mapping solving the generalization problem. The SONQL algorithm makes reinforcement learning feasible for this application.

7.3.4.1 URIS AUV

*U*nderwater *R*obotic *I*ntelligent *S*ystem is the meaning of the acronym URIS. This unmanned underwater vehicle (UUV) is the result of a project started in 2000 at the University of Girona. The main purpose of this project was to develop a small-sized underwater robot with which to easily experiment in different research areas. The shape of the vehicle is compounded of a spherical hull surrounded by various external elements (the thrusters and camera sensors). The hull is made of stainless steel with a diameter of 350 mm, designed to withstand pressures of 3 atm (30 m depth). On the outside of the sphere, there are two video cameras (forward and down looking) and 4 thrusters (2 in X direction and 2 in Z direction). All these components were designed to be waterproof, with all electrical connections made with protected cables and hermetic systems. Figure 7.5 shows a picture of URIS and its body-fixed coordinate frame. Referred to this frame, the 6 degrees of freedom (DOFs) in which a UUV can be moved are *surge*, *sway*, and *heave* for the motions in X, Y, and Z directions, respectively; and *roll*, *pitch*, and *yaw* for the rotations about X, Y and Z axes respectively.

7.3.4.2 Target-Following Task

One of the sensory systems developed for the experimental setup of URIS is the *target detection and tracking* system. This vision-based application has the goal of detecting an artificial target by means of the forward-looking camera. This camera provides a large underwater field of view (about 57° in width by 43° in height). This system was designed to provide the control architecture with a measurement of the

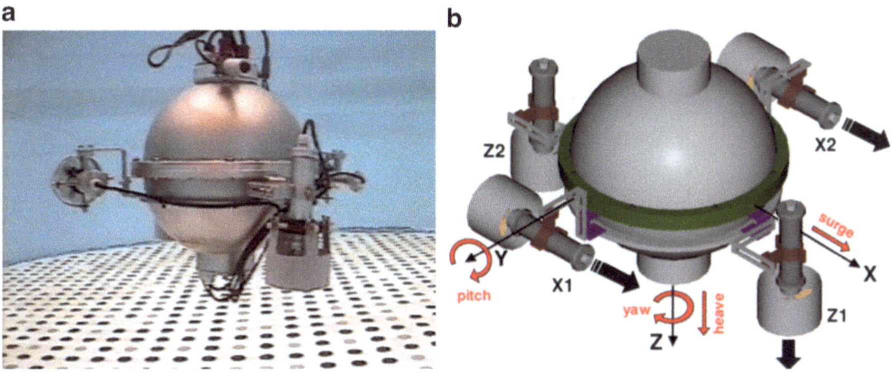

Fig. 7.5 URIS' AUV, (**a**) picture (**b**) schema

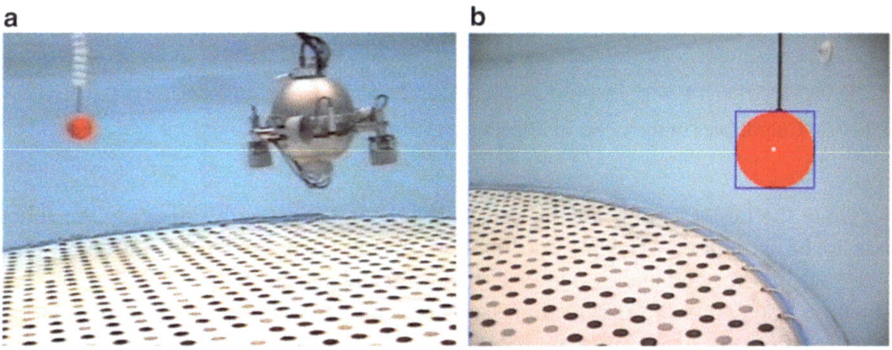

Fig. 7.6 Target detection, (**a**) URIS in front of the artificial target, (**b**) acquired image with the detected target

position of an object to be tracked autonomously. Since the goal of this work is to test control and learning systems, a very simple target was used. The shape selected for the target was a sphere because it has the same shape from whatever angle it is viewed. The color of the target was red to contrast with the blue color of the water tank. These simplifications allowed us to use simple and fast computer vision algorithms to achieve real-time (12.5 Hz) performance. Figure 7.6 shows a picture of the target being observed by URIS (a) and the acquired image with the detected target (b).

The procedure of detecting and tracking the target is based on image segmentation. The pixels which satisfy the characteristics of the target are classified as part of it. In order to express the color of an object, the *HSL* (hue, saturation, and luminance) color space is usually preferred over the standard *RGB* (red, green, and blue). Within the HSL color space, the hue and saturation values, which are extracted from the RGB values, define a particular color. Therefore, the color of an object is defined by a range of values in hue and saturation, and the segmented image is the one containing the pixels within the hue and saturation ranges.

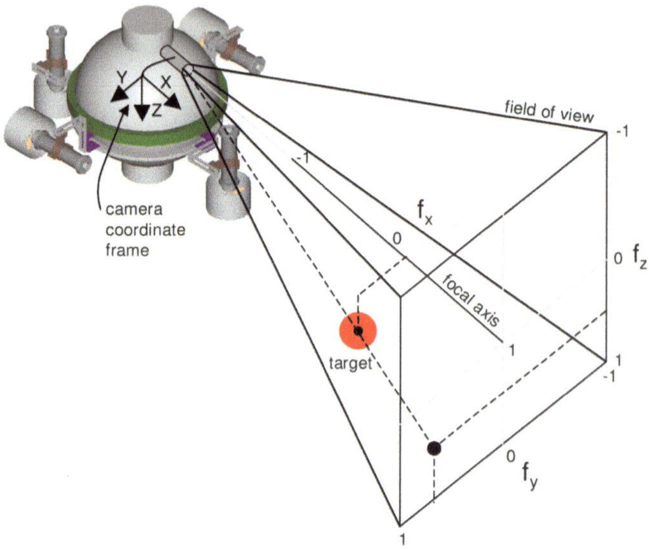

Fig. 7.7 Coordinates of the target in respect with URIS

Once the target has been detected, its relative position with respect to the robot has to be expressed. The X-coordinate of the target is related to the target size detected by the segmentation algorithm. A normalized value between -1 and 1 is linearly assigned to the range comprised between a maximum and minimum target size, respectively. It is important to note that this measure is linear with respect to the size, but nonlinear with respect to the distance between the robot and the target measured in X-axis. In Fig. 7.7 the f_x variable is used to represent the X-coordinate of the target. Similarly, the Y- and Z-coordinates of the target are related to the horizontal and vertical positions of the target in the image, respectively. However, in this case, the values represented by the f_y and f_z variables do not measure a distance, but an angle. The f_y variable measures the angle from the center of the image to the target around the Z-axis. The f_z variable measures the angle from the center of the image to the target around the Y-axis. In this case, the angles are also normalized from *-1* to *1* as it can be seen in the Fig. 7.7. Finally, the detection of the target in subsequent images is used to estimate its relative velocity.

As it has been explained, the coordinates of the target are directly extracted from the position and size found in the segmentation process. This means that the calibration of the camera has not been taken into account. Therefore, the nonlinear distortions of the camera will affect the detected position. Moreover, the measures of the f_x variable are nonlinear with the distance to the target in X-axis. These nonlinear effects have consciously not been corrected to increase the complexity with which the SONQL-based behavior will have to deal.

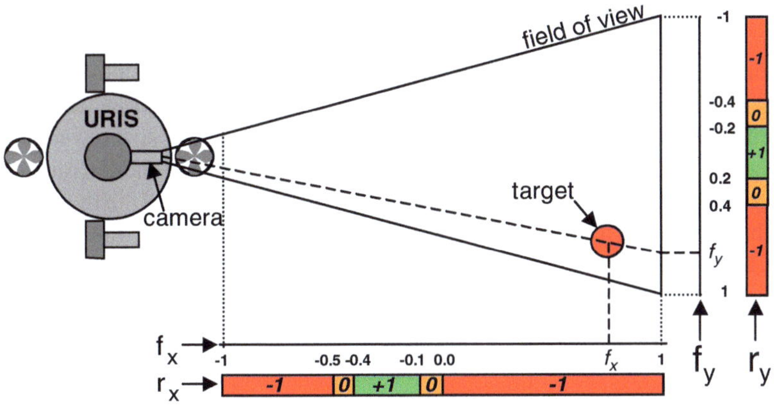

Fig. 7.8 Relative positions f_x and f_y and the reinforcement values r_x and r_y, for the "target-following" behavior

Besides the target detection, another vision-based system was used to estimate the position and velocity of the vehicle inside a water tank [14]. The system was able to localize the robot in absolute coordinates, in real time, and with high accuracy.

7.3.4.3 SONQL Behavior

The "target-following" behavior was learned using two SONQL algorithms, one for each DOF. The goal of this algorithm was to learn the state/action mapping of this behavior. This mapping determined the movements the robot had to perform in order to locate itself at a certain distance from the target and pointing directly at it. In order to generate the robot response r_f, the activation level was $a_f = 1$ whenever the target was detected and $a_f = 0$ otherwise. Moreover, the control action, v_f was generated by the SONQL algorithms. In particular, the SONQL algorithm of the X DOF generated the $v_{f,x}$ control action and the algorithm of the Yaw DOF generated the $v_{f,\text{yaw}}$. The next paragraphs will describe the state variables, the reinforcement function, and the parameters of the SONQL algorithm used in each DOF.

7.3.4.4 Environment State

The state of the environment must contain all the required measurements relating the target to the robot. The first measurement was the relative position between the target and the robot. A second necessary measurement was the relative velocity between the target and the robot. Therefore, the state for the X DOF of the SONQL algorithm was composed of f_x and fv_x, and the YAW DOF of the SONQL algorithm was composed of f_y and fv_y. Figure 7.8 shows a two-dimensional representation of the f_x- and f_y-axes.

7.3.4.5 Reinforcement Function

The reinforcement function is the only information a designer must introduce in order to determine the goal of the behavior. The reinforcement function must be a deterministic function which relates the state and action spaces to a reward value. In the case of the "target-following" behavior, the reinforcement function took only the target position as input. Therefore, according to these variables, f_x in the X DOF and f_y in the Yaw DOF, a reward value was given. Only three reward values were used: -1, 0, and 1, simplifying the implementation of this function. Other functions, such as a continuous linear function, could had also been used, obtaining similar results. Figure 7.8 shows the reward values according to the target relative position.

7.3.4.6 SONQL Parameters

After defining the states, actions, and reinforcement function, the final step to learn the "target following" behavior with the SONQL algorithm consisted of setting the parameters of the algorithm:

- NN Configuration. The neural network had three continuous variables as inputs: $\{f_x, fv_x, v_{f,x}\}$ for the X DOF and $\{f_y, fv_y, v_{f,\text{yaw}}\}$ for the Yaw DOF. The output was the estimated $Q(s,a)$ value. Only one hidden layer was used with six neurons.
- Learning Rate α. A diverse set of values were tested. The final learning rate was set to $\alpha = 0.1$, which demonstrated a fast and stable learning process.
- Discount Factor γ. The discount factor was set to $\gamma = 0.9$. Since the learning of a robot behavior is a continuous task without a final state, a discount factor smaller than 1 was required.
- Exploration Probability ε. The learning was performed with a ε-greedy policy. The exploration probability was $\varepsilon = 0.3$.
- Database Density Parameter t. The density parameter was set to $t = 0.5$. In practice, the number of learning samples resulted in being less than 30, although it depended on the state/action space exploration.

7.3.5 Results

An exhaustive set of experiments were carried out in order to test the SONQL algorithm. In these experiments, optimal parameters were found, and a fast and effective learning of the robot behavior was achieved.

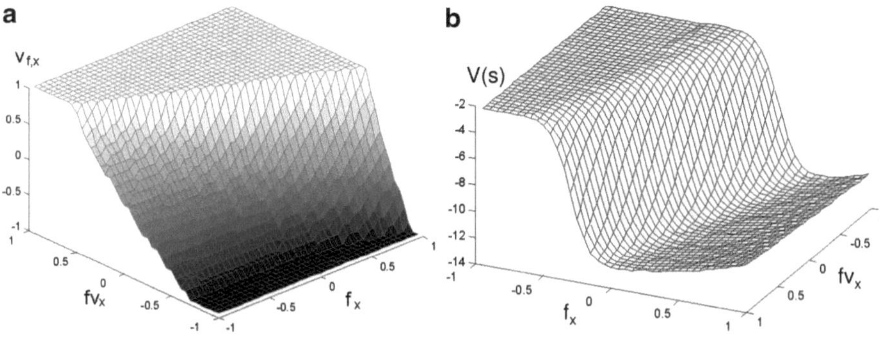

Fig. 7.9 (**a**) State/action policy learned for the X DOF. (**b**) State-value function, $V(s)$, after the learning for the X DOF

7.3.5.1 X DOF

The experiments in the X DOF were carried out by placing the target in front of the robot and starting the learning algorithm. The target was placed next to the wall of the water tank, and the robot was placed in front of it at the center of the circular tank. This positioning gave the robot the maximum space in which to explore the state. In addition, as the target was placed close to the wall of the tank, the vehicle was stopped by the "wall-avoidance" behavior when it came too close to the wall, preventing a collision with the target.

The learning of the X DOF took about 110 s, which represents about 370 learning iterations (sample time $= 0.3$ s). The policy learned after this learning can be seen in Fig. 7.9. The optimal action $v_{f,x}$ according to the state $\{f_x, fv_x\}$ can be appreciated. For example, for $f_x = 1$ and $fv_x = 0$, which means that the target is at the farthest distance and the velocity is zero, the optimal action is $v_{f,x} = 1$, that is, "go forward." This is a trivial case, however, but the policy also shows situations in which intermediate action values are given.

The Q-function learned in one of the experiments can also be seen. Figure 7.9 shows the maximum Q value for each state $s = (f_x, fv_x)$ which is also the state value $V(s)$. It can be appreciated that the maximum values are in the target positions (f_x) where the reinforcement function was maximum. However, according to the velocity variable, fv_x, the value of zones with the same position f_x changes. This is due to the prediction provided by the state-value function. If the target is located at the position where the reward is maximum but the velocity causes it to move away, the value of this state is lower than the states in which the future rewards will also be maximum. Finally, it must be noted that the function approximated by the NN is not only the one shown in this figure. Another dimension, corresponding to the continuous action $v_{f,x}$, is also contained in the function. This demonstrates the high function approximation capability of NN.

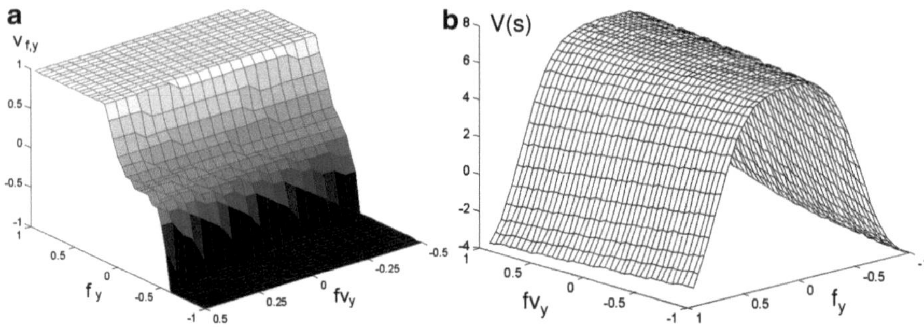

Fig. 7.10 (**a**) State/action policy learned for the Yaw DOF. (**b**) State-value function, $V(s)$, after the learning for the Yaw DOF

7.3.5.2 YAW DOF

Similar experiments were performed for the Yaw DOF. The target was located in front of the robot when the SONQL was started. During the exploration of the space, the robot frequently lost the target but the "target-recovery" behavior became active, and the target was detected again.

The learning of the Yaw DOF took about 60 s, which represents about 200 learning iterations (sample time $= 0.3$ s). The policy and state-value functions for the Yaw DOF can be seen in Fig. 7.10. The policy relates the optimal action $v_{f,\text{yaw}}$ according to the environment state $\{f_y, fv_y\}$. As far as the state-value function $V(s)$ is concerned, it can be clearly appreciated how the maximum values are in the target positions near the center ($f_y = 0$).

7.3.5.3 Convergence Test

The SONQL algorithm demonstrated that it converges to the optimal policy in a relatively small time, as commented above. The only condition to assure the convergence was to guarantee the reliability of the observed state. This means that perturbations like the influence of the umbilical cable had to be avoided. Figure 7.11 shows six consecutive learning attempts of the Yaw DOF. This figure also shows that the averaged reward increased, demonstrating that the behavior was being learned. It can be seen that the algorithm started exploring the state in order to find the maximum reward. Once the whole state had been explored, the algorithm exploited the learned Q-function and obtained the maximum reward. Although its seems that there were two independent phases, first exploration and then exploitation, the change is continuous. At the beginning, there is not valid information in the Q-function, and therefore the algorithm behaves as if it was generating random actions (exploration) 100% of the time. As the algorithm is obtaining more and more experience, greedy actions are more and more exploitation actions. This effect was obtained maintaining during all the runs the same ε-greedy policy, without changing the ε parameter.

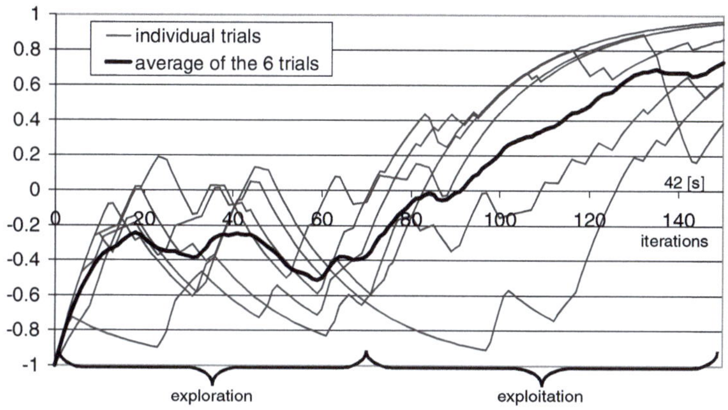

Fig. 7.11 Behavior convergence for different attempts. The results of six different learning trials in the Yaw DOF are shown. Each graph represents the mean of the last 20 rewards. This representation is useful to verify that the algorithm is receiving the maximum rewards. The minimum and maximum values correspond to the minimum ($r_y = -1$) and maximum rewards ($r_y = 1$). Also, the average of the six experiments can be seen, pointing out that the accumulated rewards increased, while the behavior was learned

Fig. 7.12 Trajectory of URIS while following the target in the water tank

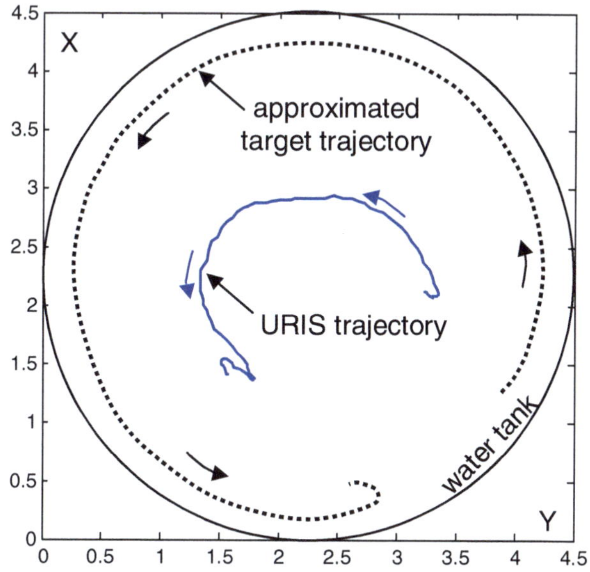

In Fig. 7.12, it can be seen how the robot followed the target at a certain distance. The target was moved manually around the water tank, and, therefore, the robot trajectory was also a circumference. Note that the position of the target is approximate since there is no system to measure it.

7.4 Policy Gradient Algorithms: Cable Tracking Behavior

7.4.1 Natural Actor-Critic

Fast convergence and adaptation to an unknown changing environment is a must when dealing with real applications, and a PG seems not able to do so alone. With the objective of finding a faster algorithm, the attention of this research work moves to actor-critic (AC) methodologies. Actor-Critic (AC) methods try to combine the advantages of PG algorithms with VF methods. The actor's PG gives convergence guarantees, while the critic's VF reduces variance of the policy update improving the convergence rate of the algorithm. Also, AC methods are on-policy algorithms and, therefore, the learning procedure benefits from convergence proofs. The main features of AC methods seem to have what real robotics is looking for.

The reinforcement learning (RL) method chosen is the natural actor-critic (NAC) algorithm [29]. The algorithm is divided into two main blocks, one concerning the critic and another the actor. The critic is represented by a VF $V^{\pi}(s)$ which is approximated by a linear function parameterization represented as a combination of the parameter vector v and a particularly designed basis function $\phi(s)$, the whole expression being $V^{\pi}(s) = \phi(s)v$. On the other hand, the actor's policy is specified by a normal distribution $\pi(a|s) = \mathcal{N}(a|Ks, \sigma^2)$. The mean μ of the actor's policy distribution is defined as a parameterized linear function of the form $a = Ks$. The variance of the distribution is described as $\sigma = 0.1 + 1/(1+\exp(\eta))$. Therefore, the whole set of the actor's parameters is represented by the vector $\theta = [K, \eta]$. The actor's policy derivative has the form $\frac{\mathrm{d}\log\pi(a|s,\theta)}{\mathrm{d}\theta}$, and in order to obtain the natural gradient, this function is inserted into a parameterized compatible function approximation, $f(s,a)^{\pi}_{w} = \frac{\mathrm{d}\log\pi(a|s,\theta)}{\mathrm{d}\theta} w$, with the parameter vector w as the true gradient.

At every iteration, action a_t is drawn from the current policy π_t, generating a new state s_{t+1} and reward r_t. After updating the basis function and the critic's statistics by means of least squares temporal difference (LSTD)-Q(λ), the VF parameters v and the natural gradient w are obtained. The actor's policy parameters are updated only if the angle between two consecutive natural gradients is small enough compared to an ε term. The learning rate of the update is controlled by the α parameter. Next, the critic has to forget part of its accumulated statistics using a forgetting factor, $\beta \in [0,1]$. The current policy is directly modified by the new parameters becoming the new policy to be followed in the next iteration, getting closer to a final policy that represents a correct solution to the problem.

7.4.2 Methods to Speeding Up Reinforcement Learning

Speeding up Reinforcement Learning (RL) algorithms in real continuous high-dimensional domains represents a key factor. The idea of providing initial high-level information to the agent is a topic that has received significant attention in the real robotic field over the last decade. The main objective of these techniques is to make learning faster in contrast to tedious reinforcement learning methods or trial-and-error learning. Also, it is expected that the methods, being user-friendly, would enhance the application of robots in human daily environments.

One way of reusing past experiences, which was investigated in the DYNA architecture [39], is to use experiences to build an *action model* and use it for planning and learning. An action model can be understood as a computer simulator which mimics the behaviors of the environment for a particular Reinforcement Learning Problem (RLP). The action model contains the dynamics of the environment together with its reward function. Instead of direct interaction with the environment, the agent can experience the consequences of actions without physically being in the real world. As a result, the robot will learn faster as the iteration steps with the simulator would be faster than those with the real world, and more importantly, fewer mistakes will be made when the robot starts running in the real world with a previously trained policy. Approximate policies learned in simulation represent a good start-up for the real learning process, meaning that the agent will face the challenge of the real world with some initial knowledge of the environment, avoiding initial gradient plateaus and local maxima dead ends [24]. The help of computer simulators has been successfully applied in [4] where a PG algorithm designed to control a RC helicopter is first trained off-line by means of a computer simulator. Once the agent's policy is considered to be *safe* enough, it is transferred to the real vehicle where the learning process continues. Also, in [23], a technique called *time hopping* is proposed for speeding up reinforcement learning algorithms in simulation. Experiments on a simulated biped crawling robot confirm that this technique can greatly accelerate the learning process.

The final proposal is the application in a real underwater robotic task of RL-based behavior in a two-step learning process. For this purpose, the learning algorithm is first trained in a simulated environment where it can quickly build an initial policy. An approximated model of the environment emulates a robot with the same number of degree of freedoms (DoFs) of the real one. Once the simulated results are accurate enough and the algorithm has acquired enough knowledge from the simulation to build an approximated policy, in a second step, the policy is transferred to the real robot to continue the learning process online in a real environment.

7.4.3 Two-Step NAC Learning with Barycentric Interpolators

A parameterized basis function represented by a barycentric interpolator is another approximator chosen to represent the policy $\pi(a_t|s_t,\theta_t)$ together with the gradient

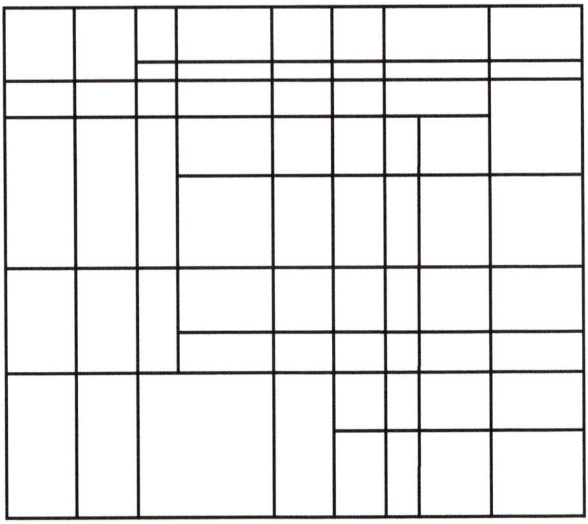

Fig. 7.13 A more typical-looking mesh, where an initial grid has been refined through iterative splitting of some of the elements

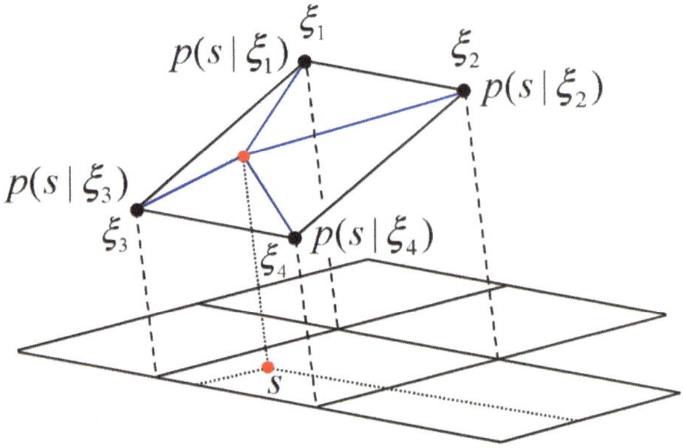

Fig. 7.14 Graphic representation of the *barycentric coordinates* given a state s in a two-dimensional mesh case

partially observable Markov decision process (GPOMDP) approach. Barycentric interpolators are a specific class of basis function approximators. These basis functions are a popular representation for VFs because they provide a natural mechanism for variable resolution discretization of the function, and the barycentric coordinates allow the interpolators to be used directly by value iteration algorithms. Another advantage in choosing linear function approximators is their convergence guarantees when used together with an on-policy method such as the Gradient Partially Observable Markov Decision Process (GPOMDP) algorithm. Barycentric interpolators apply an interpolation process based on a finite set of discrete points that form a mesh. This mesh does not need to be regular, but the method outlined here assumes that the state space is divided into a set of rectangular boxes. An example of an appropriate mesh is shown in Fig. 7.13. The key is that any particular point is enclosed in a rectangular box that can be defined by the 2^N nodes in our N-dimensional state space (Fig. 7.14).

$$V(s) = \sum_{i=1..4} p(s|\xi_i)V(\xi_i).$$

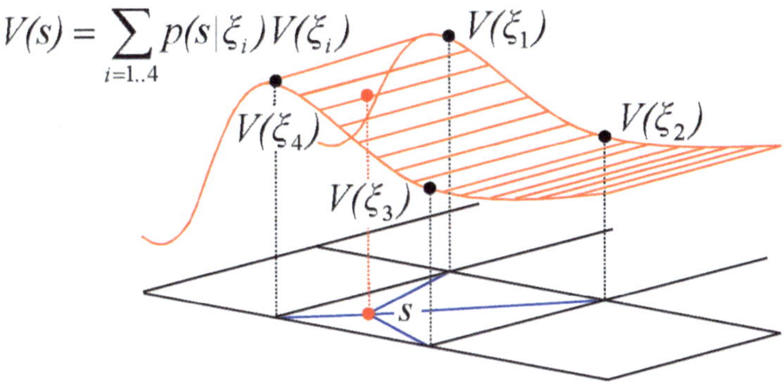

Fig. 7.15 Graphic representation of the *barycentric coordinates* given a state s in a two-dimensional mesh case

As shown in Fig. 7.15, ξ_i being a set of nodes distributed in a rectangular mesh for any state s. Once it is enclosed in a particular box $(\xi_1, \ldots, \xi_4)$ of the mesh, the state s becomes the *barycenter* inside this box with positive coefficients $p(s|\xi_i)$ of sum 1 called the *barycentric coordinates* and, therefore,

$$s = \sum_{i=1...4} p(s|\xi_i)\xi_i. \tag{7.13}$$

Thus, $V(\xi_i)$ is set as the value of the function at the nodes previously defined as ξ_i. Having defined the value of the function at the nodes of the mesh, Eq. 7.14 shows how to compute the value of the function at state s, called the *barycentric interpolator* of state s and defined as $V(s)$. The value of state s is calculated as a function of the value $V(\xi_i)$ at the nodes of the box in which the state is enclosed and its barycentric coordinates $p(s|\xi_i)$. As depicted in Fig. 7.16, we give

$$V(s) = \sum_{i=1...4} p(s|\xi_i)V(\xi_i). \tag{7.14}$$

The policy is approximated using a barycentric interpolator function where the nodes of the mesh $V(\xi_i)$ represent the policy parameters θ to be updated at each learning iteration. Therefore, given an input state s_t, the policy will compute a control action $V(s_t) = a_t$. As defined in Eq. 7.15, in the barycentric approximator context, the gradient of the policy with respect to each parameter is equal to the derivative of the approximator with respect to the parameter multiplied by an error function e_t; hence,

$$\frac{d\pi(a_t|s_t,\theta_t)}{d\theta_t} = \frac{dV(s_t)}{dV(\xi_i)}e_t. \tag{7.15}$$

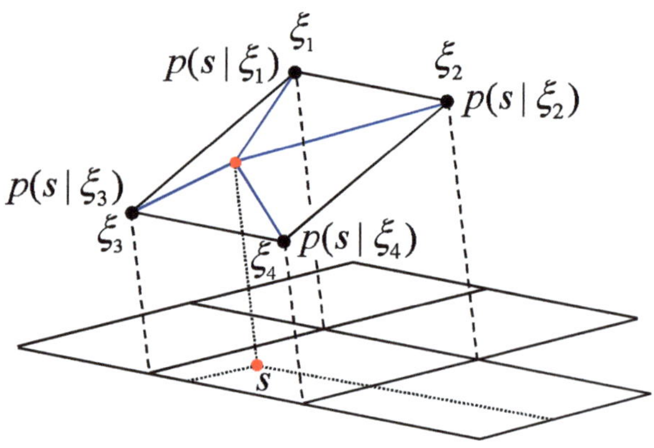

Fig. 7.16 Calculation of the function approximator output given a particular state s

Here the error is given by

$$e_t = V(s_t)_{\text{desired}} - V(s_t). \tag{7.16}$$

The eligibility trace z_{t+1} is updated following Eq. 7.17. Eligibility's decay factor λ controls the memory on past actions. The gradient of the policy with respect to a particular parameter contains the barycentric coordinate of the parameter $p(s|\xi_i)$,

$$z_{t+1} = \lambda z_t + p(s_t|\xi_i)e_t. \tag{7.17}$$

Finally, the new parameter vector is given by

$$V(\xi_i)_{t+1} = V(\xi_i)_t + \alpha r_{t+1} z_{t+1}. \tag{7.18}$$

The vector $V(\xi_i)$ represents the policy parameters to be updated, r_{t+1} is the reward given to the agent at every time step, and α as the learning rate of the algorithm.

The learning process of the proposed method is split into two steps. First, the learner interacts with the simulator, building an initial policy. During this phase, the NAC algorithm is trained in the simulator. The simulated experiments begin by setting parameterized initial basis functions for both the actor and the critic. An approximated model of the environment emulates a robot with the same Degree of Freedoms (DoFs) as the real one. The sampling time of the simulator is set equal to the one offered by the real robot. Actions, forces, accelerations, and velocities are scaled to match the real ones. Once the simulated results are accurate enough and the algorithm has acquired enough knowledge from the simulation to build a secure policy, in the second step, the learned-in-simulation policies are transferred to the robot, and step two starts. The actor and critic parameters are transferred to the real robot to continue the learning process online in a real environment, improving the initial policy. During the second step, the NAC algorithm continues its execution, this time, in the software architecture of the robot. Since the simulated model and environment are just approximations, the algorithm will have to adapt to the real world. An important idea is that, once the algorithm converges in the real environment, the learning system will continue working forever, being able to adapt to any future change.

7.4.4 Cable Tracking in Underwater Environments

After having analyzed the two-step NAC algorithm, this section shows the application for which it was designed. The goal of the algorithm is to learn the state/action mapping of a reactive behavior. An example of a reactive behavior is the cable-tracking behavior. In this case, the goal of the behavior is to generate the control actions which, according to the position of the cable, make the robot follow it. The Ictineu autonomous underwater vehicle (AUV) was equipped with a downward-looking camera that was used for detecting the cable. For the experiments, a vision-based system developed at the University of the Balearic Islands has been chosen to track a submerged cable in a controlled environment. All the tests were carried out in the Computer Vision and Robotics Group water tank facility at the University of Girona.

7.4.4.1 Ictineu AUV

The Ictineu Autonomous Underwater Vehicle (AUV) [33] was originally conceived as an entry for the student autonomous underwater challenge Europe (SAUCE) competition, a European-wide competition for students to foster research and development in underwater technology. Although the competition determined many of the vehicle's specifications, Ictineu was also designed keeping in mind its posterior use as an experimental platform for various research projects in our laboratory. The experience obtained from the development of previous vehicles by the group made it possible to build a low-cost vehicle of reduced weight (52 Kg) and dimensions ($74\times$ 46.5 $\times$ 52.4 cm) with remarkable sensorial capabilities and easy maintenance (Fig.7.17).

The Ictineu is propelled by six thrusters that allow it to be fully actuated in *Surge* (movement along the X-axis), *Sway* (movement along the Y-axis), *Heave* (movement along the Z-axis), and *Yaw* (rotation around the Z-axis), achieving maximum speeds of 3 knots. The Ictineu AUV is passively stable in both *pitch* and *roll* DoFs as its center of buoyancy is above the center of gravity. This stability is the result of an accurate distribution of the heavier elements in the lower part of the chassis combined with the effect of technical foam placed on the top, which provides a slightly positive buoyancy.

Fig. 7.17 Ictineu AUV during a mission in a real environment

7.4.4.2 The Cable-Tracking Vision-Based System

The vision-based algorithm used to locate the cable was first proposed in [27]. Once the cable has been detected, its location and orientation in the next image are predicted by means of a Kalman filter, which allows reducing the pixels to be processed to a small region of interest (ROI). In this way, the computation time is considerably lowered together with the probability of misinterpretations of similar features appearing in the image. If low or nil evidence of the pipe/cable presence in the Region of Interest (ROI) is obtained, a transient failure counter is increased after discarding the image. If this anomalous situation continues throughout too many images, then it is attributed to a failure in the prediction of the ROI, resulting in two special actions: the Kalman filter is reset and the ROI is widened to the whole image. In other words, when faced with a persistent error, the system will no longer make use of the knowledge acquired from the last processed images.

As can be seen in Fig. 7.18, the algorithm computes the polar coordinates $(\rho,\ \Theta)$ of the straight line corresponding to the detected cable in the image plane, $(\rho,\ \Theta)$; being the parameters of the cable line, the Cartesian coordinates (x,y) of any point along the line must satisfy

$$\rho = x\cos(\Theta) + y\sin(\Theta). \tag{7.19}$$

Equation (7.19) allows us to obtain the coordinates of the cable intersections with the image boundaries (X_u,Y_u) and (X_L,Y_L); thus, the midpoint of the straight line (x_g,y_g) can be easily computed $(x_g, y_g = \frac{X_L+X_u}{2}, \frac{Y_u+Y_L}{2})$. The computed parameters (ρ,Θ,x_g,y_g) together with its derivatives are sent to the control module in order to be used by the various behaviors.

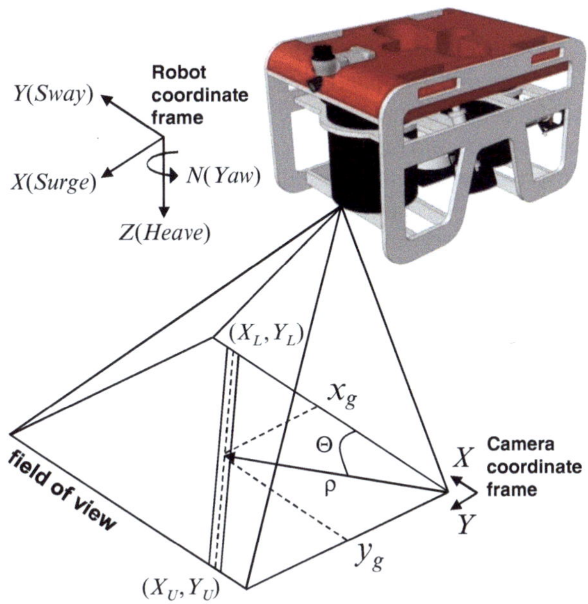

Fig. 7.18 Coordinates of the target cable with respect to the Ictineu AUV

7.4.5 Results

7.4.5.1 NAC Algorithm Configuration

For both steps, the simulated and the real one, the state of the environment is represented by a four-dimensional vector containing four continuous variables, $(\Theta, \frac{\delta\Theta}{\delta t}, x_g, \frac{\delta x_g}{\delta t})$ where Θ is the angle between the Y-axis of the image plane and the cable, x_g, is the X-coordinate of the midpoint of the cable in the image plane and, finally, $\frac{\delta\Theta}{\delta t}$ and $\frac{\delta x_g}{\delta t}$ are Θ and x_g derivatives, respectively. The bounds of these variables are $\frac{-\pi}{2}\text{rads} \leq \Theta \leq \frac{\pi}{2}\text{rads}$; $-1\text{rads}/s \leq \frac{\delta\Theta}{\delta t} \leq +1\text{rads}/s$; $-0.539 \leq x_g \leq 0.539$ (adimensional bound scaled from the size of the image in pixels), and $-0.5 \leq \frac{\delta x_g}{\delta t} \leq +0.5$ (adimensional bound derived from x_g). From this input state, the robot makes decisions concerning two DoFs: the Y movement (Sway) and the Z-axis rotation (Yaw). Therefore, the continuous action vector is defined as $(a_{\text{sway}}, a_{\text{yaw}})$ with boundaries $-1 \leq a_{\text{sway}} \leq +1$ and $-1 \leq a_{\text{yaw}} \leq +1$. The X movement or *surge* of the vehicle is not learned. A simple controller has been implemented to control the X DoF; whenever the cable is centered in the image plane (the cable is located inside the 0 reward boundaries), the robot automatically moves forward $a_{\text{surge}} = 0.3$ m/s. If the cable moves outside the *good* limits or the robot misses the cable, it stops moving forward $a_{\text{surge}} = 0$ m/s.

The main policy has been split into two subpolicies. a_{sway} actions cause Y displacements on the robot and, therefore, X displacements on the image plane. Then, it can be easily noticed that a_{sway} actions directly affect the position of x_g

along the X-axis of the image plane together with its derivative. In the same way, a_{yaw} causes rotation around the Z-axis and, therefore, Θ angle variations in the image plane. Although learning uncoupled policies certainly reduces the overall performance of the robot, this method greatly reduces the total number of stored parameters, making it easier to implement. The actor policies for Yaw and Sway are described as normal Gaussian distributions of the form $\pi(x,a,K) = N(a|Kx,\sigma^2)$, where the variance σ^2 of the distribution is fixed at 0.1. The basis function approximator combines the state vector values and the learned parameters vector K, following the expression $a = k_1x_1 + k_2x_2$. The policy parameter vectors K for Sway and Yaw policies are randomly initialized at the beginning of the experiment. Additional basis function for Sway and Yaw critic parameterization are chosen as $\phi(x) = [x_1^2, x_1x_2, x_2^2, 1]$, which experimentally have proven to be good enough to represent both functions.

A continuous function has been designed to compute specific rewards for each learned policy. Rewards are given by $r(x_t,a_t) = x_t^T Qx_t + a_t^T Ra_t$ with $Q_{\text{sway}} = \text{diag}(1.1, 0.2)$ and $R_{\text{sway}} = 0.01$ for Sway DoF and $Q_{\text{Yaw}} = \text{diag}(2, 0.25)$ and $R_{\text{Yaw}} = 0.01$ for the Yaw DoF. $Q's$ and $R's$ values have been tuned experimentally. The learning parameters have the same value for both policies. The learning rate has been fixed to $\alpha = 0.01$, and the decay factor of the eligibility has been set to $\lambda = 0.8$. A discount factor for the averaged reward is set to $\gamma = 0.95$, and a forgetting factor for the matrixes A and b has been set to $\beta = 0.9$. For every iteration, the policy is updated only when the angle between two consecutive gradient vectors accomplishes $\angle(w_{t+1}, w_t) \leq \varepsilon = \pi/180$.

7.4.5.2 Step 1, Simulated Learning Results

The computed model of the underwater robot Ictineu AUV navigates a two-dimensional world at a 0.8 m height above the seafloor. The simulated cable is placed on the bottom in a fixed position. The robot has been trained in an episodic task. An episode ends either after 150 iterations or when the robot misses the cable in the image plane, whatever comes first. When the trial ends, the robot position is reset to a random position and orientation around the cable's location, allowing any location of the cable within the image plane at the beginning of each trial. According to the value of the state and action taken, a scalar immediate reward is given at each iteration step.

The experiments in the simulation step have been repeated in 100 independent runs. At the beginning of each run, the policy parameters are randomly initialized for each one of the policies, and the learning procedure starts. The effectiveness of each episode is measured as the total reward per episode perceived by the current policy. Once the learning process is considered finished, resultant policies obtained with the NAC algorithm with its correspondent parameters are transferred to the underwater robot Ictineu AUV, ready to be tested in the real world.

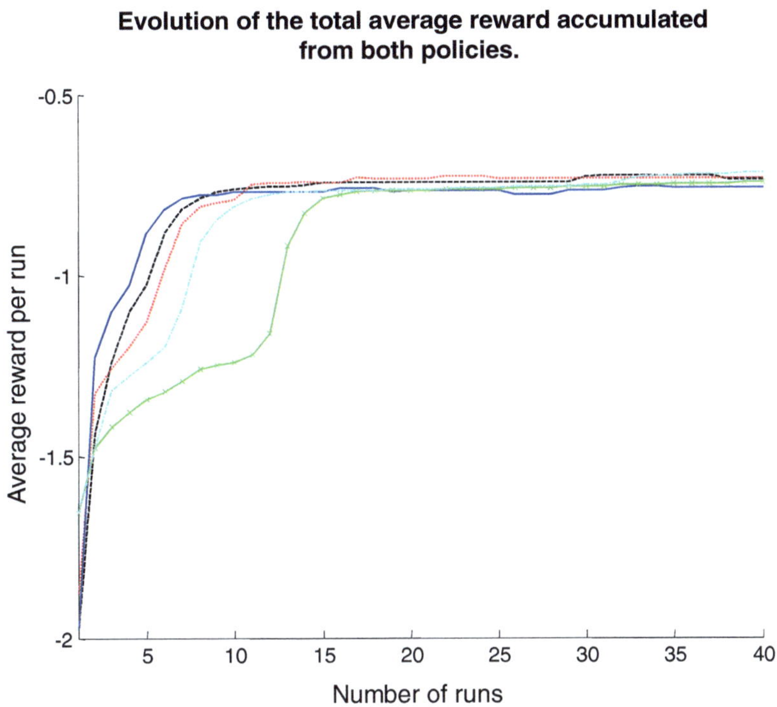

Fig. 7.19 Accumulated reward progression for five independent tests. Reward shown is the result of adding the accumulated reward for both Sway and Yaw policies

7.4.5.3 Step 2, Real Learning Results

At the beginning of step 2, the learned-in-simulation policies are evaluated on the real robot. These policies, one for the Sway DoF and another for the Yaw DoF, are transferred to the Ictineu AUV and before continuing with the policy improvement online. The results obtained with the learned-in-simulation policies on the real robot show a good performance of both simulated policies, allowing the robot to drive *safely* into the underwater scenario. Since this basic knowledge acquired in simulation has passed the tests, the transferred policies are ready to start the real learning on the robot. The main objective of the online real learning step is to improve the learned-in-simulation policies while interacting with the real environment.

Figure 7.19 shows the accumulated reward evolution during the real learning phase. This figure is the result of adding the total accumulated reward per trials of both policies, Sway and Yaw, over five independent tests of 40 runs each, remembering that each run lasts approximately 90 s, and represents a full length run from one side of the cable to the other, navigating along 9 m. It can be observed that in all five tests, the algorithm converges to a better policy as shown by the increasing evolution of the averaged rewards.

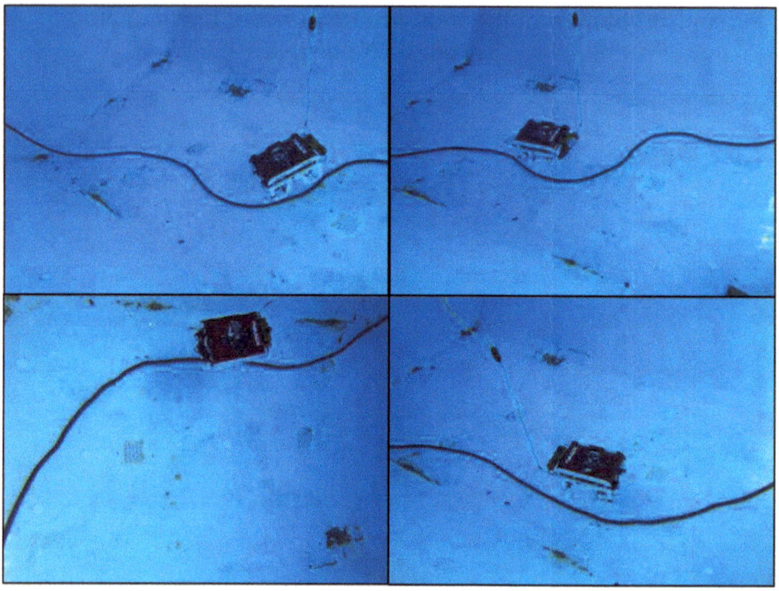

Fig. 7.20 A particular learned policy confronts different cable configurations. *upper-left image*: configuration 1, *upper-right image*: configuration 2, *bottom-left image*: configuration 3, *bottom-right image*: configuration 4

7.4.5.4 Algorithm Adaptation Capabilities

The last section of these experimental results looks into the adaptation capabilities of the NAC algorithm. This section aims to test the response of the algorithm to sudden changes in the environmental conditions once a particular policy has been learned. These changes will require a fast policy readaptation in order to successfully continue developing the task. The first adaptation experiment consists of modifying the cable's position and trajectory along the pool floor. Therefore, once the online real learning period is considered finished and a particular policy for both the Sway and the Yaw has been learned, the cable position is changed four different times. Figure 7.20 shows the four different cable configurations adopted. As can be seen in the figure, the top images correspond to a chicane configuration with close cable turns; the upper-left configuration starts with a left turn, whereas the upper-right one starts with a right turn. The bottom images correspond to a cable configuration with a big wide turn; the bottom-left configuration makes a wide right turn, whereas the bottom-right configuration makes a wide left turn.

Figure 7.21 shows the evolution of the state variable Θ for each of the four different configurations described above. Figure 7.22 shows the evolution of the state variable x_g for the same configurations. Results show that the policy learned for a particular situation is also a good one for tracking a cable positioned in a wide range of different curvature angles that may differ from the original one.

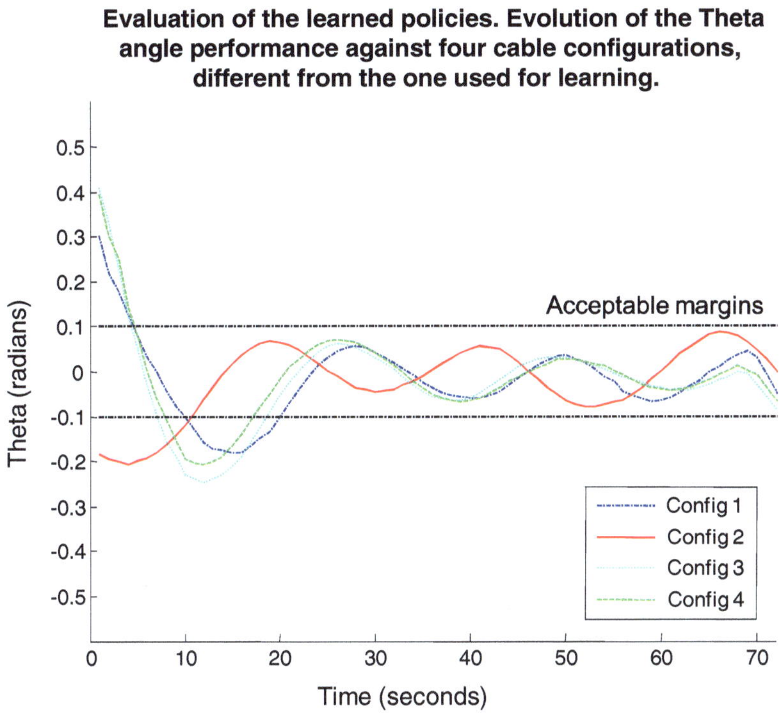

Fig. 7.21 A particular Θ policy performance over the different cable configurations showed in Fig. 7.20

A second set of experiments aims to evaluate if the NAC algorithm is able to adapt itself to unexpected changes in the environmental conditions and continue the learning process. In order to do so, a variation in the altitude with respect to the cable has been introduced in the middle of the learning process. As detailed in previous sections, the underwater robot autonomously navigates at a predefined constant height above the pool floor. For this set of tests, during the learning process, instead of tracking the cable at 0.8 m with respect to the pool floor, the altitude will suddenly change to 1.2 m. Since the altitude is not a part of the state input of the algorithm, a variation of this term causes a variation of the RL dynamics, and, therefore, current policies are no longer optimal.

Figure 7.23 shows the accumulated reward evolution of the five sample tests initially presented in Fig. 7.19. As can be seen in this figure, once the learning curves are stabilized, a sudden change in the robot's altitude is introduced (at the 40th run, the robot's altitude is changed from 0.8 m to 1.2 m). This change can be seen in the graph as a sudden change in the reward value. At this point, rewards become better because the robot navigates at a higher altitude, and tracking the cable becomes easier, as it appears smaller in the image plane. From run 40 to run 50, the robot readapts itself to the new environmental conditions, and the policies improve.

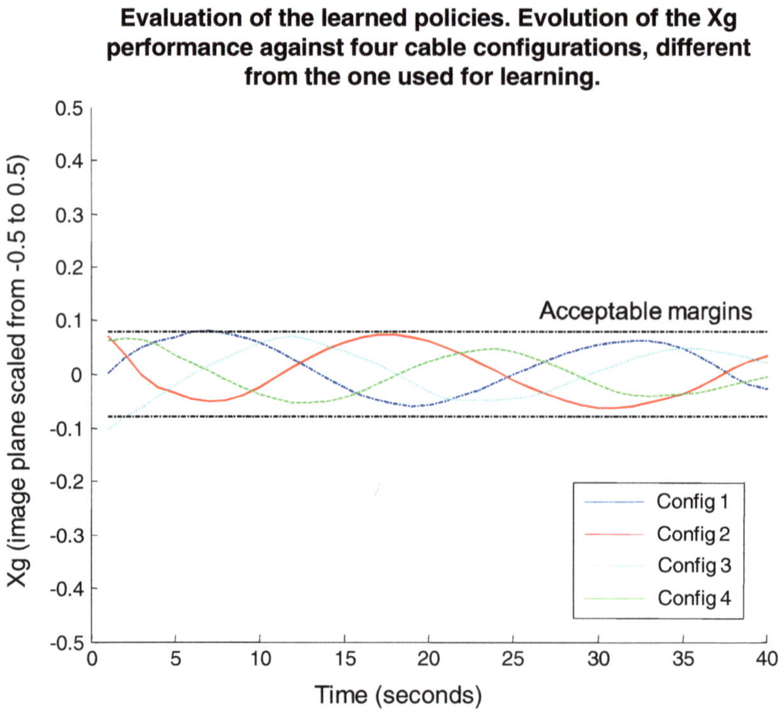

Fig. 7.22 A particular x_g policy performance over the different cable configurations showed in Fig. 7.20

From run 50 to run 70, the learning has stabilized itself, and robot navigates at 1.2 m altitude with its new, improved policies. It can be observed how the five independent tests converged to a better policy and achieved similar accumulated rewards. At around run 70, the altitude changes again, recovering its initial value of 0.8 m. Fifteen runs later, the averaged rewards stabilize at the same values before the altitude changes were introduced, denoting good online adaptation capabilities of the algorithm.

7.5 Conclusions

This chapter has presented the result of a research line of the University of Girona (Spain) in robot learning which has the final goal of making AUVs more adaptive and robust in future applications. In particular, the chapter has summarized the use of two well-known methodologies in reinforcement learning (value-function Q-learning and policy gradient-natural actor-critic) for learning a reactive AUV behavior (target following and cable tracking). All experiments have been performed

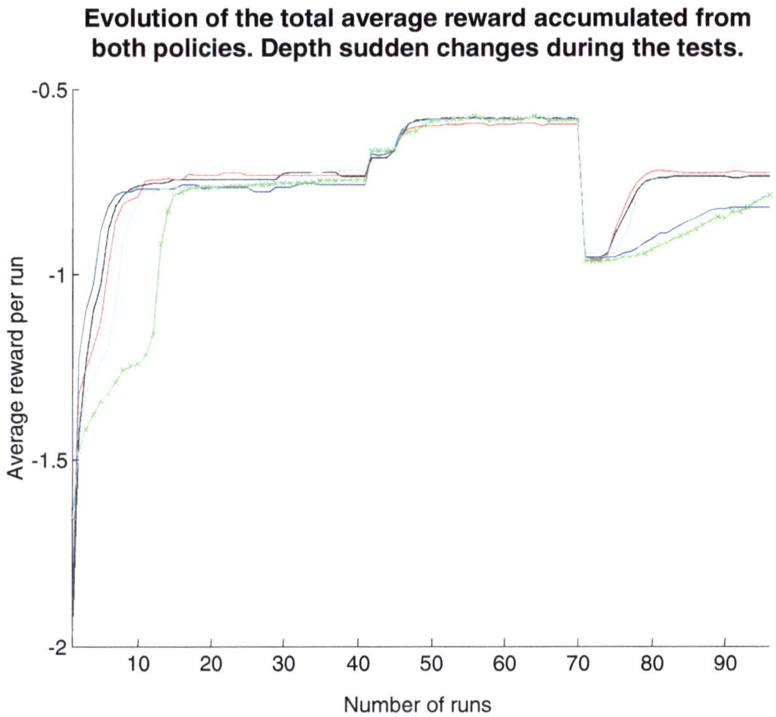

Fig. 7.23 Total averaged reward evolution over five independent tests. Initially the set point is 0.8 m. At run 40, the set point changes to 1.2 m before finally recovering its original value of 0.8 at run 70

with a real robot in a water tank. In both cases, results have pointed out the suitability of the methods, and therefore, it is required to continue the work in more challenging tasks and in more real conditions.

Nowadays, there exist quite a few underwater robots that develop some simple tasks autonomously. This work wants to introduce the application of learning techniques to the development of more challenging underwater tasks. Having a robot continuously learning to perform a task can be very valuable, specially for complex operations difficult to reprogram with control algorithms when adapting to new scenarios. The possibilities of learning in such a changing environment like underwater domains are huge. Also, the utilization of RL in robotics is very common nowadays. However, there are not many approaches which perform online learning, and, to the best of the author's knowledge, this work represents one of the first online robot behavior learning in the underwater domain. It is, therefore, an important contribution to demonstrate the feasibility of RL methods in a real-time task, especially in a complex domain such as underwater robotics. As future work, the learning algorithms used should be tested in real scenarios, extending the learning to more degrees of freedom and in different underwater applications. Operating in real scenarios like coastal waters will introduce new challenges: unpredictable water currents, irregular seafloor, and changing light conditions are only some of them. Also, there is an increasing interest on intervention AUVs which will benefit from these learning capabilities, since intervention is a clear problem in which the uncertainty of the task claims for algorithms that can adapt. Also, new techniques in robot learning such as learning by demonstration can be integrated to decrease the convergence time problem that most reinforcement learning algorithms suffer.

Acknowledgements This research was sponsored by the Spanish government (DPI2008-06548-C03-03, DPI2011-27977-C03-02) and the PANDORA EU FP7-Project under the grant agreement No: ICT-288273.

References

1. Aberdeen DA (2003) Policy-gradient algorithms for partially observable Markov decision processes. PhD thesis, Australian National University, April 2003
2. Amari S (1998) Natural gradient works efficiently in learning. Neural Comput 10:251–276
3. Anderson C (2000) Approximating a policy can be easier than approximating a value function. Computer science technical report, University of Colorado State
4. Bagnell JA, Schneider JG (2001) Autonomous helicopter control using reinforcement learning policy search methods. In: IEEE international conference on robotics and automation, ICRA. Korea
5. Baird K (1995) Residual algorithms: reinforcement learning with function approximation. In: 12th international conference on machine learning, ICML. San Francisco, USA
6. Baird LC, Klopf AH (1993) Reinforcement learning with high-dimensional, continuous action. Technical report WL-TR-93-1147, Wright Laboratory
7. Barto AG, Sutton RS, Anderson CW (1983) Neuronlike elements that can solve difficult learning control problems. IEEE Trans Syst Man Cybern 13:835–846
8. Baxter J, Bartlett PL (2000) Direct gradient-based reinforcement learning. In: International symposium on circuits and systems. Geneva, Switzerland, May 2000
9. Bellman RE (1957) Dynamic programming. Princeton University Press, Princeton
10. Benbrahim H, Doleac J, Franklin J, Selfridge O (1992) Real-time learning: a ball on a beam. In: International joint conference on neural networks (IJCNN). Baltimore, MD, USA
11. Benbrahim H, Franklin J (1997) Biped dynamic walking using reinforcement. Robot Auton Syst 22:283–302
12. Bertsekas DP, Tsitsiklis JN (1996) Neuro-dynamic programming. Athena Scientific, Belmont
13. Carreras M, Ridao P, El-Fakdi A (2003) Semi-online neural-Q-learning for real-time robot learning. In: IEEE/RSJ international conference on intelligent robots and systems, IROS. Las Vegas, USA, October 27–31 2003
14. Carreras M, Ridao P, Garcia R, Nicosevici T (2003) Vision-based localization of an underwater robot in a structured environment. In: IEEE international conference on robotics and automation, ICRA. Taipei, Taiwan
15. Carreras M, Yuh J, Batlle J, Ridao P (2007) Application of sonql for real-time learning of robot behaviors. Robot Auton Syst 55(8):628–642
16. Crites RH, Barto AG (1996) Improving elevator performance using reinforcement learning. In: Advances in neural information processing systems, NIPS. MIT Press, Cambridge, MA
17. Dayan P (1992) The convergence of TD(λ) for general λ. J Mach Learn 8:341–362
18. Gordon GJ (1999) Approximate solutions to Markov decision processes. PhD thesis, Carnegie-Mellon University

19. Gullapalli V, Franklin JJ, Benbrahim H (1994) Acquiring robot skills via reinforcement. IEEE Control Syst J, Spec Issue on Robot: Capturing Nat Motion 4(1):13–24
20. Haykin S (1999) Neural networks, a comprehensive foundation, 2nd edn. Prentice Hall, Englewood Cliffs
21. Hernandez N, Mahadevan S (2000) Hierarchical memory-based reinforcement learning. In: Advances in neural information processing systems, NIPS. Denver, USA
22. Konda VR, Tsitsiklis JN (2003) On actor-critic algorithms. SIAM J Control Optim 42(4): 1143–1166
23. Kormushev P, Nomoto K, Dong F, Hirota K (2011) Time hopping technique for faster reinforcement learning in simulations. International Journal of Cybernetics and Information Technologies 11(3):42–59
24. Lin LJ (1992) Self-improving reactive agents based on reinforcement learning, planning and teaching. J Mach Learn 8(3/4):293–321
25. Marbach P, Tsitsiklis JN (2000) Gradient-based optimization of Markov reward processes: practical variants. Technical report, Center for Communications Systems Research, University of Cambridge, March 2000
26. Meuleau N, Peshkin L, Kim K (2001) Exploration in gradient based reinforcement learning. Technical report, Massachusetts Institute of Technology, AI Memo 2001–2003, April 2001
27. Ortiz A, Simo M, Oliver G (2002) A vision system for an underwater cable tracker. Int J Mach Vis Appl 13(3):129–140
28. Peters J (2007) Machine learning of motor skills for robotics. PhD thesis, Department of Computer Science, University of Southern California
29. Peters J, Schaal S (2006) Policy gradient methods for robotics. In: IEEE/RSJ International Conference on Intelligent Robots and Systems, IROS. Beijing, China, October 9–15 2006
30. Peters J, Vijayakumar S, Schaal S (2005) Natural actor-critic. In: ECML, pp 280–291. Austria
31. Precup D, Sutton RS, Dasgupta S (2001) Off-policy temporal-difference learning with function approximation. In: 18th international conference on machine learning, ICML. San Francisco
32. Pyeatt LD, Howe AE (1998) Decision tree function approximation in reinforcement learning. Technical Report TR CS-98-112, Colorado State University
33. Ribas D, Palomeras N, Ridao P, Carreras M, Hernandez E (2007) Ictineu AUV wins the first SAUC-E competition. In: IEEE international conference on robotics and automation, ICRA. Roma
34. Richter S, Aberdeen D, Yu J (2006) Natural actor-critic for road traffic optimisation. In: Neural information processing systems, NIPS, pp 1169–1176. Cambridge
35. Singh SP, Jaakkola T, Jordan MI (1994) Learning without state-estimation in partially observable markovian decision processes. In: 11th international conference on machine learning, ICML, pp 284–292. Morgan Kaufmann, New Jersey, USA
36. Singh SP, Jaakkola T, Jordan MI (1995) Reinforcement learning with soft state aggregation. Advances in neural information processing systems, NIPS, 7. Cambridge
37. Smart WD, Kaelbling LP (2000) Practical reinforcement learning in continuous spaces. In: International conference on machine learning, ICML. San Francisco
38. Sutton RS (1988) Learning to predict by the method of temporal differences. J Mach Learn 3:9–44
39. Sutton RS (1990) Integrated architectures for learning, planning, and reacting based on approximating dynamic programming. In: 7th international workshop on machine learning, pp 216–224. San Francisco
40. Sutton RS (1996) Generalization in reinforcement learning: Successful examples using sparse coarse coding. Advances in Neural Information Processing Systems, NIPS 9:1038–1044
41. Sutton RS, Barto A (1998) Reinforcement Learning, an introduction. MIT Press, Cambridge
42. Sutton RS, McAllester D, Singh S, Mansour Y (2000) Policy gradient methods for reinforcement learning with function approximation. Advances in Neural Information Processing Systems, NIPS 12:1057–1063
43. Tesauro GJ (1992) Practical issues in temporal difference learning. J Mach Learn 8(3/4): 257–277

44. Tsitsiklis JN, Van Roy B (1996) Feature-based methods for large scale dynamic programming. J Mach Learn 22:59–94
45. Tsitsiklis JN, Van Roy B (1997) Average cost temporal-difference learning. IEEE Trans Autom Control 42:674–690
46. Ueno T, Nakamura Y, Shibata T, Hosoda K, Ishii S (2006) Stable learning of quasi-passive dynamic walking by an unstable biped robot based on off-policy natural actor-critic. In: IEEE/RSJ International Conference on Intelligent Robots and Systems, IROS
47. Watkins CJCH, Dayan P (1992) Q-learning. J Mach Learn 8:279–292
48. Weaver S, Baird L, Polycarpou M (1998) An analytical framework for local feedforward networks. IEEE Trans Neural Netw 9(3)
49. Zhang W, Dietterich TG (1995) A reinforcement learning approach to job-shop scheduling. In: International Joint Conference on Artificial Intelligence, IJCAI

Chapter 8
Simultaneous Localization and Mapping in Marine Environments

Maurice F. Fallon, Hordur Johannsson, Michael Kaess, John Folkesson, Hunter McClelland, Brendan J. Englot, Franz S. Hover, and John J. Leonard

Abstract Accurate navigation is a fundamental requirement for robotic systems—marine and terrestrial. For an intelligent autonomous system to interact effectively and safely with its environment, it needs to accurately perceive its surroundings. While traditional dead-reckoning filtering can achieve extremely low drift rates, the localization accuracy decays monotonically with distance traveled. Other approaches (such as external beacons) can help; nonetheless, the typical prerogative is to remain at a safe distance and to avoid engaging with the environment. In this chapter we discuss alternative approaches which utilize onboard sensors so that the robot can estimate the location of sensed objects and use these observations to improve its own navigation as well as its perception of the environment. This approach allows for meaningful interaction and autonomy. Three motivating autonomous underwater vehicle (AUV) applications are outlined herein. The first fuses external range sensing with relative sonar measurements. The second application localizes

M.F. Fallon (✉) • H. Johannsson • M. Kaess • J.J. Leonard
Computer Science and Artificial Intelligence Laboratory, Massachusetts Institute of Technology, Cambridge, MA, USA
e-mail: mfallon@mit.edu; hordurj@mit.edu; kaess@mit.edu; jleonard@mit.edu

J. Folkesson
Center for Autonomous Systems, Kungliga Tekniska Högskolan (KTH), Stockholm, Sweden
e-mail: johnf@kth.se

H. McClelland
Robotics and Mechanisms Laboratory, Virginia Polytechnic Institute of Technology and State University (Virginia Tech), Blacksburg, VA, USA
e-mail: hgm@vt.edu

B.J. Englot • F.S. Hover
Department of Mechanical Engineering, Massachusetts Institute of Technology, Cambridge, MA, USA
e-mail: benglot@alum.mit.edu; hover@mit.edu

M.L. Seto (ed.), *Marine Robot Autonomy*, DOI 10.1007/978-1-4614-5659-9_8,

relative to a prior map so as to revisit a specific feature, while the third builds an accurate model of an underwater structure which is consistent and complete. In particular we demonstrate that each approach can be abstracted to a core problem of incremental estimation within a sparse graph of the AUV's trajectory and the locations of features of interest which can be updated and optimized in real time on board the AUV.

8.1 Introduction

In this chapter we consider the problem of simultaneous localization and mapping (SLAM) from a marine perspective. Through three motivating applications, we demonstrate that a large class of autonomous underwater vehicle (AUV) missions can be generalized to an underlying set of measurement constraints which can then be solved using a core pose graph SLAM optimization algorithm known as incremental smoothing and mapping (iSAM) [39].

Good positioning information is essential for the safe execution of an AUV mission and for effective interpretation of the data acquired by the AUV [25, 46]. Traditional methods for AUV navigation suffer several shortcomings. Dead reckoning and inertial navigation systems (INS) are subject to external disturbances and uncorrectable drift. Measurements from Doppler velocity loggers can be used to achieve higher precision, but position error still grows without bound. To achieve bounded errors, current AUV systems rely on networks of acoustic transponders or surfacing for GPS resets, which can be impractical or undesirable for many missions of interest.

The goal of SLAM is to enable an AUV to build a map of an unknown environment and concurrently use that map for positioning. SLAM has the potential to enable long-term missions with bounded navigation errors without reliance on acoustic beacons, *a priori* maps, or surfacing for GPS resets. Autonomous mapping and navigation is difficult in the marine environment because of the combination of sensor noise, data association ambiguity, navigation error, and modeling uncertainty. Considerable progress has been made in the past 10 years, with new insights into the structure of the problem and new approaches that have provided compelling experimental demonstrations.

To perform many AUV missions of interest, such as mine neutralization and ship hull inspection, it is not sufficient to determine the vehicle's trajectory in post-processing after the mission has been completed. Instead, mission requirements dictate that a solution is computed in real time to enable closed-loop position control of the vehicle. This requires solving an ever-growing optimization problem incrementally by only updating quantities that actually change instead of recomputing the full solution—a task for which iSAM is well suited.

Each application presents a different aspect of smoothing-based SLAM:

- Smoothing as an alternative to filtering: the use of nontraditional acoustic range measurements to improve AUV navigation [18]

- Relocalizing in an existing map: localizing and controlling an AUV using natural features using a forward looking sonar [23]
- Loop closure used to bound error and uncertainty: combining AUV motion estimates with observations of features on a ship's hull to produce accurate hull reconstructions [34]

A common theme for all three applications is the use of pose graph representations and associated estimation algorithms that exploit the graphical model structure of the underlying problem.

First we will overview the evolution of the SLAM problem in the following section.

8.2 Simultaneous Localization and Mapping

The earliest work which envisaged robotic mapping within a probabilistic framework was the seminal paper by Smith et al. [68]. This work proposed using an extended Kalman filter (EKF) to estimate the first and second moments of the probability distribution of spatial relations derived from sensor measurements. Moutarlier and Chatila provided the first implementation of this type of algorithm with real data [54], using data from a scanning laser range finder mounted on a wheeled mobile robot operating indoors. The authors noted that the size of the state vector would need to grow linearly with the number of landmarks and that it was necessary to maintain the full correlation between all the variables being estimated; thus, the algorithm scales quadratically with the number of landmarks [11].

The scalability problem was addressed by a number of authors. The sparse extended information filter (SEIF) by Thrun et al. [73] uses the information form of the EKF in combination with a sparsification method. One of the downfalls of that approach was that it resulted in overconfident estimates. These issues were addressed in the exactly sparse delayed-state filters (ESDFs) by Eustice et al. [14,15] and later with the exactly sparse extended information filter (ESEIF) by Walter et al. [78].

Particle filters have also been used to address both the complexity and the data association problem. The estimates of the landmark locations become independent when conditioned on the vehicle trajectory. This fact was used by Montemerlo et al. [53] to implement FastSLAM. The main drawback of particle filters applied to the high-dimensional trajectory estimation is particle depletion. In particular, when a robot completes a large exploration loop and decides upon a loop closure, only a small number of particles with independent tracks will be retained after any subsequent resampling step.

In purely localization tasks (with static prior maps) particle filters have been successful. Monte Carlo localization allowed the Minerva robotic museum guide to operate for 44 km over 2 weeks [71]. More recently it has been used by Nuske et al.

to localize an AUV relative to a marine structure using a camera [59], exploiting GPU-accelerated image formation to facilitate large particle sets.

Filtering approaches have some inherent disadvantages when applied to the SLAM problem: measurements are linearized only once based on the current state estimate—at the time the measurement is added. Further, it is difficult to apply delayed measurements or to revert a measurement once it has been applied to the filter. The Atlas framework by Bosse et al. [6] addresses these issues by combining local submaps and a nonlinear optimization to globally align the submaps. Each submap has its own local coordinate frame, so the linearization point cannot deviate as far from the true value as in the case of global parameterization.

8.2.1 Pose Graph Optimization Using Smoothing and Mapping

As the field has evolved, full SLAM solutions [47, 72] have been explored to overcome the linearization errors that are the major source of suboptimality of filtering-based approaches. Full SLAM includes the complete trajectory into the estimation problem rather than just the most recent state. This has led to the SLAM problem being modeled as a graph where the nodes represent the vehicle poses and optionally also landmarks. The edges in this graph are measurements that put constraints between these variables. By associating probability distributions to the constraints, the graph can be interpreted as a Bayes network.

Under the assumption that measurements are corrupted by zero-mean Gaussian noise, the maximum likelihood solution of the joint probability distribution is found by solving a nonlinear least squares problem. Many iterative solutions to the SLAM problem have been presented, such as stochastic gradient descent [28,60], relaxation [10], preconditioned conjugate gradient [43], and loopy belief propagation [63].

Faster convergence is provided by direct methods that are based on matrix factorization. Dellaert and Kaess [9] introduced the square root smoothing and mapping (SAM) algorithm using matrix factorization to solve the normal equations of the sparse least squares problem. Efficiency is achieved by relating the graphical model to a sparse matrix in combination with variable reordering for maintaining sparsity. Similar methods are used by [44, 45], and more efficient approximate solutions include [27].

The aforementioned incremental smoothing and mapping algorithm provides an efficient incremental solution [39]. In iSAM the matrix factorization is incrementally updated using Givens rotations, making the method better suited for online operations. In addition they developed an efficient algorithm for recovering parts of the covariance matrix [36], which is useful for online data association decisions.

Recently, further exploration of the connection between graphical models and linear algebra allowed a fully incremental formulation of iSAM. The Bayes tree data structure [37] can be considered as an intermediate representation between the Cholesky factor and a junction tree. While not obvious in the matrix formulation, the Bayes tree allows a fully incremental algorithm, with incremental variable

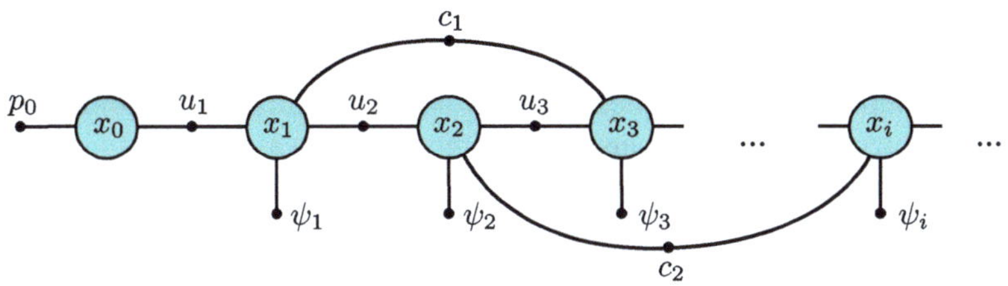

Fig. 8.1 Factor graph for the pose graph formulation of the SLAM problem. The *large circles* are variable nodes, here the AUV states x_i. The *small solid circles* are factor nodes: relative pose measurements u_i, absolute pose measurements ψ_i, a prior on the first pose p_0, and loop closure constraints c_j

reordering and fluid relinearization. The resulting sparse nonlinear least squares solver is called iSAM2 [38].

Using a nonlinear solver for the full SLAM problem overcomes the problems caused by linearization errors in filtering methods, and it is also the case that estimation of the full trajectory results in a sparse estimation problem [9]. It is not necessary to explicitly store the correlation between all the landmarks, making these methods very efficient. One downside is that the problem grows with time (or at least distance traveled) instead of the size of the environment, although the rate of growth is not significant for the applications discussed in this chapter.

8.2.2 Mathematical Summary

In this section, we will briefly present the mathematical formulation of the full SLAM problem as a nonlinear least squares optimization. The full SLAM problem can be described as a constantly growing factor graph. A factor graph is a bipartite graph consisting of variable nodes and factor nodes, connected by edges. The factor graph represents a factorization of a function $f(X)$ over some variables $X = \{x_i\}_{i=0}^{N}$:

$$f(X) = \prod_{k=1}^{K} f_k(X_k) \tag{8.1}$$

where X_k denotes the subset of variables involved in the kth factor. The factor nodes $F = \{f_k\}_{k=1}^{K}$ represent constraints involving one or more variables. Each edge connects one factor node with one variable node.

For our navigation setting, consider the simple factor graph example in Fig. 8.1, where the variable nodes $x_1 \ldots x_N$ represent the vehicle states sampled at discrete times, together forming the vehicle trajectory. Here, the factor nodes F are partitioned into multiple types that represent relative pose constraints u_i between consecutive poses, absolute pose constraints ψ_i on individual poses, and loop closure constraints c_j on arbitrary pairs of poses form these measurements.

When assuming Gaussian measurement noise, we arrive at a *nonlinear least squares* problem. Under the Gaussian assumption, a measurement z_k is predicted based on the current estimate X_k through a deterministic function h_k and with added zero-mean Gaussian measurement noise v_k with covariance Λ_k:

$$z_k = h_k(X_k) + v_k \qquad v_k \sim \mathcal{N}(0, \Lambda_k) \tag{8.2}$$

Hence, the factor f_k to encode the actual measurement z_k is defined as

$$f_k(X_k) \propto \exp\left(-\frac{1}{2}\|h_k(X_k) - z_k\|^2_{\Lambda_k}\right) \tag{8.3}$$

where $\|x\|^2_\Sigma := x^\top \Sigma^{-1} x$. To find the nonlinear least squares solution $\hat{X}$ we make use of the monotonicity of the logarithm function for converting the factorization into a sum of terms:

$$\hat{X} = \arg\max_X \prod_{k=1}^{K} f_k(X_k) \tag{8.4}$$

$$= \arg\min_X - \log \prod_{k=1}^{K} f_k(X_k) \tag{8.5}$$

$$= \arg\min_X \sum_{k=1}^{K} - \log f_k(X_k) \tag{8.6}$$

$$= \arg\min_X \sum_{i=k}^{K} \|h_k(X_k) - z_k\|^2_{\Lambda_k} \tag{8.7}$$

Standard Gauss-Newton [26]-based solutions, such as Levenberg-Marquardt or Powell's dog leg, repeatedly linearize and solve this sparse nonlinear least squares problem. By stacking the linearized equations, a sparse matrix A is obtained whose block structure mirrors the structure of the factor graph

$$\delta\hat{X} = \arg\min_{\delta X} \|A\delta X - b\|^2 \tag{8.8}$$

The vector b contains the measurements and residuals; details are given in [9]. This linear system can be solved by matrix factorization and forward and back substitution. After each iteration the current estimate is updated by $\hat{X} \leftarrow \hat{X} + \delta\hat{X}$. The new estimate is then used as new linearization point, and the process is iterated until convergence.

iSAM [38, 39] provides an incremental solution to Gauss-Newton style methods, in particular Powell's dog leg [65]. When new measurements are received, this approach updates the existing matrix factorization rather than recalculating the nonlinear least squares system anew each iteration. For a detailed account of this process, the reader is referred to the original papers.

8.2.3 Data Association

A fundamental problem in feature-based SLAM is the correct association of point measurements from different time steps to one another. Given a series of raw laser, camera, or sonar measurements, the challenge is to identify the observed features which originated from the same physical entity. Knowledge of this data association provides a set of valid measurement constraints. As explained previously, these constraints can be optimized efficiently; however, this data association problem must first be solved.

Data association in its most generalized form is a well-studied problem, for example [58]. Where the measurements are indistinct, noisy, or contradictory, there remains the possibility of association errors. A core weakness of current SLAM approaches is brittleness and suboptimality resulting from these errors becoming "baked into" the optimization problem. Currently, the predominant approach is to avoid adding such associations if not absolutely confident in their correctness—instead assuming access to informative sensor data at a later time. That is the approach we are taking for ship hull inspection in Sect. 8.6, where navigation uncertainty of the onboard sensors is low, allowing for many minutes of open-loop navigation without significant loss of accuracy.

Discarding uninformative sensor information unfortunately is not a luxury available in many AUV applications in which interesting features are often rare. While approaches which maintain multiple data association hypotheses for an extended time have been proposed, the exponential growth in the size of a hypothesis tree cannot be supported indefinitely. In Sect. 8.5 we present an application which tackles this problem in a typical marine environment for a low-cost AUV with significant navigation uncertainty. Data association decisions are taken just after a feature has left the field of view so as to have access to all available observations of a particular feature before making the critical association decision.

While a detailed discussion of the field of data association is outside the scope of this work, it remains a problem specific to each problem or application.

8.3 Navigation in Marine Environments

In the following sections we will motivate the use of the smoothing and mapping approach by way of three separate autonomous marine applications. In particular, we will demonstrate that the estimation problem at the heart of each application can be reduced to a set of navigation and perception constraints which can be optimally, incrementally, and efficiently solved using the iSAM algorithm.

First we will give a more general overview of SLAM in marine environments. The modern AUV contains proprioceptive sensors such as compasses, fiber-optic gyroscopes (FOG), and Doppler velocity loggers (DVL) [82]. The sensor output of these sensors is fused together using navigation filters, such as the EKF, to produce a high-quality estimate of the AUV position and uncertainty. This estimate is then used by the AUV to inform onboard decision-making logic and to adaptively complete complex survey and security missions. Kinsey et al. provides a survey of state-of-the-art approaches to AUV navigation [42].

Acoustic ranging has been widely used to contribute to AUV navigation [84, 85]. Long baseline (LBL) navigation was initially developed in the 1970s [30, 33] and is commonly used by industrial practitioners [51]. It requires the installation of stationary beacons at known locations surrounding the area of interest which measure round-trip acoustic time of flight before triangulating for 3-D position estimation. Operating areas are typically restricted to a few square kilometers.

Ultrashort baseline (USBL) navigation [48] is an alternative method which is typically used for tracking an underwater vehicle's position from a surface ship. Range is measured via time of flight to a single beacon, while bearing is estimated using an array of hydrophones on the surface vehicle transducer. Overall position accuracy is dependent on many factors, including the range of the vehicle from the surface ship, the motion of the surface ship, and acoustic propagation conditions.

In addition, many modern AUVs have multiple exteroceptive sensors. Side-scan sonar, initially developed by the US Navy, has been widely used for ship, ROV, and AUV survey since its invention in the 1950s. More recently, forward looking sonars, with the ability to accurately position a field of features in two dimensions, have also been deployed for a variety of applications such as 3-D reconstruction [31] and harbor security [13, 40, 49, 64]. In scenarios in which water turbidity is not excessively high, cameras have been used to produce accurate maps of shipwrecks and underwater historical structures, for example, the mapping of RMS Titanic [16] and of Iron Age shipwrecks [3].

These more recent applications have a common aspect; to maintain consistency of sensor measurements over the duration of an experiment, smoothing online of an AUV's trajectory and the location of measured features is necessary. We will now demonstrate how SLAM smoothing in a marine environment is applied in practice.

8.4 Smoothing: Cooperative Acoustic Navigation

The first application we will consider is that of cooperative acoustic navigation. In this application non-traditional sources of acoustic range measurements can be used to improve the navigation performance of a group of AUVs with the aim of achieving bounded error or at the least reducing the frequency of GPS fix surfacings.

Within the context of the data association discussion in Sect. 8.2.3, this application is much simpler in that the acoustic range measurements are paired with the location of the surface beacon originating them—by design. This avoids data association entirely.

Following on from traditional LBL navigation, the moving long baseline (MLBL) concept proposed two mobile autonomous surface vehicles (ASVs) aiding an AUV using acoustic modem ranging. This was proposed by Vaganay et al. [75] and extended by Bahr et al. [1, 2]. This concept envisaged the ASVs transmitting acoustic modem messages containing their GPS positions paired with a modem-estimated range to the AUV which could then uniquely fix its position while maintaining full mobility—which is not afforded by typical LBL positioning.

More recent research has focused on utilizing only a single surface vehicle to support an AUV using a recursive state estimator such as the extended Kalman filter [19] or the distributed extended information filter (DEIF) [81].

For many robotic applications, however, estimating the vehicle's entire trajectory as well as the location of any observed features is important (e.g., in survey missions). As mentioned previously, the EKF has been shown to provide an inconsistent SLAM solution due to information lost during the linearization step [35]. Furthermore, our previous work, [22], demonstrated (off-line) the superior performance of NLS methods in the acoustic ranging problem domain versus both an EKF and a particle filtering implementation—although requiring growing computational resources. For these reasons we present here an application in which iSAM is used for full pose trajectory estimation using acoustic range data.

Additionally we demonstrate that mapping of bottom targets (identified in side-scan sonar imagery) can be integrated within the same optimization framework. The effect of this fusion is demonstrated in Fig. 8.2. This figure demonstrates the alignment of side-scan sonar mosaics from three separate observations of the same feature. Without optimizing the entire global set of constraints, the resultant data reprojection would be inconsistent.

As an extension, we demonstrate the ability to combine relative constraints across successive missions, enabling multi-session AUV navigation and mapping, in which data collected in previous missions is seamlessly integrated online with data from the current mission on board the AUV.

8.4.1 Problem Formulation

The full vehicle state is defined in three Cartesian and three rotation dimensions, $[x, y, z, \phi, \theta, \psi]$. Absolute measurements of the depth z, roll ϕ, and pitch θ are measured using a water pressure sensor and inertial sensors. This leaves three dimensions of the vehicle to be estimated in the horizontal plane: x, y, ψ.

The heading is instrumented directly using a compass, and this information is integrated with inertial velocity measurements to propagate estimates of the x and y

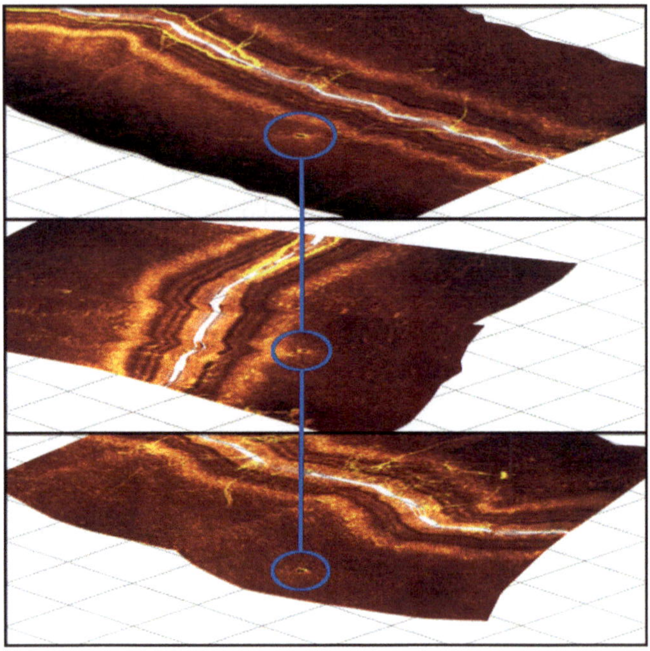

Fig. 8.2 Optimizing the entire set of vehicle poses and target observations facilitates explicit alignment of sonar mosaics and understanding of the motion of the AUV during the mission. This allows for reactive decision making in the water—as opposed to post-processing which is common currently. In this figure this optimization allows three different observations of a single target to be explicitly aligned

position.[1] This integration is carried out at a high frequency ($\sim$ 10 Hz) compared to the exteroceptive range and sonar measurements ($\sim$ 1 Hz).

The motion of the vehicle at time step i is described by a Gaussian process model as follows:

$$u_i = h_u(x_{i-1}, x_i) + w_i \quad w_i \sim N(0, \Sigma_i) \tag{8.9}$$

where x_i represents the 3-D vehicle state (as distinct from the dimension x above). Note that while the heading is directly estimated using a compass, we use this estimate only as a prior within the smoothing framework. In this way the smoothed result will produce a more consistent combined solution.

[1]In our case this integration is carried out on a separate proprietary vehicle control computer, and the result is passed to the payload computer.

8.4.1.1 Acoustic Ranging

Instead of either LBL or USBL, our work aims to utilize acoustic modems, such as the WHOI Micro-Modem [24], which arc alrcady installed on the majority of AUVs for command and control. The most accurate inter-vehicle ranging is through one-way travel-time ranging with precisely synchronized clocks, for example, using the design by Eustice [17], which also allows for broadcast ranging to any number of vehicles in the vicinity of the transmitting vehicle. An alternative is round-trip ranging, which, while resulting in more complexity during operation and higher variance, requires no modification of existing vehicles.

Regardless of the ranging method, the range measurement $r_{j,3D}$, a 2-D estimate of the position of the transmitting beacon, $g_j = [x_{gj}, y_{gj}]$, and associated covariances will be made known to the AUV at intervals on the order of 10–120 seconds. Having transformed the range to a 2-D range over ground r_j (using the directly instrumented depth), a measurement model can be defined as follows:

$$r_j = h_r(x_j, b_j) + \mu_j \quad \mu_j \sim N(0, \Xi_j) \tag{8.10}$$

where x_j represents the position of AUV state at that time. GPS measurements g_j of the beacon position b_j are assumed to be distributed via a zero-mean normal distribution with covariance Ξ_j.

Comparing the onboard position estimates of the AUV and the ASV in the experiments in Sect. 8.4.2, round-trip ranging is estimated to have a variance of approximately 7 m, compared with a variance of 3 m for one-way ranging reported in [22]. An additional issue is that with the ranging measurement occurring as much as 10 s before the position and range are transmitted to the AUV, an acausal update of the vehicle position estimate is required.

The operational framework used by Webster et al. [80, 81] is quite similar to ours. Their approach is based on a decentralized estimation algorithm that jointly estimates both the AUV position and that of a supporting research vessel using a distributed extended information filter. Incremental updates of the surface vehicle's position are integrated into the AUV-based portion of the filter via a simple and compact addition which, it is assumed, can be packaged within a single modem data packet.

This precise approach hypothesizes the use of a surface vehicle equipped with a high accuracy gyrocompass and a survey-grade GPS (order of 0.5 m accuracy). Furthermore, as described in [80], the approach can be vulnerable to packet loss, resulting in missing incremental updates which would cause the navigation algorithm to fail. While rebroadcasting strategies to correct for such a failure could be envisaged, it is likely that significant (scarce) bandwidth would be sacrificed, making multi-vehicle operations difficult.

Our approach instead aims to provide independent surface measurements to the AUV in a manner that is robust to inevitable acoustic modem packet loss. The goal is a flexible and scalable approach that fully exploits the one-way travel-time

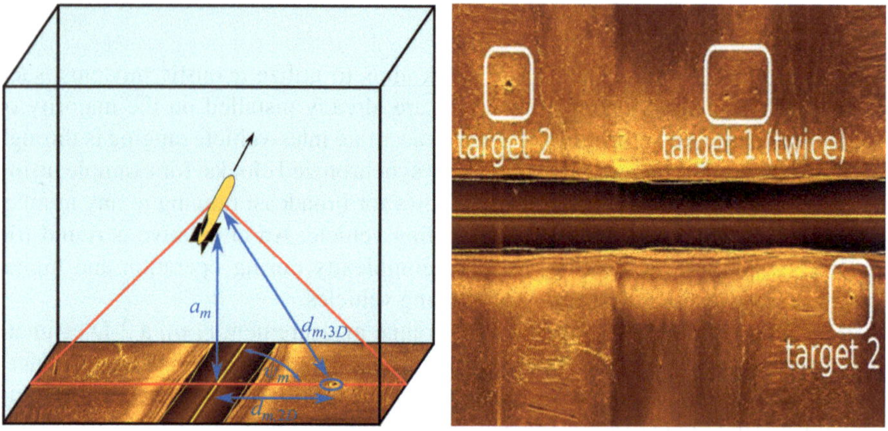

Fig. 8.3 **(a)** As the AUV travels through the water, the side-scan sonar images laterally with objects on the ocean floor giving strong returns. **(b)** A top-down projection of the side-scan sonar for a 120 m of vehicle motion *(left to right)*. The lateral scale is 30 m in each direction which yields a 1:1 aspect ratio. Note that in this case targets 1 and 2 have been observed twice each after a turn

ranging data that the acoustic modems enable. The solution should be applicable to situations in which only low-cost GPS sensors are available on the ASVs or gateway buoys to provide maximum flexibility.

8.4.1.2 Side-Scan Sonar

To demonstrate the compatibility of this approach with typical side-scan sonar surveys, the algorithm was extended to support relative observations from the sonar in a SLAM framework.

Side-scan sonar is a common sonar sensor often used for ocean sea-floor mapping. As the name suggests, the sonar transducer device scans laterally when towed behind a ship or flown attached to an AUV through the water column. A series of acoustic pings are transmitted, and the amplitude and timing of the returns combined with speed of sound in water are used to determine the existence of features located perpendicular to the direction of motion.

By the motion of the transducer through the water column, two-dimensional images can be produced which survey the ocean floor and features on it. See Fig. 8.3 for an example side-scan sonar image. These images, while seemingly indicative of what exists on the ocean floor, contain no localization information to register them with either a relative or global position. Also it is often difficult to repeatedly detect and recognize features of interest; for example, Fig. 8.3 illustrates two observations each of two different targets of interest. Target 1 (a metallic icosahedron) appears differently in its two observations. Also targets are typically not identified using the returned echoes from the target itself but by the shadow cast by the target [12].

For these reasons we must be careful in choosing side-scan sonar features for loop closure. Appearance-based matching techniques, such as FABMAP [8], would most likely encounter difficulties with acoustic imagery. Metric-based feature matching requires access to accurate, fully optimized position and uncertainty estimates of the new target relative to all previously observed candidate features. For these reasons, we will use iSAM to optimize the position and uncertainty of the entire vehicle trajectory, the sonar target positions, as well as all the beacon range estimates mentioned in Sect. 8.4.1.1.

The geometry of the side-scan sonar target positioning is illustrated in Fig. 8.3. Distance from the side-scan sonar to a feature corresponds to the *slant range*, $d_{m,3\mathrm{D}}$, while the distance of the AUV off the ocean floor (*altitude*, a_m) can be instrumented. We will assume the ocean floor to be locally flat which allows the slant range to be converted into the horizontal range, resulting in the following relative position measurement:

$$d_{m,2\mathrm{D}} = \sqrt{d_{m,3\mathrm{D}}^2 - a_m^2} \tag{8.11}$$

$$\rho_m = \pm\pi/2 \tag{8.12}$$

where ρ_m is the relative bearing to the target defined from the front of the vehicle anticlockwise. These two measurements paired together give a relative position constraint, $z_m = [d_{m,2\mathrm{D}}, \rho_m]$ for an observation of target s_m. This target can either be a new, previously unseen target or a reobservation of an older target. In the experiments in Sect. 8.4.2 this data association is done manually, while in future work we will aim to do this automatically as in [69]. The resultant measurement model will be as follows:

$$z_m = h_z(x_m, s_m) + v_m \quad v_m \sim N(0, \Lambda_m) \tag{8.13}$$

where x_m is the pose of the AUV at that time. In effect, repeated observations of the same sonar target correspond to loop closures. Such repeated observations of the same location allow uncertainty to be bounded for the navigation between the observations.

8.4.1.3 Integration into the SAM Framework

Using the set of J acoustic ranges, M side-scan sonar constraints as well as the N incremental inertial navigation constraints, the optimization problem is formulated as follows:

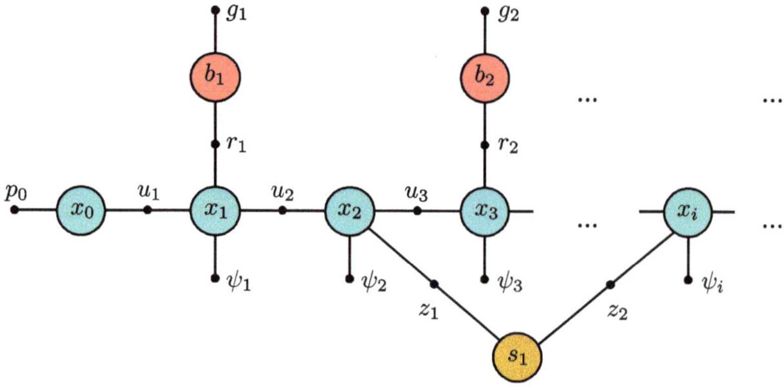

Fig. 8.4 Factor graph formulation of the measurement system showing vehicle states x_i, surface beacons b_j, and sonar targets s_k. Also illustrated are the respective constraints: range r_j in the case of the surface beacons and range and relative bearing z_m in the case of sonar targets. Ranges are paired with surface beacon measurements, while multiple observations of a particular sonar target is in effect a loop closure. The initial pose is constrained using an initial prior p_0 using the GPS position estimate when the AUV dived

$$\hat{X} = \arg\min_X \sum_{i=1}^{N} ||h_u(x_{i-1}, x_i) - u_i||^2_{\Sigma_i}$$
$$+ \sum_{j=1}^{J} ||b_j - g_j||^2_{\Phi_j} + \sum_{j=1}^{J} ||h_r(x_j, b_j) - r_j||^2_{\Xi_j}$$
$$+ \sum_{m=1}^{M} ||h_z(x_m, s_m) - z_m||^2_{\Lambda_m} \tag{8.14}$$

In summary, x_j represents the vehicle pose when measuring the range r_j to beacon b_j, x_m is the pose when observing sonar target s_m at relative position z_m, and finally u_i is the control input between poses x_{i-1} and x_i. Unlike the simple derivation outlined in Sect. 8.2.2, the beacon and target positions require explicit insertion into the problem factor graph. The corresponding factor graph is illustrated by Fig. 8.4.

8.4.2 *Experiments*

A series of experiments were carried out in St. Andrews Bay in Panama City, Florida to demonstrate this proposed approach. A Hydroid REMUS 100 AUV carried out four different missions while collecting side-scan sonar data (using a Marine Sonic transducer) as well as range and GPS position information transmitted from either

Fig. 8.5 The vehicles used in our experiments: the Hydroid REMUS 100 AUV was supported by the MIT Scout ASV or by the research vessel—the *Steel Slinger*

the Scout ASV (Fig. 8.5) or a deck box on the 10 m support vessel. In each case, a low-cost Garmin 18x GPS sensor was used to provide GPS position estimates.

The Kearfott T16 INS, connected to the REMUS front-seat computer, fused its FOG measurements with those of a Teledyne RDI DVL, an accelerometer and a GPS sensor to produce excellent navigation performance. For example after a 40 min mission the AUV surfaced with a 2 m GPS correction—drift of the order of 0.1% of the distance traveled.

The AUV did not have the ability to carry out one-way ranging, and as a result, two-way ranging was used instead. The navigation estimate was made available to a *backseat* computer which ran an implementation of the algorithm in Sect. 8.2.2 (less the sonar portion).

Given the variance of two-way ranging ($\sim$7 m) and the accuracy of the vehicle INS, it would be ambitious to expect to demonstrate significant improvement using cooperative ranging-assisted navigation in this case. For this reason these missions primarily present an opportunity to validate and demonstrate the system with combined sensor input and multiple mission operation.

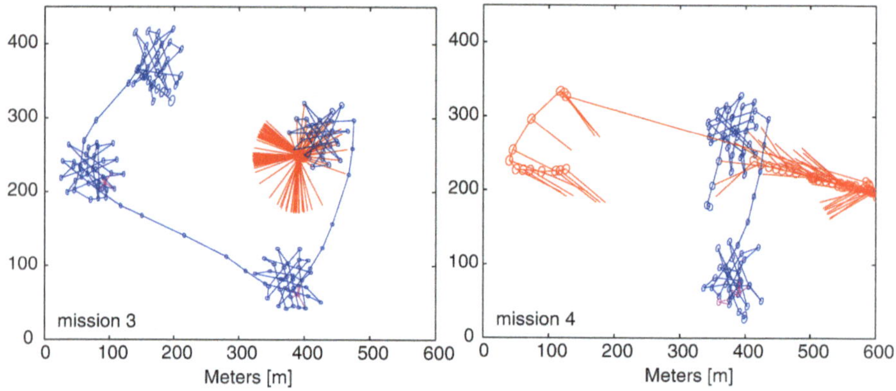

Fig. 8.6 An overview of the optimized trajectory estimates of the AUV (*blue*) and the surface vehicle (*red*), as well as the estimated position of three sonar targets (*magenta*) for two of the missions. The mutually observed feature in the southeast allows for the joint optimization of the two missions. This corresponds to target 3 in Fig. 8.7. The *red lines* indicate the relative vehicle positions during ranging, while the *ellipses* indicate position uncertainty

For simplicity, we will primarily focus on the longest mission—Mission 3 in Fig. 8.7—before discussing the extension to successive missions in Sect. 8.4.2.3. The missions are numbered chronologically.

8.4.2.1 Single Mission

During Mission 3, the AUV navigation data was combined with the acoustic range/position pairs and optimized online on board the AUV using iSAM to produce a real-time estimate of its position and uncertainty. After the experiments, sonar targets were manually extracted from the Marine Sonic data file and used in combination with the other navigation data to produce the combined optimization illustrated in Fig. 8.6. (The two remaining applications of this chapter describe online algorithms for sonar processing.)

An overview of the mission is presented in Fig. 8.7 as well as quantitative results from the optimization where 3σ uncertainty was determined using $3\sqrt{\sigma_x^2 + \sigma_y^2}$.

Starting at (400, 250), the vehicle carried out a set of four *re-identification* (RID) patterns. These overlapping patterns are designed to provide multiple opportunities to observe objects on the ocean floor using the side-scan sonar. Typically this mission is carried out after having first coarsely surveyed the entire ocean floor. In this case two artificial targets were placed at the center of patterns 2 and 3 and were detected between 15–24 min (6 times) and 27–36 min (7 times), respectively. The surface beacon, in this case the support vessel on anchor at (400, 250), transmitted round-trip ranges to the AUV on a 20-second cycle.

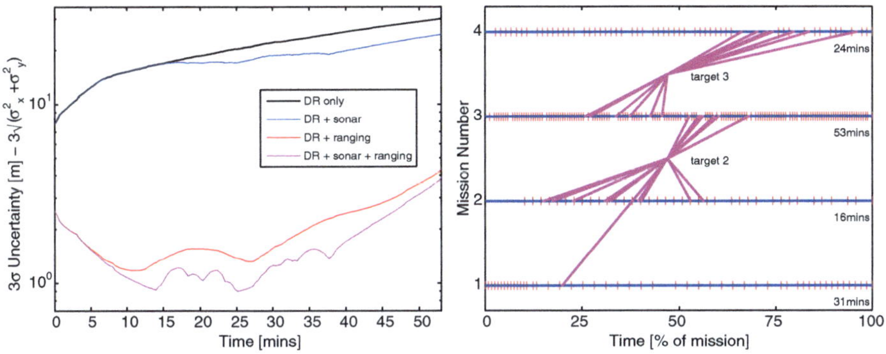

Fig. 8.7 (**a**) Navigation uncertainty for Mission 3 for four different algorithm configurations. Acoustic ranging alone can bound error growth—subject to observability (*red*), while the full sonar and acoustic fusion produces the solution with minimum uncertainty (*magenta*). (**b**) During the four (consecutive) missions, range measurements (represented by the *red lines*) were frequently received from the ASV (Mission 1 and 2) or the research vessel (Mission 3 and 4). Occasionally targets were detected in the side-scan sonar data. Repeated observations of the same target (illustrated in magenta) allow for a SLAM loop closure and for interloop uncertainty to be bounded

8.4.2.2 Analysis

A quantitative analysis of the approach is presented in Fig. 8.7. The typical case (black) of using only dead reckoning for navigation results in ever-increasing uncertainty. The second approach (blue) utilizes target re-identifications in the sonar data but not acoustic range measurements. This temporarily halts the growth of uncertainty, but monotonic growth continues in their absence.

Acoustic ranging by comparison (red) can achieve bounded error navigation—in this case with a 3σ-bound of about 2 m. As the AUV's mission encircled the support vessel, sufficient observability was achieved to properly estimate the AUV's state—which results in the changing alignment of the uncertainty function. However performance deteriorates when the relative positions of the vehicles do not vary significantly (such as during patterns 3–4; 40–53 min).

Finally, the best performance is observed when the sonar and acoustic ranging data are fully fused. Interestingly, the two modalities complement each other: during *re-identification* patterns 2 and 3, sonar target observations bound the uncertainty while the AUV does not move relative to the support vessel. Later the vehicle transits between patterns—allowing for the range observability to improve.

In summary, the combination of the onboard, sonar, and ranging sensor measurements allows for online navigation to be both globally bounded and locally drift-free.

8.4.2.3 Multiple Missions

In this section we will describe how the algorithm has been extended to combine the maps produced by multiple successive AUV missions within a single optimization framework. As mentioned in previous sections, it is advantageous to provide a robot with as much prior information of its environment before it begins its mission, which it can then improve on as it navigates.

Space considerations do not permit a full analysis of this feature, but briefly: during Missions 1 and 2, surface information was transmitted from an autonomous surface vehicle, MIT's Scout kayak (shown in Fig. 8.5), which moved around the AUV so as to improve the observability of the AUV, as previously demonstrated in [22]. In Mission 4, as in Mission 3, the support vessel was instead used—although in this case, the support vessel moved from a location due east of the AUV to another location due west of the AUV, as illustrated in Fig. 8.6. This demonstrates that a basic maneuver by the support vessel is sufficient to ensure mission observability. The mission started at (350, 200).

Figure 8.7 illustrates the intermission connectivity. This demonstrates that the two targets were observed numerous times during the missions, which allows us to combine the navigation across all of the missions into a single fully optimized estimate of the entire operation area.

While such an approach could possibly be carried out for several vehicles operating simultaneously, sharing minimal versions of their respective maps [21], it is unclear if the acoustic bandwidth available would be sufficient to share sonar target observation thumbnails to verify loop closure.

8.4.3 Discussion

In this section we presented a method for the fusion of onboard proprioceptive navigation and relative sonar observations with acoustic ranges transmitted from an autonomous surface vehicle. It allows for operation for many hours in real time for missions of the type described above. Factors resulting in a reduction in performance of this approach are as follows: (1) infrequent ranging, (2) ranging from the same relative direction, and (3) sonar targets not being present or being infrequently observed. We estimate that the bounded error for a non-FOG enabled AUV with several percent drift would be of the order of 3–5 m (depending on the relative geometry and frequency of the one-way travel-time range measurements).

The specific acoustic ranging problem defined above is one problem in a wider class of problems each of which is defined by the connectivity of the fleet of vehicles and the direction of information flow (which result in inter-vehicle correlations being created). A recent overview of the various subproblems is presented in [77].

Fig. 8.8 iRobot Ranger—a low-cost single-man portable AUV

8.5 Localization Using a Prior Map

In this second application we consider the challenge of using a prior map (generated using techniques described above) as part of a greater mission to neutralize mines in very shallow water—a task that has traditionally been carried out by human divers. The potential for casualties associated with this method of mine countermeasures (MCM) motivates the use of unmanned systems to replace human divers. While tethered robotic vehicle could be remotely controlled to perform MCM, a solution using untethered AUVs offers numerous advantages.

When mission requirements dictate that vehicle cost must be extremely low, the navigation problem for target reacquisition is quite challenging. The crux of the problem is to achieve good navigation performance despite the use of sensors with very low cost.

Resultantly the application unfolds within the context of a multiple-step effort, involving a variety of vehicles and technologies. The mission assumes a target field of moored and bottom mines along a shoreline. In this scenario, a remote environmental measuring unit (REMUS) AUV [76] (Fig. 8.5) performs a survey of the operating area, scouting the operating area, and collecting data using its side-scan sonar. The REMUS data are used to create an a priori map of the underwater environment via processing software developed by SeeByte, Ltd. This a priori map consists of the locations of any strong sonar features in the target field.

This map and the location of the feature of interest (FOI) acts as input to a second *low-cost* relocalization vehicle. In the mission scenario we aim to release this vehicle at a distance of 100 to 1,500 m from the center of the prior map and have it swim at the surface close to the feature field before diving to the seabed. Upon reentering this feature field, the vehicle will extract features from its sonar and use these features to build a map of the features.

Having reobserved a sufficient number of features, the AUV will localize relative to the a priori map and attach itself to the FOI. If successful, the AUV will self-detonate or place an explosive charge. Because of this the vehicle is not intended to be recoverable. For these reasons a low-cost vehicle design requirement has had significant impact on the SLAM algorithms mentioned here.

Overview of Vehicles Used

The vehicle used in development has been the iRobot Ranger AUV [66]. This vehicle was equipped with a depth sensor, altimeter, GPS receiver, a 3-D compass, an acoustic pinger, and a Blueview Proviewer 900 kHz forward looking sonar. The vehicle's design was intended to be low cost and light weight. As indicated by Fig. 8.8, it is single-man portable and deployable.

The design of the vehicle incorporates a propeller which is entirely servoed. This allows the vehicle to be highly maneuverable with a very tight turning radius of 0.5 m (compared with 10 m for the REMUS 100). This is of particular importance for the target homing at the end of the mission. The cruising speed of the AUV is quite low at about 0.6 m/s—comparable with typical surface currents. Thus, the dead-reckoning error due to the current can be quite significant. Given the small diameter of the vehicle, a processor smaller than the typical PC104 generation with limited capability was used. This resulted in severe processing restrictions which are mentioned in subsequent sections.

The vehicle specifically did not have a DVL, used for precise velocity estimation due to cost regions. It would be remiss for us not to mention that the current range of FLS devices are comparable in price to a typical DVL; however, a significant proportion of this price represents the overhead cost of research and development. The manufacturer expects that mass production can reduce cost by an order of magnitude. Nonetheless the utility of the capabilities outlined herein go far beyond this particular application.

While the Hydroid REMUS 100 was primarily used as a survey vehicle (as discussed in Sect. 8.5.2), it was also used in several experiments demonstrated in Sect. 8.5.5.

Marine Vehicle Proprioception

At high-frequency depth estimates, altimeter altitudes, GPS fixes and compass estimates of roll, pitch and heading are fused with actuation values (orientation of the servoed propeller and the estimated propeller RPM) using a typical EKF prediction filter to produce an estimate of the vehicle position and uncertainty at each time. In benign current-free conditions, with careful tuning and excellent compass calibration, this procedure produced a dead-reckoning estimate with about 1% error per distance traveled.

However as we transitioned to more challenging current-prone conditions in later stages of the project (as discussed in Sec. 8.4.2), a current estimation model was developed so as to reject the vehicle's drift in this situation. (Because of the nature of this project, it is not possible to use the aforementioned DVL-enabled vehicle's estimate of the current profile.) This module is designed to be run immediately prior to the mission as the vehicle approaches the target field.

This simplistic model performed reasonably well in smaller currents (below 0.3 m/s) and allowed the AUV to enter the field. After this, success was primarily due to the sonar-based SLAM algorithm (outlined in Sect. 8.5.2). In this current

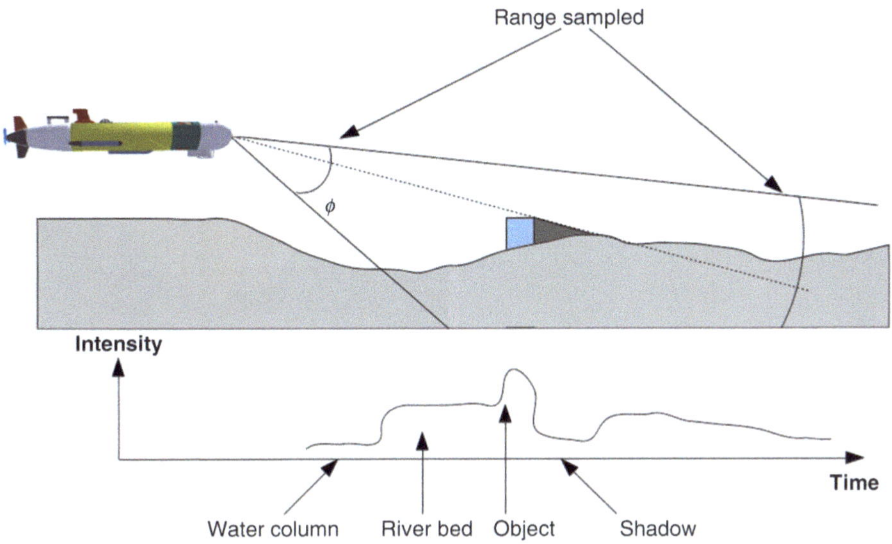

Fig. 8.9 The sonar image generation process for a single sonar beam. Each beam return is an vector of intensities of the returned sonar signal with objects of high density resulting in high returns and shadows resulting in lower intensities

regime, we were able to enter the field approximately 85% of the time using this model, and we estimate the error as about 5% per distance traveled.

8.5.1 Forward Looking Sonar Processing

The sonar is our most important sensor allowing the AUV to perceive its environment. During the project a series of Blueview Proviewer FLS sonars were used. In this section we will give an overview of the sensor technology before presenting our sonar processing algorithms in Sect. 8.5.1.1.

The Proviewer FLS operates using Blazed Array technology [70]. Typically the sonar consisted to two transducer heads (horizontal and vertical) each with a field of view of 45°, although 90° and 135° units were later used.

An outgoing ensonifying signal (colloquially known as a "ping") reflects off of objects of incidence (in particular metal and rock), and the phase, amplitude, and delay of the returned signals are processed to produce a pattern as indicated in Fig. 8.9 (by the manufacturer BlueView). This return is evaluated for each array element with one-degree resolution in the plane of the head, and the output is then fused together via digital signal processing to produce the image in Fig. 8.10.

The outgoing sonar signal also has a significant lobe width, $\phi \sim 20°$, which means that there is significant ambiguity as to the location of the returning object

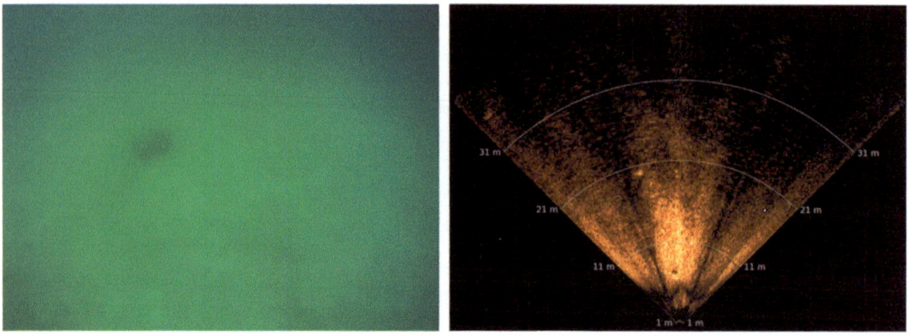

Fig. 8.10 Typical underwater camera and sonar images. The clear water and well-lit scenario represents some of the best possible optical conditions; nonetheless, visibility is only a few meters. This 90° Blazed Array sonar horizontal image indicates 3 features (one at 5 m in front; one at 20 m and 5° to the left and one at 35 m and 40° to the left)—which is more than typical

in the axis off of the return. This distribution was used to estimate the elevation of detections using only the horizontal image.

8.5.1.1 Sonar Feature Detection

In this section we will outline our algorithms which extract point features from regions of high contrast. Forward looking sonar has previously been used for obstacle detection and path planning as in [62]; in this application the feature extraction is focused on conservative estimation of all detected objects given the very noisy output of the FLS systems. Finally, [7] carried out multi-target tracking of multiple features from a FLS using a PHD filter.

Most normal objects are visible at 20 m, while very brightly reflective objects are detectable to 40 m. Adaptability to bottom reflective brightness was achieved by the online estimation of an average background image immediately after the vehicle leveled out at its cruising depth. Estimating this background noise image was essential for us to achieve excellent performance in both sandy and muddy bottom types. Having done this, the features are detected based on gradients of the sonar image in each of four different sonar regions. The specific details of our feature detector and a quantitative analysis of its performance are available in [20, 23].

In terms of processing, the feature detector uses negligible processing power. The formation of the input image (using the Blueview SDK)—the input to this process—requires a substantial 180 ms, per image. The feature detector requires about 18 ms while the remaining CPU power is used to fuse the measurements, to make high level mission decisions and to control the AUV.

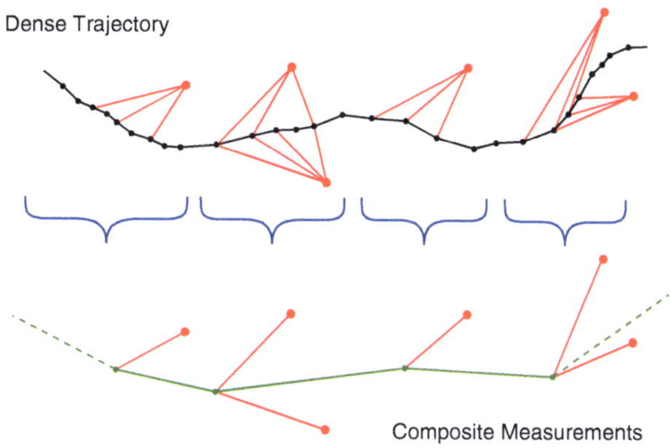

Fig. 8.11 As the robot explores, a pose (*black*) is added to the graph at each iteration, while feature detections (*red*) are also added to produce a dense trajectory. This dense trajectory is very large, so we periodically marginalize portions of the trajectory and the feature observations into composite measurements (*green*) at a much lower rate

8.5.2 Marine Mapping and Localization

As in the case of cooperative acoustic navigation, this application results in a series of constraints which can be optimized to best inform the AUV of its location relative to the map.

The complexity of the system of equations is tied to the sparseness of the A matrix which is itself dependent on the fill-in caused by loops in the graph structure. We explicitly avoid carrying out loop closures in this filter so as to maintain sparsity. All of this ensures that the matrices remain sparse and computation complexity predictable. Decomposition will not grow in complexity at each iteration, while the computational cost of back substitution will grow, but it is linear.

So as to avoid computational growth due to an ever-increasing graph size and also to produce an input to the next estimation stage, we periodically rationalize the oldest measurements from this graph to form a *composite measurement*. To do this we marginalize out all the poses that have occurred during the previous (approximately) 10 s period to produce a single node for the relative motion for that period as well as nodes for fully detected features and the associated covariances. This approach is very similar in concept to key frames in vision SLAM and is illustrated in Fig. 8.11.

We time this marginalization step to occur after a feature has left the sonar field of view as this allows us to optimally estimate its relative location given all available information. This composite measurement is then added to a lower-frequency higher-level graph. This low-frequency graph is used as input to the prior map matching algorithm in Sect. 8.5.3. Meanwhile the high-frequency graph begins to grow again by the insertion of newer constraints into A_i.

An alternative approach would be to maintain the dense trajectory of the robot pose at all times. This is the approach taken by iSAM [39]; however, given the size of the resultant graph, we are not certain that such an approach would have been able to yield a computationally constant solution required for our low-powered embedded CPU.

Additionally and unlike most land-based systems, the underwater domain is characterized by extended periods where the seabed is featureless for long distances and the resultant composite measurement is simply the relative trajectory of the distance traveled.

8.5.2.1 Feature Tracking

While the section above explains how the graph of the trajectory and sonar observations is optimized and efficiently solved, we have not discussed the way in which sonar targets are proposed.

The sonar detector passes point extractions to a target nursery which maintains a vector of all recent detections. The nursery feature projects the detections into a local coordinate frame using the recent vehicle dead reckoning and uses a probabilistic distance threshold to associate them with one another. Should a sufficiently large number of detections be clustered together (approximately 7–8 but dependent on the spread and intensity of detections), it is inferred that a consistent physical feature is present.

At this stage this nursery feature is added to the square root smoother. All of the relative AUV-to-point constraints for that feature are then optimized which results in improved estimation of the feature and the AUV trajectory. Subsequent point detections, inserted directly into the pose graph, result in an improved estimate via further square root smoothing. This approach also estimates the height/altitude of the sonar target using the sonar intensities measured at each iteration.

Finally it should be noted that the input to this feature tracker are point features characterized only by their location and covariance (due to the poor resolution of the sensor). This makes it difficult to robustly infer SLAM loop closures on the graph structure.

8.5.3 Global Estimation and Map Matching

Given this high-level graph of the robot trajectory and observed feature locations, it still remains for the autonomous system to make a critical judgment of where it is relative to the *a priori* map and to decide if this relative match is certain enough to be declared convincingly. To do this we maintain a set of match hypotheses in parallel. We compare them probabilistically so as to quantify the quality of the map match.

This comparison is implemented using a bank of estimators—each tracking a different match hypothesis in parallel. The relative likelihood of one match hypothesis over another is computed using positive information (of prior features detected by the sonar) as well as negative information (of prior features that were expected but undetected by the sonar), and in this way matching can be done in a probabilistically rigorous manner. Simply put, *if one expects to detect features predicted to lie along the trajectory of a robot and these features were not seen, then the trajectory must be less likely.*

The inclusion of this extra information is motivated by the regular rows of feature in the field and the inability of positive information metrics to estimate the relative position of the AUV along these lines. The incorporation of negative information in this way is to, our knowledge, a novel contribution and was by motivated information not captured by algorithms such as joint compatibility branch and bound (JCBB) algorithm [58].

8.5.3.1 Negative and Positive Scoring

In SLAM, multi-hypothesis comparison can typically be reduced to a scoring algorithm of the relative probabilities of candidate solutions. Here we propose an algorithm for multi-hypothesis scoring which uses both positive as well as negative information which we name the negative and positive scale (NAPS). An early version of this concept was introduced in [23]. More details are provided in [20].

At time t, we define NAPS for hypothesis i as the ratio of the probability of its map matching hypothesis, $h_{i,t}$, compared to a null hypothesis, h_{null}, when both are conditioned on the measurements $z_{1:t}$

$$NAPS(h_{i,t}) = \ln\left(\frac{p(h_{i,t}|z_{1:t})}{p(h_{null,t}|z_{1:t})}\right) \tag{8.15}$$

We define a hypothesis as the combination of an estimate of the graph structure of the SLAM problem x_h (the vehicle trajectory and all detected features) as well as all data association matches of these features to map features in the prior map. The null hypothesis is a special version of this hypothesis in which no data associations exist and in which it is proposed that each detected features is a new feature independent of the map. We use it as normalization for maps of growing size.

Dropping reference to i for simplicity and using Bayes' rule gives

$$NAPS(h_t) = \ln\left(\frac{p(z_t|h_t)p(h_t)}{p(z_t|h_{null})p(h_{null})}\right) \tag{8.16}$$

We split $p(z_t|h)$ into two terms representing both negative and positive information

$$p(z_t|h) = \eta p(z_{pos}|h)p(z_{neg}|h) \tag{8.17}$$

Positive information is then defined, in the same way as for JCBB, as the likelihood of the measurements given the hypothesis

$$\begin{aligned} p(z_{t,pos}|h) &= \eta_{z,pos} e^{-\frac{1}{2}(x_h - z_t)^{\mathrm{T}}\Sigma^{-1}(x_h - z_t)} \\ &= \eta_{z,pos} e^{-D_h} \end{aligned}$$

where Σ represents the covariance, $\eta_{z,pos}$ is a normalization constant, and D_h is the Mahalanobis distance.

The term $p(h)$ represents a prior probability of a particular map hypothesis being created by the robot which we propose is directly related to the number of features N_f matched to the prior map is given by

$$p(h) = \eta_x \mathrm{e}^{\lambda N_f} \tag{8.18}$$

where η_x is a normalization constant, λ is a free parameter, and N_f is an integer between zero and the total number of features in the prior map. While this formulation does not take into account aspects such as a target's measured visibility or other such specific terms, it does give us a measure of the confidence of a map match.

Combining these terms and canceling where possible give the following expressions for NAPS and as well as more common positive-only scoring (POS) metrics:

$$NAPS_t(h) = -D_h + \lambda N_f + C_{h,neg} \tag{8.19}$$

$$POS_t(h) = -D_h + \lambda N_f \tag{8.20}$$

This specifically indicates the contribution of negative information, $C_{h,neg}$, that we believe is neglected in typical multi-hypothesis scoring algorithms. POS algorithms implicitly assume $C_{h,neg} = 0$ and do not account for it in scoring the hypotheses. Most approaches assume very high λ: essentially selecting the hypotheses that match the most total features and then ordering those by Mahalanobis distance—as in the case of JCBB. A overview of such algorithms is presented in [57, 61].

8.5.4 Evaluating Negative Information

We define negative information as

$$\begin{aligned} C_{h,neg} &= \ln\left(\frac{p(z_{t,neg}|h)}{p(z_{t,neg}|h_{null})}\right) \\ &= \ln(p(z_{t,neg}|h)) - \ln(p(z_{t,neg}|h_{null})) \end{aligned} \tag{8.21}$$

As each hypothesis NAPS score will eventually be compared to one another, the second term need not be calculated.

For a particular hypothesis, consider an entire vehicle trajectory and the sonar footprint that it traced out (such as in Fig. 8.12). Also consider a prior map feature which is located within this footprint but was not detected. We wish to measure the number of times that this feature ought to have been detected, given that trajectory. NI is formed as the product of the probability of each undetected feature given the

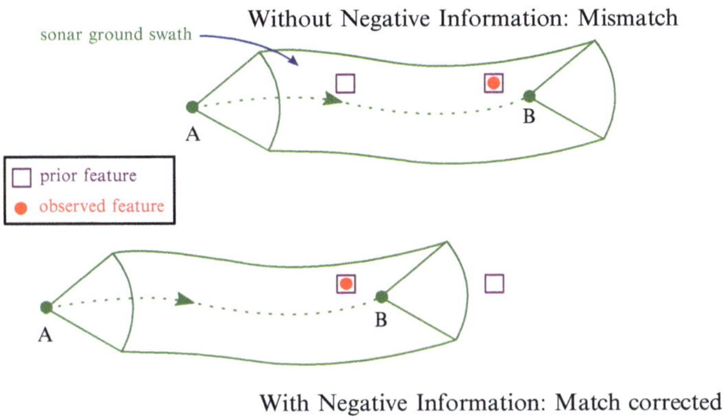

Fig. 8.12 Illustration of the effect of negative and positive scoring (NAPS). Consider the AUV trajectory from A to B with the sonar sensor footprint enclosed in *green*. If the AUV observes the *red* feature, how do we match its trajectory to the prior map (*purple squares*)? Using JCBB, the observed feature will be matched equally well to either prior feature. However, using negative information, NAPS indicates that the match in the *lower figure* is more likely. The *upper figure* is less likely because we would have expected to have observed both features—but only observed one

hypothesized vehicle trajectory

$$\begin{aligned} p(z_{t,neg}|h) &= p(z_{t,neg,f_1} \cap \ldots \cap z_{t,neg,f_{nu}}|h) \\ &= \prod_{f \in N_u} p(z_{t,neg,f}|h) \\ &= \prod_{f \in N_u} (1 - p(z_{t,pos,f}|h)) \end{aligned} \tag{8.22}$$

where s_t is whole area sensed during measurement z_t; thus,

$$p(z_{t,pos,f}|h) = \int_{p(f) \cap p(s_t)} v_f p(f) p(s_t) \mathrm{d}A \tag{8.23}$$

where v_f is the visibility of feature f and $p(f)$ is the prior probability of that feature.

In words, the probability of not detecting each conditionally independent feature is the product of one minus the probability of detecting each feature, integrated across the intersection of the PDF of each feature and the PDF of the scanned sensor area. This formulation is subject to the following assumptions: (1) the sensor occlusion model is well-defined and accurate, (2) all features are static, (3) feature detections are independent, and (4) feature visibility can be accurately modeled. This calculation, often intractable due to complicated integration limits, theoretically defines the probability of a set of negative measurements $z_{t,neg}$ given sensed area s_t.

More information about its precise evaluation is presented in [20]. The result of the metric is a positive value which scores a particular hypothesis more likely when its observations do not contradict the prior map.

In particular, combining negative information with the other (positive-only) metrics in Eq. 8.19 allowed us to disambiguate similar locations along a row of otherwise indistinguishable features, as indicated in Fig. 8.12.

While the AUV operated in the field this metric is evaluated for each hypothesis. The vehicle controls itself off of the most likely hypothesis: giving heading, speed, and depth commands to the low level vehicle controller so as to travel to a set of preprogrammed waypoints in the field. When the metric for a particular hypothesis exceeds a threshold, it is decided that the AUV is matched to the prior map and switches to a final target capture mode.

When it approaches this location, the FOI should be observed in the sonar imagery. The mission controller then transitions to directly controlling using the sonar detections using a PID—which we call sonar servoing. It opens a pair of tines with a tip separation of approximately 1m and drives onto the mooring line of the FOI.

8.5.5 Field Experiments

The system has undergone extensive testing and evolution over a number of years. Starting in November, 2006, we have conducted approximately 14 sea trials, each lasting 2 to 3 weeks. Our experiments began in fairly benign environments, using highly reflective moored objected as features, and progressed to more challenging conditions such as natural sea bottom targets and strong currents. After each trial we have refined and improved the system. In the following we summarize the progress of the development of the algorithms and the vehicle platform.

Typically the ingress point/direction to the field was varied for each mission, while the choice of feature of interest was taken at random just before placing the AUV in the water. After reaching the field, the vehicle typically traveled along the rows of features indicated in Fig. 8.14. This was so as to keep the number of map match hypotheses low (to about 4–5). The typical mission duration was 15–25 min, although the mission planner could be programmed to repeat the mission if the AUV failed to find the feature field. A typical water depth was 15 m.

Detailed comparison of mission parameters is difficult as the effect of the vehicle's control decisions is that different paths and observations follow. For this reason, this section focuses on the progression of our core map matching algorithm.

St. Andrews Bay, Florida, June 2007: The NAPS and joint compatibility branch and bound (JCBB) criteria were alternately used over 18 trials on a field of strongly reflective moored targets. The JCBB implementation uses a threshold on the Mahalanobis distance for multiple pair matching and chooses the most compatible pairs. The results of this live test and selected other tests are summarized in Table 8.1. The difference in the frequencies is 1.41 standard deviations which

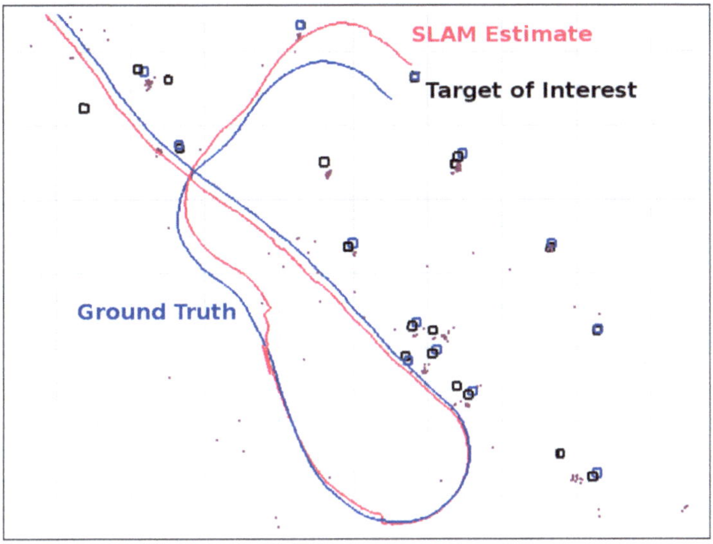

Fig. 8.13 A top-down overview of a successful mission using the accurate *REMUS 100* vehicle. The vehicle approached from the northwest and extracted feature points (*purple dots*). Using these points and the prior map (*blue squares*), the SLAM map (*black squares*) and the vehicle trajectory estimate (*magenta line*) were formed. Having matched against the map, the vehicle homed to the feature of interest. The abrupt position changes are the result of the square root smoother. The scale of the grid is 10 m. It is important to note that using only the DVL-INS-based position estimate the AUV would have failed to reacquire the FOI without using sonar as the map itself was only accurate to 5 m (*blue line*)

gives a 91% significance. We believe this demonstrates that the NAPS outperforms the simpler JCBB matching criteria in our application.

Narragansett Bay, Rhode Island, June 2008: Using the data from June 2007, significant improvements to our sonar processing algorithms allowed for improved detection of man-made and natural bottom features. This includes the addition of an adaptive noise floor model discussed in Sect. 8.5.1.1 and a reimplementation in integer logic for increased efficiency. The field for these tests consisted of various man-made and naturally occurring objects on the sea bottom as well as moored targets. The bay had a significant tidal current comparable to the 0.5 m/s velocity of the vehicle, which gave us substantial dead-reckoning errors.

Of the nine runs, we attached to the target once and had two mechanical failures. In both cases the tine mechanism broke upon hitting the mine mooring line. Thus the overall success rate of the sonar navigation system was 33%. After these tests the current model mentioned in Sect. 8.5 was developed.

Gulf of Mexico, near Panama City, Florida, June 2009: The entire system was tested on a field of 12 bottom objects and 3 moored objects over a two-week period. These experiments tested an improved model for current estimation along

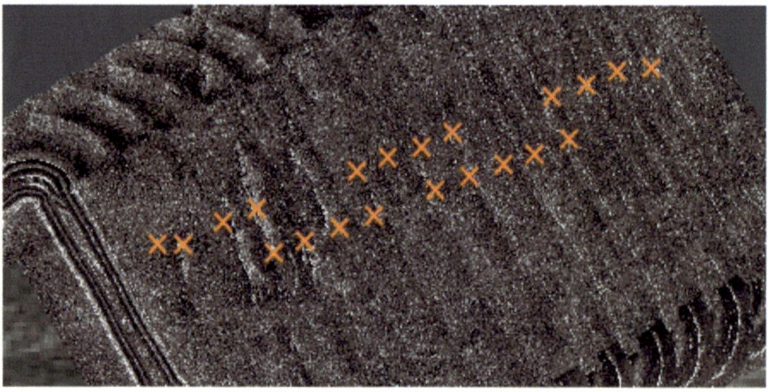

Fig. 8.14 Typical prior map generated using a REMUS 100 equipped with a Marine Sonic side-scan sonar. A series of features were extracted by trained human operates from the side-scan sonar imagery to produce an a priori map for the target reacquisition mission. The distance between the features is approximately 20 m (Fig. courtesy of SeeByte, Ltd.)

Table 8.1 Selected results in different conditions, with and without use of NAPS

Match criteria	No. of runs	Successes	Frequency (%)	$\sqrt{s_n^2/n}$ (%)
Bright targets - June 2007				
NAPS	9	6	67	17
JCBB	9	3	33	17
NAPS and JCBB			33	24
NAPS and multi-hypothesis	18	14	78	10
Normal targets - June 2008				
NAPS multi-hypothesis	9	3	33	17
Normal targets, low currents - June 2009				
NAPS multi-hypothesis	26	17	65	10
Normal targets, high currents - June 2010				
NAPS multi-hypothesis	42	13	31	7
As above, having reached the field - June 2010				
NAPS multi-hypothesis	18	13	72	11

with minor adjustments to the feature modeling. The current during this period was estimated as typically being 0.2 m/s using GPS surfaces. We had 17 successful target attachments in 26 runs.

Gulf of Mexico, July 2010: An additional set of experiments were carried out. In this circumstance we observed much higher currents which changed significantly. These currents varied from day to day but were estimated to be greater than the vehicle velocity (greater than 0.5 m/s) on certain days meaning that the vehicle could not make any headway against the current when it found itself down field from the features.

Presented in Table 8.1 are two different results for this experiment. One result gives the overall percent success when including all of the 42 runs carried out: 31%. Filtering the runs to the 18 runs in which the AUV was able to enter the field (as defined by at least a single feature detection in the sonar) produced a success percentage of 72% which we believe is more in fitting with the performance of the SLAM algorithm and comparable to the previous year's results. Nonetheless, this demonstrates the limitation of this particular vehicle platform as well as current estimation without a direct sensor.

8.5.6 Discussion

This section described target reacquisition system for small low-cost AUVs, based on forward looking sonar-based SLAM aided by a prior map. Our results indicate that when the AUV correctly matches to the prior feature map, it is regularly able to revisit a designated feature of interest.

The main failure mode of the algorithm is failing to enter the feature field, due to disturbances that exceed the vehicle's control authority. For small to moderate ocean currents, we developed an online current estimation procedure which allows the vehicle to avoid being driven off course during the initial vehicle dive. Room exists to improve this estimation procedure by estimation of the known current features mentioned in Sect. 8.5. Unsurprisingly in currents of more than 50–70% of the vehicle's velocity, successful performance was limited. This presented an obvious engineering limitation for this technology.

While more research is necessary to understand the many variables that can effect the system performance, such as the density and complexity of environmental features, the project has shown the viability of the viability of the FBN concept for feature reacquisition with low-cost vehicles.

8.6 Loop Closure: Ship Hull Inspection

The third application we consider is autonomous ship hull inspection. Hull inspections of large ships are frequently performed for safety and security purposes. It is not feasible to put the ships into dry dock every time an inspection is required. Currently, this inspection is primarily carried out by divers. This is a time-consuming and dangerous task for the divers. To address these risks, Bluefin Robotics and MIT built a ship hull inspection vehicle (see Fig. 8.15) called the hovering autonomous underwater vehicle (HAUV) [74]. The HAUV is equipped with a DVL to measure velocity relative to a surface, an IMU with ring laser gyro for attitude measurements and a dual-frequency identification sonar (DIDSON) [4] for imaging the structures being inspected.

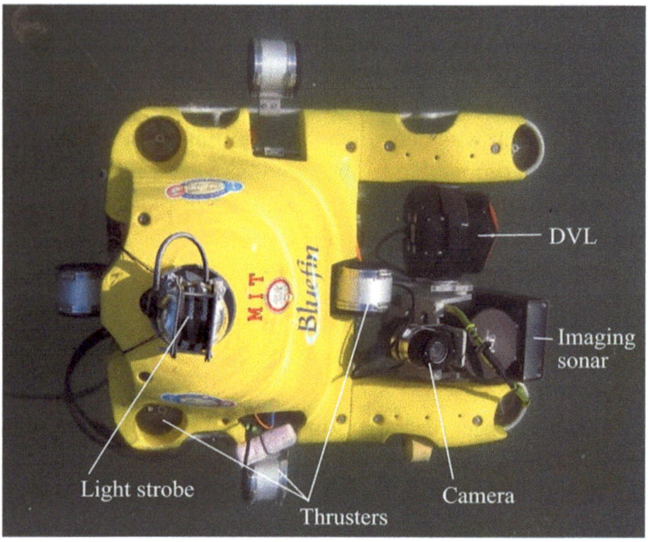

Fig. 8.15 Top view of the Bluefin-MIT hovering autonomous underwater vehicle (HAUV). The vehicle is equipped with a Doppler velocity log (DVL), an imaging sonar, an optical camera, and a light strobe. The sonar and DVL can be actuated independently

For autonomous ship hull inspection, it is crucially important, but difficult, to accurately track the vehicle position during mission execution. Accurate position information is essential for ensuring full coverage of the area being inspected. The ship hull inspection task further requires reporting the location of potential targets, so they can later be identified and removed. It is difficult, however, to obtain the global position estimate underwater from an external source. GPS is only available at the surface, so acoustic beacons would need to be deployed. Employing only rate gyros and odometry, over time sensor errors accumulate, and the position estimate will drift.

Using time-of-flight measurements with acoustic beacons has been commonly used in underwater navigation [83, 84, 86] to obtain a global position estimate; it has also proved successful in various applications like underwater archaeology[52] and ship hull inspection [29]. Here, we want to avoid the need for external infrastructure and instead are interested in achieving drift-free navigation by using the onboard imaging sonar. In particular, registering current data with previously observed sonar frames provides the necessary constraints to eliminate long-term drift.

Augmenting vehicle localization using sonars has been undertaken in a number of prior works. Walter et al. [79] used manually extracted landmarks and later automatic feature detection [78] with the ESEIF to produce a map of the environment. An automatic feature detector and a landmark formulation using an EKF filter were used in [13]. Sekkati et al. used extracted corner features from DIDSON frames to estimate vehicle motion over several frames [67]. In related work Negahdaripour et al. combined the DIDSON with an optical camera for

3-D target reconstruction using opti-acoustic stereo [56]. Eustice et al. [16] used constraints from overlapping camera frames within a SLAM information filter to estimate vehicle pose. A full 360-degree sonar scanner has been used in partially structured underwater environments [64] for localization by tracking line features in the environment using an EKF for the estimation process. Mallios et al. recently showed promising results in [50] using an mechanical scanning sonar and scan matching in an EKF framework.

Here, we use the pose graph formulation from Sect. 8.2.2 to combine onboard navigation information with sonar registration based on automated dense feature extraction [34]. We focus on imaged areas that are locally flat, such as the open areas of ship hulls and the seafloor. Our system allows for drift-free navigation without depending on any external infrastructure.

8.6.1 Drift-Free Navigation Using Imaging Sonar

The goal of this application is to correct drift in the vehicle state estimate over time using the imaging sonar—we begin by defining the quantities to be estimated. The vehicle pose consists of position and attitude. The vehicle position in 3-D is specified by Cartesian coordinates x, y, z with respect to some arbitrary reference frame, such as the starting point of the mission or a GPS frame acquired before diving. The attitude of the vehicle is specified by the standard Euler angles ϕ, θ, ψ or roll, pitch, and heading, respectively.

Without the imaging sonar information, only three out of the six degrees of freedom can be estimated without long-term drift. The ring laser gyro is used for heading estimation by integrating the measured angular rotations. A magnetic compass is not a viable option in close vicinity to a ship hull. The DVL provides velocities that are used for dead reckoning. In addition to the relative measurements, absolute measurements of depth from pressure and roll and pitch from the IMU are available. These absolute measurements are integrated into the estimation of the pose graph shown in Fig. 8.1. Nonetheless, no global information is available to limit long-term drift in the heading and x, y position.

Adding loop closure constraints from imaging sonar into the optimization problem eliminates long-term drift in the remaining three dimensions. The loop closure constraints are obtained by registering current sonar images to previously observed ones. Next, we describe the imaging sonar geometry, followed by our approaches to feature extraction and sonar registration.

8.6.1.1 Imaging Sonar Geometry

Following the formulation in [55, 56, 67], we define the geometry of the imaging sonar and derive a model that describes how the image is formed. To generate an

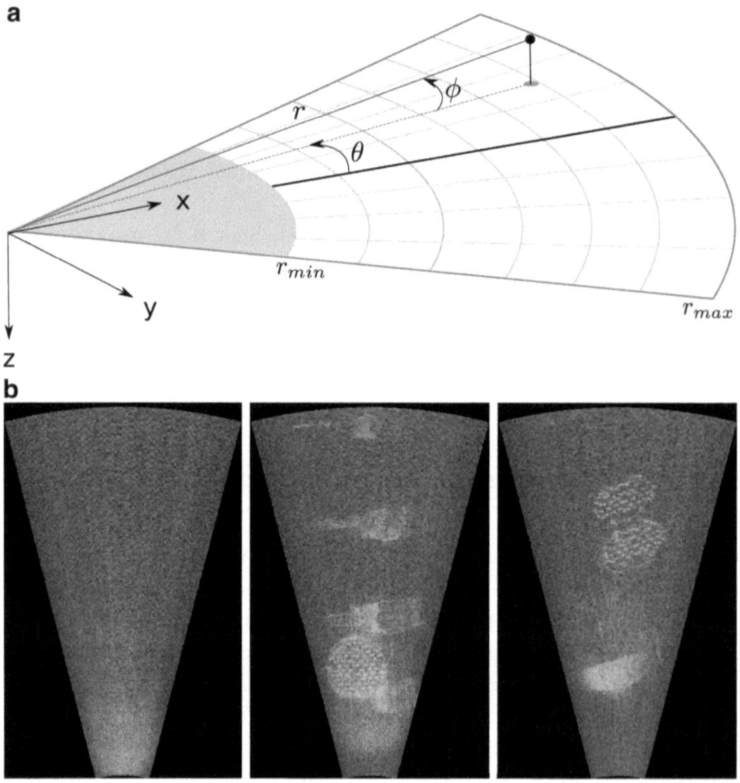

Fig. 8.16 **(a)** DIDSON imaging sonar geometry. **(b)** Sample images showing a clean hull (*left*) and several views of intakes and other structures

image, the sonar emits a narrow-beam sound wave and then listens to the returns, sampling the acoustic energy returned from different directions. The sonar provides time of flight and intensity for each azimuth angle. Combining the returns from all the elements provides an image of the reflective surfaces in front of the sonar. We use an imaging sonar with vertical beam width of 28°, covering 96 beams over a 29-degree horizontal field of view. Note that for a given point in the image, it can lie anywhere on an arc at a fixed range, spanning the vertical beam width.

Mathematically, the imaging process can be described as follows. We define the coordinate system for the sonar as shown in Fig. 8.16a. Let us consider a point $\mathbf{p} = [x\ y\ z]^\top$ in the sensor coordinate frame, and let $\mathbf{s} = [r\ \theta\ \phi]^\top$ be the same point in spherical coordinates, where r is the range, θ is the azimuth, and ϕ is the elevation of the point. We can relate the spherical and Cartesian coordinates with the following equations:

$$\mathbf{p} = \begin{bmatrix} x \\ y \\ z \end{bmatrix} = \begin{bmatrix} r\cos\phi\cos\theta \\ r\cos\phi\sin\theta \\ r\sin\phi \end{bmatrix} \tag{8.24}$$

$$\mathbf{s} = \begin{bmatrix} r \\ \theta \\ \phi \end{bmatrix} = \begin{bmatrix} \sqrt{x^2+y^2+z^2} \\ \arctan 2(y,x) \\ \arctan 2\left(z, \sqrt{x^2+y^2}\right) \end{bmatrix} \tag{8.25}$$

The sonar does not provide azimuth ϕ, so we measure point $\mathbf{p}$ as $I(\mathbf{p}) = [r\ \theta]^\top$, and the Cartesian projection of this point is

$$\hat{I}(\mathbf{p}) = \begin{bmatrix} u \\ v \end{bmatrix} = \begin{bmatrix} r\cos\theta \\ r\sin\theta \end{bmatrix} \tag{8.26}$$

For a small vertical beam width, this can be viewed as an approximation to an orthographic projection.

8.6.1.2 Feature Extraction

The imaging sonar returns intensity measurements at a number of ranges along each azimuth beam. The example sonar image in Fig. 8.17a shows some features on a flat surface. A strong return followed by a shadow likely indicates an object standing above the imaged surface, while a shadow on its own indicates a hole or a depression in the surface. Variations in the returned signal are also caused by changes in material properties, the strength of the transmitted signal, receiver sensitivity, distance to target, and the grazing angle, among other factors.

Stable features are extracted from sharp transitions in image intensity that mark the boundary between a strong return and a shadow. The main steps of the algorithm are as follows:

1. Smooth the image.
2. Calculate gradient.
3. Threshold a top fraction as features.
4. Cluster points and discard small clusters.

First, the image is smoothed using a median filter, significantly reducing noise, while still preserving edges, as shown in Fig. 8.17b. Next, the gradient is calculated by computing the difference between the local value and the mean of the n_p previous values along the beam (Fig. 8.17c). The number of previous values n_p used to calculate the mean around the current values affect the type of objects that are detected. Then points with gradient exceeding a given threshold are marked as candidate features (Fig. 8.17d). The threshold is adaptively chosen, such that a fixed fraction of the features is retained. Note that strong positive gradients are ignored because these correspond to the ends of shadows and are not as stable as negative

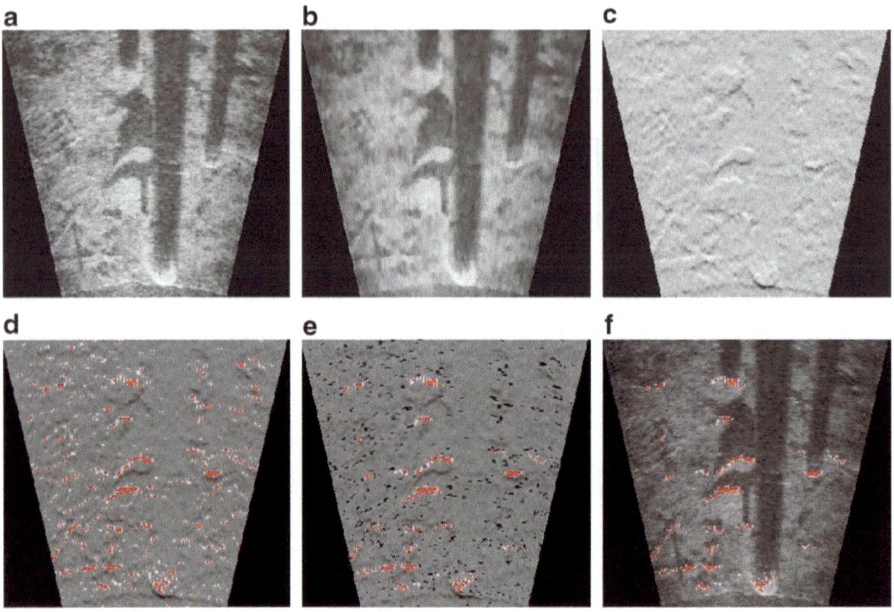

Fig. 8.17 Intermediate steps of the feature extraction process. The extracted features are shown in red (**a**) Initial sonar image (**b**) Smoothed (**c**) Gradient (**d**) Threshold (**e**) Clustering (**f**) Extracted features

gradients, which are closer to the sensor. Next, spurious features are eliminated by clustering the points and eliminating small clusters (Fig. 8.17e). The remaining extracted features are shown in Fig. 8.17f, typically containing on the order of one thousand points.

Assuming a locally flat surface, the Cartesian error associated with a successful registration arises mostly from the vertical beam width. The inclination angle between sensor plan and imaged surface is typically around twelve degrees, with a perpendicular distance between one and two meters. For a vertical beam width of 28°, the error can therefore reach 15 cm but is typically much smaller.

8.6.1.3 Registration

We align two overlapping sonar images by registration of the extracted features using the normal distribution transform (NDT) algorithm [5]. The NDT algorithm assigns the feature points of a scan to cells of a regular grid spanning the covered area. For each cell we calculate the mean and variance of its assigned points. This is done for four overlapping grids, where each grid is shifted by half a cell width along each axis. Using multiple shifted grids alleviates the effect of discontinuities resulting from the discretization of space. Two of the benefits using the NDT are that it provides a compact representation of the scan and no exact correspondences

Fig. 8.18 HAUV trajectory along a ship hull with sonar footprint visualized against a 3-D ship model. The current sonar image with detected features is shown on the *right*

between points are needed for alignment. This is useful here, because the movement of the HAUV causes variations in the returns from surfaces, causing some points to drop in and out of the extracted feature set.

The NDT of a scan serves as our model for registration. Given a new scan, a score is calculated for each point by evaluating the Gaussian of the NDT cell that receives the point. This provides a measure of the likelihood that a given point is observed based on the model. We define a cost function as the sum of the negative scores of all the points in the current view. Minimizing the cost with respect to the change in x, y position and heading ψ of the sonar provides the transformation between the scans. Because the main goal of the registration method is to close loops after drift has accumulated, we do not use the current estimate of the vehicle location to initialize the search. Instead, we repeat optimization from several initial values in an attempt to find the global minimum. To avoid incorrect matches, acceptance is based on a conservative threshold of a normalized score and also requires a minimum number of points to be matched. A successful registration is added to the pose graph as a loop-closing constraint.

8.6.2 *Experiments and Results*

Initial experiments were performed in a testing tank to verify that the system runs online and can stay localized over an extended period of time. Further tests were performed inspecting a patch of the bottom of the Charles river near the MIT Sailing Pavilion and on a small vessel with flat bottom in Boston harbor. While initial experiments focused on flat surfaces, the open areas of larger ship hulls are only locally flat, requiring a full six degrees of freedom state estimation. A model view of a ship from an actual experiment is shown in Fig. 8.18, with the trajectory in

Fig. 8.19 SS Curtiss in San Diego, with the submerged HAUV visible in the foreground as the yellow object in the water

cyan and the sonar viewing cone in blue. Note the large-scale difference between the vehicle and the ships this system is targeted for.

The HAUV is tethered during experiments but uses onboard power. The onboard battery allows operation for several hours, while the vehicle moves at 0.25 meters per second. An onboard computer controls the vehicle. Our software runs on a laptop on shore, connected by a fiber tether to the vehicle. The laptop serves for development purposes as well as for visualization during missions. Without the tether, the trajectory as well as selected data can be sent by acoustic communication to the shore. The DVL is locked to the hull to allow operation at a constant distance from the hull.

Recently we demonstrated online ship hull inspection on various large ships. One experiment was performed in early 2011 at the US Naval Station in San Diego on the 183 m-long SS Curtiss, a roll-on/roll-off container ship shown in Fig. 8.19. Another experiment was performed in mid-2011 in Boston, on a 82 m medium endurance US Coast Guard cutter, the USCGC Seneca. One trajectory segment is shown in Fig. 8.20, with loop-closing constraints obtained from sonar registration. The track lines are spaced approximately four meters apart, providing redundant coverage of the hull with sonar for the inspection mission.

We demonstrated the accuracy of our SLAM-derived state estimate by revisiting waypoints under closed-loop control. During inspection, the operator selected interesting waypoints along the hull using our real-time visualization. The waypoints were selected based on human-recognizable features, which were saved for later comparison. Later in the mission, the vehicle was commanded back to various waypoints, and the current and recorded images were shown for comparison. This served as verification of the consistency and accuracy of our SLAM system. The uncorrected coordinates estimated by the vehicle significantly differed after longer operation, showing the effectiveness of our navigation system in eliminating long-term drift.

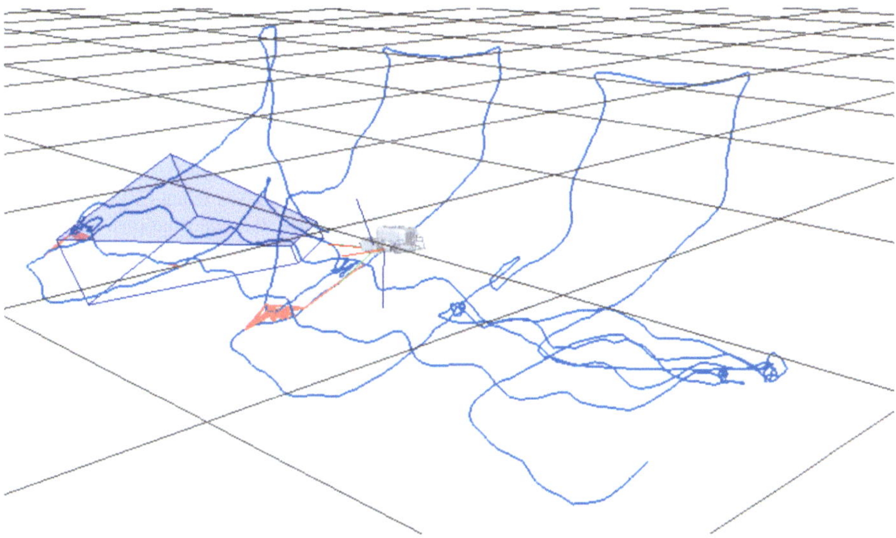

Fig. 8.20 Estimated vehicle trajectory along the ship hull of the USCGC Seneca in Boston—detected loop closures are shown in *pink*. The vehicle is shown partially under the ship hull, with the sonar viewing cone in *blue* indicating the part of the hull visible at this instance

A fuller and more complete overview of this particular project can be found in [32], which also presents more detailed experimental results and demonstrations. In addition to the sonar-based measurements discussed in this section, the wider project also incorporated information extracted from the optical camera illustrated in Fig. 8.15. Combining sonar and visual constants within the same SLAM estimation problem allows for sensor redundancy as well, taking advantage of complementary information. The specific application of visual SLAM to this problem is described in [41].

8.7 Conclusions

We have outlined three very different applications of simultaneous localization and mapping in the marine environment. SLAM is increasingly mature and is contributing to ever more complex marine problems which move to closer interaction with the underwater environment. In particular we have demonstrated that pose graph optimization methods for SLAM can operate onboard modern AUVs in real time, enabling closed-loop autonomous operation for many missions of interest.

Despite the substantial progress in SLAM for AUVs over the past decade, there are numerous important topics for future research. Future challenges for the applications outlined here include (1) robust long-term operation incorporating recovery from failures and detection of environmental changes, (2) cooperative mapping by multiple AUVs using undersea acoustic modems, and (3) integration of SLAM with motion planning and task control to enable close-range subsea inspection and intervention tasks.

Acknowledgements The work described in this chapter was partially supported by the Office of Naval Research under grants N00014-05-10244, N00014-11-10119, N00014-07-11102, N00014-06-10043, N00014-10-10936, and N00014-12-10020, and by the MIT Sea Grant College Program under research grant 2007-R/RCM-20. We wish to acknowledge the contributions of Michael Benjamin, Joseph Curcio, Georgios Papadopoulos, and Andrew Patrikalakis at MIT. Each project had a number of commercial and academic partners which we also wish to acknowledge including Ryan Eustice and Ayoung Kim (University of Michigan); Doug Horner and Tad Mazek (Naval Postgraduate School); Ed Matson and Bryan Schulz (iRobot); Jerome Vaganay, Kim Shurn, and Mike Elkins (Bluefin Robotics); Scott Reed, Alastair Cormack, and Jose Vasquez (SeeByte); and Dan Kucik and Paul Carroll (Naval Surface Warfare Center-Panama City).

References

1. Bahr A (2009) Cooperative Localization for Autonomous Underwater Vehicles. PhD thesis, Massachusetts Institute of Technology, Cambridge, MA, USA.
2. Bahr A, Leonard J, Fallon M (2009) Cooperative localization for autonomous underwater vehicles. Intl. J. Robotics. Res. 28(6):714–728
3. Ballard RD, Stager LE, Master D, Yoerger D, Mindell D, Whitcomb, LL, Singh H, Piechota D (2002) Iron Age shipwrecks in deep water off Ashkelon, Israel. Am J Archaeol 106(2):151–168
4. Belcher E, Hanot W, Burch J (2002) Dual-frequency identification sonar (DIDSON). In: Underwater Technology, Proceedings of the International Symposium on, p 187–192
5. Biber P, Strasser W (2003) The normal distributions transform: a new approach to laser scan matching. In: IEEE/RSJ Intl. Conf. on Intelligent Robots and Systems (IROS), vol 3, pp 2743–2748
6. Bosse M, Newman P, Leonard J, Teller S (2004) Simultaneous localization and map building in large-scale cyclic environments using the Atlas framework. Intl J Robotics Res 23(12): 1113–1139
7. Clark DE, Bell J (2005) Bayesian multiple target tracking in forward scan sonar images using the PHD filter. IEE Radar, Sonar and Navigation 152:327–334
8. Cummins M, Newman P (2007) Probabilistic appearance based navigation and loop closing. In: IEEE Intl. Conf. on Robotics and Automation (ICRA), pp 2042–2048
9. Dellaert F, Kaess M (2006) Square Root SAM: Simultaneous localization and mapping via square root information smoothing. Intl J Robotics Res 25(12):1181–1203
10. Duckett T, Marsland S, Shapiro J (2002) Fast, on-line learning of globally consistent maps. Auton Robotics 12(3):287–300
11. Durrant-Whyte H, Bailey T (2006) Simultaneous localisation and mapping (SLAM): Part I. Robotics Automat Mag 13(2):99 –110
12. Edgerton HE (1986) Sonar images. Prentice-Hall, Englewood Cliffs
13. Englot B, Johannsson H, Hover F (2009) Perception, stability analysis, and motion planning for autonomous ship hull inspection. In: Proceedings of the Int. Symp. on Unmanned Untethered Submersible Technology (UUST)
14. Eustice R (2005) Large-Area Visually Augmented Navigation for Autonomous Underwater Vehicles. PhD thesis, Massachusetts Institute of Technology, Cambridge, MA, USA.
15. Eustice R, Singh H, Leonard J (2006). Exactly sparse delayed-state filters for view-based SLAM. IEEE Trans. Robotics., 22(6):1100–1114

16. Eustice R, Singh H, Leonard J, Walter M, Ballard R (2005) Visually navigating the RMS Titanic with SLAM information filters. In: Robotics: Science and Systems (RSS), pp 3281–3288
17. Eustice RM, Whitcomb LL, Singh H, Grund M (2007) Experimental results in synchronous-clock one-way-travel-time acoustic navigation for autonomous underwater vehicles. In: IEEE Intl. Conf. on Robotics and Automation (ICRA), pp 4257–4264, Rome, Italy
18. Fallon M, Kaess M, Johannsson H, Leonard J (2011) Efficient AUV navigation fusing acoustic ranging and side-scan sonar. In: IEEE Intl. Conf. on Robotics and Automation (ICRA), Shanghai, China
19. Fallon M, Papadopoulos G, Leonard J (2009) Cooperative AUV navigation using a single surface craft. In: Field and Service Robotics
20. Fallon MF, Folkesson J, McClelland H, Leonard JJ (2012). Relocating underwater features autonomously using sonar-based SLAM. IEEE J. Ocean Engineering. To Appear
21. Fallon MF, Papadopoulos G, Leonard, JJ (2010) A measurement distribution framework for cooperative navigation using multiple AUVs. In: IEEE Intl. Conf. on Robotics and Automation (ICRA), pp 4803–4808
22. Fallon MF, Papadopoulos G, Leonard JJ, Patrikalakis NM (2010) Cooperative AUV navigation using a single maneuvering surface craft. Intl J Robotics Res 29(12):1461–1474
23. Folkesson J, Leonard J (2009) Autonomy through SLAM for an underwater robot. In: Proc. of the Intl. Symp. of Robotics Research (ISRR) 70:55–70
24. Freitag L, Grund M, Singh S, Partan J, Koski P, Ball K (2005) The WHOI micro-modem: An acoustic communications and navigation system for multiple platforms. In: Proceedings of the IEEE/MTS OCEANS Conference, vol 1, pp 1086–1092
25. Geyer E, Creamer P, D'Appolito J, Gains R (1987). Characteristics and capabilities of navigation systems for unmanned untethered submersibles. In: Proc. Int. Symp. on Unmanned Untethered Submersible Technology, pp 320–347
26. Golub G, Loan CV (1996) Matrix computations, 3rd edn. Johns Hopkins University Press, Baltimore
27. Grisetti G, Kümmerle R, Stachniss C, Frese U, Hertzberg C (2010) Hierarchical optimization on manifolds for online 2D and 3D mapping. In :IEEE Intl. Conf. on Robotics and Automation (ICRA), Anchorage, Alaska
28. Grisetti G, Stachniss C, Grzonka S, Burgard W (2007) A tree parameterization for efficiently computing maximum likelihood maps using gradient descent. In: Robotics: Science and Systems (RSS).
29. Harris S, Slate E (1999) Lamp Ray: Ship hull assessment for value, safety and readiness. In: Proceedings of the IEEE/MTS OCEANS Conference, vol 1, pp 493 –500
30. Heckman DB, Abbott RC (1973) An acoustic navigation technique. In: Proceedings of the IEEE/MTS OCEANS Conference, pp 591–595
31. Horner DP, McChesney N, Masek T, Kragelund SP (2009) 3D reconstruction with an AUV mounted forward looking sonar. In: Proc. Int. Symp. on Unmanned Untethered Submersible Technology, pp 1464–1470
32. Hover FS, Eustice RM, Kim A, Englot B, Johannsson H, Kaess M, Leonard JJ (2012) Advanced perception, navigation and planning for autonomous in-water ship hull inspection. Intl J Robotics Res. To Appear
33. Hunt M, Marquet W, Moller D, Peal K, Smith W, Spindel R (1974) An acoustic navigation system. Technical Report WHOI-74-6, Woods Hole Oceanographic Institution.
34. Johannsson H, Kaess M, Englot B, Hover F, Leonard J (2010) Imaging sonar-aided navigation for autonomous underwater harbor surveillance. In: IEEE/RSJ Intl. Conf. on Intelligent Robots and Systems (IROS), Taipei, Taiwan
35. Julier S, Uhlmann J (2001) A counter example to the theory of simultaneous localization and map building. In: IEEE Intl. Conf. on Robotics and Automation (ICRA), vol 4, pp 4238–4243
36. Kaess, M Dellaert F (2009) Covariance recovery from a square root information matrix for data association. J Robotics Auton Syst 57(12):1198–1210

37. Kaess M, Ila V, Roberts R, Dellaert F (2010) The Bayes tree: An algorithmic foundation for probabilistic robot mapping. In: Intl. Workshop on the Algorithmic Foundations of Robotics, WAFR, Singapore
38. Kaess M, Johannsson H, Roberts R, Ila V, Leonard JJ, Dellaert, F (2012) iSAM2: Incremental smoothing and mapping using the Bayes tree. IntL J RoboticS Res 31:217–236
39. Kaess M, Ranganathan A, Dellaert F (2008) iSAM: Incremental smoothing and mapping. IEEE Trans Robotics 24(6):1365–1378
40. Kim A, Eustice R (2009) Pose-graph visual SLAM with geometric model selection for autonomous underwater ship hull inspection. In: IEEE/RSJ Intl. Conf. on Intelligent Robots and Systems (IROS)
41. Kim A, Eustice RM (2011) Combined visually and geometrically informative link hypothesis for pose-graph visual SLAM using bag-of-words. In: IEEE/RSJ Intl. Conf. on Intelligent Robots and Systems (IROS), San Francisco, CA, pp 1647–1654
42. Kinsey JC, Eustice RM, Whitcomb LL (2006) A survey of underwater vehicle navigation: Recent advances and new challenges. In: IFAC Conference of Manoeuvering and Control of Marine Craft, Lisbon, Portugal. Invited paper
43. Konolige K (2004) Large-scale map-making. In: AAAI'04: Proceedings of the 19th national conference on Artificial intelligence, pp 457–463. AAAI Press/The MIT Press
44. Konolige K, Grisetti G, Kummerle R, Burgard W, Limketkai B, Vincent R (2010) Efficient sparse pose adjustment for 2D mapping. In: IEEE/RSJ Intl. Conf. on Intelligent Robots and Systems (IROS), Taipei, Taiwan, pp 22–29
45. Kümmerle R, Grisetti G, Strasdat H, Konolige K, Burgard W (2011) g2o: A general framework for graph optimization. In: IEEE Intl. Conf. on Robotics and Automation (ICRA), Shanghai, China
46. Leonard JJ, Bennett AA, Smith CM, Feder HJS (1998) Autonomous underwater vehicle navigation. Technical Report Marine Robotics Laboratory Technical Memorandum 98–1, MIT
47. Lu F, Milios E (1997) Globally consistent range scan alignment for environmental mapping. Auton Robotics 4:333–349
48. Mandt M, Gade K, Jalving, B (2001) Integrating DGPS-USBL positioning measurements with inertial navigation in the HUGIN 3000 AUV. In: International Conference on Integrated Navigation Systems, Saint Petersburg, Russia
49. Mahon I, Williams S, Pizarro O, Johnson-Roberson M (2008) Efficient view-based SLAM using visual loop closures. IEEE Trans Robotic 24(5):1002–1014
50. Mallios A, Ridao P, Hernandez E, Ribas D, Maurelli F, Petillot Y (2009) Pose-based SLAM with probabilistic scan matching algorithm using a mechanical scanned imaging sonar. In: Proceedings of the IEEE/MTS OCEANS Conference
51. Milne PH (1983) Underwater acoustic positioning systems. E. F. N. Spon, London
52. Mindell D (2007) Precision navigation and remote sensing for underwater archaeology. Remote sensing in archaeology, p 499
53. Montemerlo M, Thrun S, Roller D, Wegbreit B (2003) FastSLAM 2.0: An improved particle filtering algorithm for simultaneous localization and mapping that provably converges. In: Intl. Joint Conf. on Artificial Intelligence, Morgan Kaufmann, pp 1151–1156
54. Moutarlier P, Chatila R (1990) An experimental system for incremental environment modeling by an autonomous mobile robot. In: Intl. Sym. on Experimental Robotics (ISER).
55. Negahdaripour S, Firoozfam P, Sabzmeydani P (2005) On processing and registration of forward-scan acoustic video imagery. In: Computer and Robot Vision, 2005. Proceedings. The 2nd Canadian Conference on, pp 452–459
56. Negahdaripour S, Sekkati H, Pirsiavash H (2009) Opti-acoustic stereo imaging: On system calibration and 3-D target reconstruction. IEEE Transactions on Image Processing:
57. Neira J, Tardos J, Castellanos J (2003) Linear time vehicle relocation in SLAM. In: IEEE Intl. Conf. on Robotics and Automation (ICRA), pp 427–433
58. Neira J, Tardos JD (2001) Data association in stochastic mapping using the joint compatibility test. IEEE Trans Robotic Autom 17(6):890–897
59. Nuske S, Roberts J, Prasser D, Wyeth G (2010) Experiments in visual localisation around underwater structures. In: Field and Service Robotics. Springer Tracts in Advanced Robotics, vol 62. Springer, Berlin, pp 295–304

60. Olson E, Leonard J, Teller S (2006) Fast iterative alignment of pose graphs with poor initial estimates. In: IEEE Intl. Conf. on Robotics and Automation (ICRA), pp 2262–2269
61. Paz L, Guivant J, Tardós J, Neira J (2007) Data association in *O(n)* for divide and conquer SLAM. In: Robotics: Science and Systems, RSS, Atlanta, GA, USA
62. Petillot Y, Ruiz IT, Lane DM (2001) Underwater vehicle obstacle avoidance and path planning using a multi-beam forward looking sonar. J Ocean Eng 26:240–251
63. Ranganathan A, Kaess M, Dellaert F (2007) Loopy SAM. In: Intl. Joint Conf. on Artificial Intelligence (IJCAI), Hyderabad, India, pp 2191–2196
64. Ribas D, Ridao P, Neira J, Tardós, J (2006) SLAM using an imaging sonar for partially structured underwater environments. In: IEEE/RSJ Intl. Conf. on Intelligent Robots and Systems (IROS)
65. Rosen D, Kaess M, Leonard J (2012) An incremental trust-region method for robust online sparse least-squares estimation. In: IEEE Intl. Conf. on Robotics and Automation (ICRA), St. Paul, MN, pp 1262–1269
66. Schulz B, Hughes R, Matson R, Moody R, Hobson B (2005) The development of a free-swimming UUV for mine neutralization. In: Proceedings of the IEEE/MTS OCEANS Conference, IEEE, pp 1443–1447
67. Sekkati H, Negahdaripour S (2007) 3-D motion estimation for positioning from 2-D acoustic video imagery. Lecture Notes in Computer Science, 4478:80
68. Smith R, Self M, Cheeseman P (1990) Estimating uncertain spatial relationships in robotics. In: Autonomous Robot Vehicles. Springer, pp 167–193
69. Tena Ruiz I, de Raucourt S, Petillot Y, Lane D (2004) Concurrent mapping and localization using sidescan sonar. J Ocean Eng 29(2):442–456
70. Thompson R, Zehner W (1999) Frequency-steered acoustic beam forming system and process. US Patent 5,923,617
71. Thrun S, Beetz M, Bennewitz M, Burgard W, Cremers A, Dellaert F, Fox D, Hahnel D, Rosenberg C, Roy N, Schulte J, Schulz D (2000) Probabilistic algorithms and the interactive museum tour-guide robot minerva. Intl J Robotics Res 19(11):972–999
72. Thrun S, Burgard W, Fox D (2005) Probabilistic robotics. The MIT press, Cambridge
73. Thrun S, Liu Y, Koller D, Ng A.Y., Ghahramani Z., Durrant-Whyte H (2004) Simultaneous localization and mapping with sparse extended information filters. Intl J Robotics Res, 23(7):693–716
74. Vaganay J, Elkins M, Esposito D, O'Halloran W, Hover F, Kokko M (2007) Ship hull inspection with the HAUV: US Navy and NATO demonstrations results. In: Proceedings of the IEEE/MTS OCEANS Conference and Exhibition, vol 1, pp 761–766
75. Vaganay J, Leonard J, Curcio J, Willcox J (2004) Experimental validation of the moving long base line navigation concept. In: Autonomous Underwater Vehicles, 2004 IEEE/OES, pp 59–65
76. von Alt C, Allen B, Austin T, Stokey R (1994) Remote environmental monitoring units. In: AUV 94
77. Walls JM, Eustice RM (2011) Experimental comparison of synchronous-clock cooperative acoustic navigation algorithms. In: Proceedings of the IEEE/MTS OCEANS Conference, Kona, HI, USA.
78. Walter M (2008) Sparse Bayesian information filters for localization and mapping. PhD thesis, Massachusetts Institute of Technology.
79. Walter M, Hover F, Leonard J (2008) SLAM for ship hull inspection using exactly sparse extended information filters. In: IEEE Intl. Conf. on Robotics and Automation (ICRA), pp 1463–1470
80. Webster SE (2010) Decentralized single-beacon acoustic navigation: combined communication and navigation for underwater vehicles. PhD thesis, Johns Hopkins University, Baltimore, MD, USA.
81. Webster SE, Whitcomb LL, Eustice RM (2010) Preliminary results in decentralized estimation for single-beacon acoustic underwater navigation. In: Robotics: Science and Systems (RSS), Zaragoza, Spain.

82. Whitcomb L, Yoerger D, Singh H (1999a) Advances in doppler-based navigation of underwater robotic vehicles. In: IEEE Intl. Conf. on Robotics and Automation (ICRA), vol 1, pp 399–406
83. Whitcomb L, Yoerger D, Singh H (1999b) Doppler/LBL based navigation of underwater vehicles. In: Proceedings of the Intl Symp on Unmanned Untethered Submersible Technology (UUST)
84. Whitcomb L, Yoerger D, Singh H, Mindell D(1998) Towards precision robotic maneuvering, survey, and manipulation in unstructured undersea environments. In: Proc. of the Intl. Symp. of Robotics Research (ISRR), vol 8, Springer, London, pp 45–54
85. Yoerger D, Jakuba M, Bradley A, Bingham B (2007) Techniques for deep sea near bottom survey using an autonomous underwater vehicle. Intl J Robotics Res 416–429
86. Yoerger D, Mindell D (1992) Precise navigation and control of an ROV at 2200 meters depth. In: Proceedings of Intervention/ROV, vol 92

Index

A
Acoustic Doppler current profiler (ADCP), 31–32
Actor critic (AC) algorithms, 290
Ad hoc self-organizing process, 281
Aerial vehicle community, 271
Artificial neural networks
  neural activity landscape, 200–201
  neural activity propagation, 202
  shunting equation, 201
  topology, 202
Automated planning
  constraint satisfaction, 104
  EUROPA, 103–104
  human decision making, 102
  machine intelligence, 102
  milk-buying plan, 103
  scheduling, 104–105
  Shopping Agent problem, 105
Automated schedule and planning environment (ASPEN) system, 227
Autonomous science craft experiment (ASE), 100
Autonomous underwater vehicles (AUVs). *See also* Unmanned underwater vehicles (UUVs)
  autonomous control systems, 288
  data gathering and intervention, 225
  machine learning techniques, 288
  mission planning
    adaptive planning, 228
    ASPEN system, 227
    autonomous and semi-autonomous platforms, 226
    bio-inspired rheotaxis approach, 228
    CLEaR system, 227
    decision-making technologies, 226
    declarative, 226
    domain and inferred knowledge,226
    fault tolerant adaptive mission planning (*see* Ontology-based model)
    GPG system, 227
    plume detection and tracking, 228
    RAX, 227
    Sherpa system, 227
    space operations and Mars rovers, 227
  operability, 225
AvoidCollision behavior, IvP Helm
  approach calculation and caching, 87–88
  approach mapping, 84–85
  bravo vehicle, 88
  configuration parameter, 86
  spawning and killing, 86
  utility function, 86

B
Battlespace access for unmanned underwater vehicles (BAUUV), 231
Bayes filter, 32–33
Bistatic sensing, 281–282

C
Constraint-based planning, 107–108
Constraint reasoning engine (CRE), 116
Constraint satisfaction programming, 106–107
Controlled, agile, and novel observing network (CANON) program, 155
Cooperation, underwater vehicles
  architectures
    bottom-up approach,274–275
    composite systems,275–276
    ground robot community, 274

M.L. Seto (ed.), *Marine Robot Autonomy*, DOI 10.1007/978-1-4614-5659-9,

Cooperation, underwater vehicles (*cont.*)
planning,276–277
top-down approach,274–275
decision-making (*see* Decision-making, cooperative systems)
definitions,258–259
elements, 265
navigation and environmental estimation, 258
sample tasks
bistatic sensing,281–282
communication bandwidth, 277
coverage,281–282
FBN, 277
formation development,279–280
group navigation, 278
navigation accuracy, 277
networks, 281
rendezvous planning, 279
SLAM,277–278
shared task view
data management,265–267
goal representation,264–265
structure and content, 263
underwater environment impacts
communications, 260
navigation, 261
platform limitations,261–262
windsocks, 257
Cooperative acoustic navigation
acoustic range measurement, 336
acoustic ranging,339–340
alignment of side-scan sonar mosaics, 337
analysis,345
autonomous surface vehicle, 346
full pose trajectory estimation, 337
Garmin 18x GPS sensor, 343
Kearfott T16 INS, 343
multiple missions,343–346
recursive state estimator, 337
SAM framework,341–342
side-scan sonar,340–341
single mission,344–345
Coverage path planning (CPP), 179–180

D

Decision-making, cooperative systems
dissent sources,269–270
group/centralized controller, 267
negotiation and consensus
aerial vehicle community, 271
auctions, 272
communications limitations, 271
definition of negotiation, 270
game theory, 271
hierarchy,271–272
multi-agent systems community, 271
performance metrics,272–273
specific coordination mechanisms,273–274
task allocation,268–269
task decomposition, 267
Degrees of freedom (DOFs),311–312
Deliberative control
AI-based robotic planning, 98
architectural block diagram, AI, 97
automated planning
constraint satisfaction, 104
EUROPA, 103–104
human decision making, 102
machine intelligence, 102
milk-buying plan, 103
scheduling, 104–105
Shopping Agent problem, 105
autonomous platforms, 92
autonomy, 95
constraint-based planning, 107–108
constraint satisfaction programming, 106–107
EUROPA planner
architecture, 116–119
automated inference, 122–127
built-in object types, 113
constraints, 111
domain model, 97, 109
flaw resolution, 128–131
flexible partial plan, 109–110
modeling, 119–122
objects, 111–112
rules, 115–116
space telescope scheduling, 108
state variable representation, 110
tokens, 112–113
token state model, 113–115
variables, 109
Yo-Yo sawtooth pattern, 110
event response scenario, 167
experimental results
Dorado vehicle, 151
Gulper water sampler, 151
hidden Markov model, 153
intermediate nepheloid layer, 152
Lagrangian surveys (*see* Lagrangian surveys)
mixed-initiative frontal tracking,–154–155
replanning, 152
self-organizing maps, 153

failure recovery mechanism, 166
FDIR, 164
feature recognition, 165
MBARI Dorado AUV,94–95
"multi-agent" system, 163
planning and execution, T-REX (*see* Planning and execution, T-REX)
portable mobile observatory, 166
RAX experiment,99–101, 165
reactive control methods, 96
reactive subsumption-based approach, 93
script-based controllers, 97
sense-plan-act paradigm, 96
targeted marine science applications, 163
terrestrial robotic platform, 93
waypoint-based predefined mission, 95
Deliberative reactor
continuous planning, 147
current state flaw, 142
decision-making problem, 141
deliberation solver, 139
dispatch window, 146
embedding planning, 140
Going end condition, 145–146
inertial value assumption, 143
interleaved synchronization and execution, 147–149
partial plan failure, 149–151
planning window, 146–147
reactive planner, 140–141
reactor's deliberation latency, 145
shopping agent, 143–144
synchronization flaw, 141–142
synchronization solver, 139
Disruption-tolerant networks, 281
Doppler velocity log (DVL), 31–32
Drifter tracking application (DTA), 156
Dual-frequency identification sonar (DIDSON), 359

E

EUROPA planner
architecture
API layer, 117
chronological backtracking solver, 117
constraint reasoning engine, 116
NDDL model, 118
plan database, 116
solver module, 116–117
automated inference
built-in constraint library, 124
miscellaneous constraints, 124, 126
object hierarchy constraints, 124–125
problem domain description, 122
reservoir resources, 126
reusable resources, 126127
set constraints, 124–125
unary resources, 127
built-in object types, 113
domain model, 97, 109
flexible partial plan, 109–110
modeling
NDDL instance and goal specification, 122–123
object types, 119
predicate types, 121
shopping agent problem, 119–120
variables and constraints, 122
variable types and instances, 121
objects, 111–112
rules, 115–116
search
chronological backtracking algorithm, 129
constraint propagation, 130
flaw detection and resolution, 129
N-Queens solver, 131
ordering/filtering policy, 130
plan space planning, 128
unbound variable, 128
space telescope scheduling, 108
state variable representation, 110
tokens,–112–113
token state model, 113–115
variables, 111
Yo-Yo sawtooth pattern, 110
Explicit synchronization, 274
Explosive ordinance disposal (EOD) process, 12
Extended Kalman filter (EKF), 33–34

F

Fault diagnosis, isolation and recovery (FDIR), 164
Feature-based navigation (FBN), 277
Finite Markov decision processes (FMDP),296–297
First-in-first-out (FIFO) method, 196
Forward looking sonar processing
blazed array technology, 349
sonar feature detection,350–351
sonar image generation process,349–350
Function approximator, 299

G
Generalized Voronoi diagram (GVD), 190–192
Genetic algorithms (GA)
  Bezier curve, 203
  crossover, 207
  Darwin's theory, 199
  elitism, 200
  fitness function,204–206
  genome, 200
  multiple AUVs,208–209
  mutation, 207
  parameter setting, 207
  representation, 204
  selection mechanism,207–208
  trigonometric function-based seabed, 208
Global estimation and map matching
  field experiments,356–359
  negative and positive scoring,353–354
  negative information evaluation,354–356
  online current estimation procedure, 359
Global navigation satellite systems, 28
Gradient partially observable Markov decision process (GPOMDP) algorithm, 314
Ground robot community, 274
GVD. *See* Generalized Voronoi diagram (GVD)

H
Heuristic-based reactive algorithm
  Bug1 algorithm, 183–184
  Bug2 algorithm, 183, 185
  obstacle-laden environment, 183
  range sensor, 184
  TangentBug, 183
Hidden Markov model (HMM), 153
Hierarchical mode declarations (HMDs)
  mission planning, 66
  scripted *vs.* adaptive mission, 64
  syntax, 64–65
Hovering autonomous underwater vehicle (HAUV), 360

I
Implicit synchronization, 274
Incremental smoothing and mapping (iSAM), 330
Industrial mission, 259
Inertial measurement unit (IMU), 32
Inertial navigation system (INS), 32, 216
In-mission adaptation
  instance classification, 249
  mission performance timeline, 250
  mission planning action execution parameters,247–248
  payload system, 247
  reasoning process, 249
  starboard transducer gain,249–250
  status concepts and reusability, 249
  status monitoring, 247, 249–250
  symbols,251–252
  vehicle's track, 251
Intermediate nepheloid layer (INL), 152
IvP Helm autonomy
  behavior conditions, 66
  behavior flags and behavior messages, 62, 67–68
  behavior run states, 66–67
  decision making, 60
  HMDs
    mission planning, 64
    scripted *vs.* adaptive mission,64
    syntax, 64–65
  interval programming solution algorithm, 60
  iterate loop, 59, 62–63
  multi-objective optimization
    IvP functions, 62, 68–70
    IvP solver and behavior priority weights, 70–72
    IvP Toolbox, 69–70
  nontraditional aspects, 61–62
IvP Helm behavior
AvoidCollision behavior
    approach calculation and caching, 87–88
    approach mapping,84–85
    bravo vehicle, 88
    configuration parameter, 86
    spawning and killing,86
    utility function, 86
  Loiter behavior, 77
  MemoryTurnLimit behavior
    memory time, 82
    schematic, 84
    turn range, 83
    vehicle heading, 83
    vehicle turn radius limit, 82
  OpRegion behavior, 77
  overview,73–74
  parameters, 74–76
  waypoint behavior
    capture radius and slip radius parameters, 79, 80
    configuration parameters, 79
    lead parameter, 81
    objective function, 81, 82

points, order and repeat parameters, 80
schematic, 78–79
variables, 79

J

Joint architecture for unmanned systems (JAUS), 232

K

Kalman filter, 33

L

Lagrangian surveys
CANON program, 155
drifter cruise, 156–157
drifter tracking application, 156
drifter tracking tool, 160, 162
fluorometer, 158
genomic sensor, 157
human-in-the-loop approach, 162
microbial communities, 155
oceanographic field experiment, 159
ODSS, 160–161
offshore CANON ESP-drifter, 157–
Wirewalker instrument, 159–160
Long baseline (LBL) navigation, 35, 336

M

Marine mapping and localization,351–352
Marine vehicle proprioception, 348–349
MemoryTurnLimit behavior, IvP Helm
memory time, 82
schematic, 84
turn range, 83
vehicle heading, 83
vehicle turn radius limit, 82
Micromodem, 36
Military mission, 259
Mine countermeasures (MCM), 246, 347
Mission oriented operating suite (MOOS), 230
Mission oriented operating suite IvP (MOOS-IvP)
adaptive autonomy, 50–51
backseat driver design, 52–53
behavior-based control design, 54–55
C++, 50
collaborative autonomy, 50–51
Helm autonomy
behavior conditions, 66
behavior flags and behavior messages, 62, 67–68
behavior run states,66–67
decision making, 60
HMDs,64–65
interval programming solution algorithm, 60
iterate loop, 59, 62–63
multi-objective optimization,68–72
nontraditional aspects,61–62
Helm behavior
AvoidCollision behavior,84–88
Loiter behavior, 77
MemoryTurnLimit behavior,82–84
OpRegion behavior, 77
overview,73–74
parameters,74–76
waypoint behavior,78–82
inter-process communication with publish/subscribe, 53, 55–56
mail handling,57–58
message content,56–57
multi-objective optimization, 51
nested autonomy paradigm,53–55
open-source project, 49–50
overloaded functions,58–60
publish-subscribe middleware design, 52–54
sensor technology, 50
Monterey Bay Aquarium Research Institute (MBARI),92
MOOS-IvP. *See* Mission oriented operating suite IvP (MOOS-IvP)
Multi-hop networks, 281
Multi-objective optimization, IvP Helm autonomy
IvP functions, 62, 68–69, 71
IvP solver and behavior priority weights,70–72
IvP Toolbox,69–70

N

NAC method. *See* Natural actor-critic (NAC) method
Natural actor-critic (NAC) method
algorithm configuration,319–320
convergence and adaptation, 312
linear function parameterization, 312
two-step learning with barycentric interpolators
appropriate mesh, 314
barycentric coordinates,314–315
eligibility trace, 316

Natural actor-critic (NAC) method (*cont.*)
error, 316
function approximator output,316–317
GPOMDP approach, 314
parameterized basis function, 313
parameter vector, 316
policy gradient, 312
simulated model and environment, 317
Neural Q-learning (NQL),300–301

O

Observe-orient-decide-act (OODA),229–230
Oceanographic decision support system (ODSS), 160–161
Off-policy algorithm, 296
One-way travel time (OWTT), 35
On-policy algorithm, 296
Ontology-based model
application ontology, 241
core domain ontology, 239
fault tolerant adaptive mission
FDD systems, 233
mission plan adaptation,234–235
mission plan refinement, 236
shaft bearing failure, 233
foundation and core ontology,240–241
knowledge representation
*ABox,*237–238
components, 237
conceptualisation encoding, 237
definition of ontology,236–237
diagnostic system, 238
fault, 237
intelligent processing, 238
reasoning interactions, 238
*TBox,*237–238
levels of generality,238–239
library construction and organisation, 238
mission planning ontology,243–244
OWL language, 240
RDF and XML standard, 240
semantic world model framework, 228
situation awareness (*see* Situation awareness (SA_H))
status monitor ontology
concepts,242–243
environmental and internal data,241–242
reuse and consistency, 242
system architecture,245–246
test case scenario
in-mission adaptation (*see* In-mission adaptation)
knowledge base representation,246–247
MCM scenario, 246
procedural mission, 246, 248
REMUS 100 AUV platform, 246
vertical segmentation, 238
Ontology web language (OWL), 240

P

Path planning
artificial neural network
neural activity landscape, 200–201
neural activity propagation, 202
shunting equation, 201
topology, 202
AUV, 180–181
bug algorithm, 219
cell decomposition
connectivity graph, 194
critical slice, 195
Moorse cell decomposition, 196–197
Reeb graph, 196
start-to-goal path planning, 195
configuration space, 178
CPP, 179–180
degrees of freedom, 178–179
2D environment, 182
genetic algorithms
Bezier curve, 203
crossover, 207
Darwin's theory, 199
elitism, 200
fitness function, 204–206
genome, 200
multiple AUVs, 208–209
mutation, 207
parameter setting, 207
representation, 204
selection mechanism, 206–207
trigonometric function-based seabed, 208
heuristic-based reactive algorithm
Bug1 algorithm, 183–184
Bug2 algorithm, 183, 185
obstacle-laden environment, 183
range sensor, 184
TangentBug, 183
potential function approach
potential field, 185–188
wavefront planner, 188–189
roadmaps
GVD, 190–192
probability, 192–195
visibility graph, 191–193

robot configuration, 177
search algorithm
Dijkstra' s algorithm, 197
distance-cost heuristics, 197, 199
FIFO, 196
search tree, 198, 200
sensor-driven online path planning
information gain, 211–216
in-water trials, 216–219
sidescan sonar sensor, 210–212
uncertainty handling, 208
underwater environment, 209
waypoint-defined path, 177
PG algorithms. *See* Policy gradient (PG) algorithms
Plan database (PDB), 116
Planning and execution, T-REX
architectural concepts
directed acyclic graph, 133, 135
EUROPA state variables, 133
intertwined plan execution, 134
shared state variables, 135
token dispatching, 134
coupling, 132
deliberative reactor
continuous planning, 147
current state flaw, 142
decision-making problem, 141
deliberation solver, 139
dispatch window, 146
embedding planning, 140
Going end condition, 144–145
inertial value assumption, 143
interleaved synchronization and execution, 147–159
partial plan failure, 150–153
planning window, 146–147
reactive planner, 140–141
reactor's deliberation latency, 145
shopping agent, 143–144
synchronization flaw, 141–142
synchronization solver, 139
execution cycle
backtracking, 136
natural information flows, 137
reactor failure handling, 138–139
synchronization and deliberation, 137
low-level control, 131
partitioning, 132
Planning domain definition language, 243–244
Policy gradient (PG) algorithms
algorithm adaptation capabilities
cable configurations, 322
chicane configuration, 322
θ policy performance, 322–323
reward evolution, 323, 325
x_g policy performance, 322, 324
cable tracking, underwater environments
cable-tracking vision-based system, 318–319
Ictineu AUV, 317–318
convergence process, 290
independent function approximator, 289
Markov property, 289
NAC method (*see* natural actor-critic (NAC) method)
real learning results, 321–322
simulated learning results, 320
speeding up RL algorithms, 313

Q

Q-learning algorithm, 298–299

R

Rapidly exploring random tree (RRT), 194
Reinforcement learning (RL)
curse of dimensionality, 289
dynamics model, 288
formulation, 289
optimal state/action mapping, 288
PG algorithms (*see* Policy gradient (PG) algorithms)
RLP (*see* Reinforcement learning problem (RLP))
scalar value, 288
vs. supervised learning methods, 288
value-function algorithms (*see* Value-function (VF) algorithms)
Reinforcement learning problem (RLP)
dynamics function, 293
exploration and exploitation, 292
FMDP, 294–295
learner/environment interaction, 291–292
mapping function, 292
optimal policy, 294
policy function, 293
policy gradient algorithms, 296–297
reinforcement function, 291, 293
states, actions, and rewards sequence, 292
temporal difference learning, 296
value function, 293–294
Remote agent experiment (RAX), 165
autonomous sciencecraft experiment, 100
CIRCA, 103
constraint-based representation, 99
flexible temporal representation, 99

Remote agent experiment (RAX), 165 (*cont.*)
IDEA, 100
ReSSAC project, 100
ROAR, 101
Remote environmental monitoring units (REMUS), 12, 343, 347
Rendezvous planning, 279
Resource description framework (RDF) standard, 240
RL. *See* Reinforcement learning (RL)
RLP. *See* Reinforcement learning problem (RLP)
Roadmaps
GVD
deformation retraction, 190–191
homotopy, 190
retraction, 190
schematic, 191–192
probability
free configuration space, 193–194
RRT, 194
sampling-based method, 193
schematic, 194–195
visibility graph
free space representation, 191
schematic, 192–193
supporting and separating lines, 192
Robot operating system (ROS), 231

S
SA_H. *See* Situation awareness (SA_H)
Scientific mission, 259
Search algorithm
Dijkstra' s algorithm, 197
distance-cost heuristics, 197, 199
FIFO, 196
search tree, 198, 200
Self-organizing maps (SOM), 153
Semi-online learning, 302–303
Semi-online neural-Q-learning (SONQL)
algorithm
action selection, 303–304
neural Q-learning, 300–301
semi-online learning, 306–307
behavior,317
parameters, 318
Sensor-driven online path planning
information gain
conditional entropy, 212
mine detection probability, 214–216
mutual information, 212
objective function, 213
seabed conditions, 212–213
Shannon entropy, 212
utility quantification, 211
in-water trials
confidence map, 218–219
inertial navigation system, 216
lawn mower path, 216–217
non-convex environment test, 218
sidescan sonar sensor
AUV trajectory, 210
geo-referenced data, 210–211
"lawn mower"/"zigzag" pattern, 210
seabed conditions, 211–212
uncertainty handling, 208
underwater environment, 209
Short baseline (SBL) navigation, 35
Simultaneous localization and mapping (SLAM)
acoustic modem, 368
cooperative acoustic navigation
acoustic range measurement, 336
acoustic ranging, 339–340
analysis, 345
autonomous surface vehicle, 346
full pose trajectory estimation, 337
Garmin 18x GPS sensor, 343
Kearfott T16 INS, 343
multiple missions, 346
recursive state estimator, 337
SAM framework, 341–342
side-scan sonar, 337, 340–341
single mission, 344–345
data association, 335
forward looking sonar processing
blazed array technology, 349
sonar feature detection, 350–351
sonar image generation process, 348–349
global estimation and map matching
field experiments, 356–359
negative and positive scoring, 353–354
negative information evaluation, 354–356
online current estimation procedure, 359
Hydroid REMUS 100, 348
iRobot Ranger, 348
iSAM, 330
marine mapping and localization, 351–352
marine vehicle proprioception, 348–349
mathematical summary, 333–335
mine countermeasures, 347
navigation, marine environment, 335–336
particle filters, 331
pose graph optimization, 332–333

REMUS AUV, 343, 347
sample tasks, 277–278
SEIF, 331
ship hull inspection
Bluefin-MIT HAUV, 360
DIDSON, 359
drift-free navigation, 361–365
experiments and results, 365–368
time-of-flight measurement, 366
smoothing-based, 330
Single-hop networks, 281
Situation awareness (SA_H)
decision component, 229
definition, 229
fault detection and diagnosis, 232–233
human and AUV, 229
knowledge representation and middleware
BAUUV, 231
human-robot interaction, 232
information transfer protocols, 230
interoperability, 232
JAUS, 232
MOOS, 230
OceanSHELL libraries, 231
Player libraries, 230
robot operating system, 231
observation component, 229
orientation component, 229
RAX architecture, 230
status monitor and mission plan adapter, 229–230
SLAM. *See* Simultaneous localization and mapping (SLAM)
SONQL. *See* Semi-online neural-Q-learning (SONQL)
Sparse extended information filter (SEIF), 331
Stoplight coordination, 273
STRIPS language, 243

T
Teleo-reactive executive (T-REX)
MBARI's *Dorado* AUV, 94
planning and execution (*see* Planning and execution, T-REX)
Temporal difference (TD) learning, 296
Time hopping, 313
Two-way travel time (TWTT), 35

U
Ultra-short baseline (USBL) navigation, 35, 339
Underwater communications
acoustic propagation
absorption loss, 25–26
ambient noise, 27
channel impulse response, 24
geometric spreading loss, 25
low sound speed, 25
multi-path, 26–27
ocean waveguide, 24
scattering loss, 26
surface scattering, 27
detectability, 23
electromagnetic propagation underwater, 23–24
Underwater navigation
acoustic navigation
positioning systems, 35–36
underwater modems, 36
aided dead reckoning
DVL, 31–32
inertial navigation systems, 32
state estimation techniques, 32–35
dead reckoning, 28, 31
geophysical navigation, 37
GPS, 28
navigation accuracy, 28
and payload sensors, 28–30
positional accuracy, 28
Underwater Robotic Intelligent System (URIS), 304–305
Unmanned surface vehicle (USV), 2
Unmanned underwater vehicles (UUVs)
adaptation, 3
adaptive sampling, 18–19
autonomous agents, 4
autonomous behaviours, 2
autonomy levels, 7–9
deliberative architectures, 5
energy
common sources, 39–40
endurance missions, 37
fuel cells, 38, 40
high energy density power sources, 40
on-board power, 37–38
payload computer, 38
propulsion power, 38
specific energy density, 38
hybrid architectures, 5–6
intelligent autonomy
decision-making, 3–4
deliberation and reasoning, 3
fault tolerance, 4
learning and adaptation, 10
military and scientific communities, 10
mission planning/re-planning, 4

Unmanned underwater vehicles (UUVs) (*cont.*)
monitoring and diagnosis, 9–10
networking and cooperation, 10
planning and decision, 9
sensing and perception, 9
intelligent behaviours, 3
military mission requirements
EOD process, 12
MCM mission, 11–12
military autonomy requirements, 13–15
prioritized missions, 11
REMUS, 12
simultaneous localization and mapping, 13
warfare platform, 11
ocean observation
coastal ocean observation, 18
mesoscale variability, 16
moored arrays, 16–17
ocean role, climate, 17–18
opportunistic measurements, 15–16
scientific research missions, 15
temporal and spatial attributes, 16
under-sampling, 16
oceanographic and military tasks, 1
oceanographic research autonomy
covert and non-covert communications, 21
data collection, independent human operator, 20
denied/degraded GPS operation, 21
force multiplication, multiple vehicles operation and coordination, 20
high-resolution information, 20
intelligent tasking, scientific sensors, 22
long distances reach and transit, 19–20
mesoscale variability, 22
monitoring and diagnosis, 22
opportunistic measurements, 21
persistence, 20
planning and decision, 22
potential threats, 21
tracking moving objects, 21
payload autonomy implementation, 6–7
reactive architectures, 5
requirements, 4–5
robotic autonomy, 5
structure, 5
surfaced explorer, 2
surfaced underwater glider, 2–3
underwater communications (*see* Underwater communications)
underwater navigation (*see* Underwater navigation)
URIS AUV, 304

V

Value-function (VF) algorithms
deterministic policies, 289
discrete lookup table parameterization, 289
generalization problem, 299–300
Q-learning algorithm, 298–299
SONQL algorithm (*see* Semi-online neural-Q-learning (SONQL))
target-following behavior
convergence test, 310–311
environment state, 307
HSL color, 305
image segmentation, 305
nonlinear effects, 306
reinforcement function, 308
SONQL behavior, 307
SONQL parameters, 308
target coordinates, 306
target detection and tracking system, 304–305
URIS AUV, 304–305
X DOF, 309
Yaw DOF, 310

W

Waypoint behavior, IvP Helm
capture radius and slip radius parameters, 78, 80
configuration parameters, 79
lead parameter, 81–82
objective function, 82–83
points, order and repeat parameters, 80
schematic, 78
variables, 79

MIX
Papier aus verantwortungsvollen Quellen
Paper from responsible sources
FSC® C105338

If you have any concerns about our products,
you can contact us on
ProductSafety@springernature.com

In case Publisher is established outside the EU,
the EU authorized representative is:
Springer Nature Customer Service Center GmbH
Europaplatz 3, 69115 Heidelberg, Germany

Printed by Libri Plureos GmbH
in Hamburg, Germany